75

MANAGING THE PLANET

Edited by

Peter Albertson
&
Margery Barnett

A SPECTRUM BOOK

PRENTICE-HALL, INC. Englewood Cliffs, N.J.

PRENTICE-HALL, INC. Englewood Cliffs, N.J.

This book is an abridgment of *Environment and Society in Transition,*
a symposium of The New York Academy of Sciences, © 1971 by The
New York Academy of Sciences, and is published by arrangement with
the Academy.

Printed in the United States of America

Library of Congress Catalog Card Number: 72–3949

ISBN   0–13–550715–4
ISBN   0–13–550707–3   (pbk)

10   9   8   7   6   5   4   3   2   1

PRENTICE-HALL INTERNATIONAL, INC. (*London*)
PRENTICE-HALL OF AUSTRALIA, PTY. LTD. (*Sydney*)
PRENTICE-HALL OF CANADA, LTD. (*Toronto*)
PRENTICE-HALL OF INDIA PRIVATE LIMITED (*New Delhi*)
PRENTICE-HALL OF JAPAN, INC. (*Tokyo*)

CONTENTS

II. DISCUSSIONS

III. CONFERENCE DINNER

IV. A CONFERENCE SUMMARY

CONFERENCE OFFICERS

Chairman:

Boris Pregel
President, American Division, World Academy of Art and Science
Vice-President, American Geographical Society
Past President, The New York Academy of Sciences

Conference Cochairmen:

Hugo Boyko
President, World Academy of Art and Science

Carl-Göran Hedén
Department of Microbiological Engineering
Karolinska Institute
Stockholm, Sweden

Serge A. Korff
President, American Geographical Society
President-Elect, The New York Academy of Sciences
Professor of Physics
New York University

Harold D. Lasswell
Professor of Law and the Social Sciences
Yale University

John McHale
Director, Center for Integrative Studies
State University of New York
Binghamton, N.Y.

Stuart Mudd
Vice-President, World Academy of Art and Science
Professor Emeritus
University of Pennsylvania

Arne Tiselius
Head, Institute of Biochemistry
The Nobel Institute of the Royal Swedish Academy of Sciences

Session Chairmen:

Lloyd L. Motz, President
The New York Academy of Sciences
Professor of Astronomy
Columbia University

Detlev W. Bronk
President Emeritus
The Rockefeller University

Walter A. Rosenblith
Provost
Massachusetts Institute of Technology

Margaret Mead
Curator Emeritus of Ethnology
The American Museum of Natural History

Harold D. Lasswell
Professor of Law and the Social Sciences
Yale University

Jan Tinbergen,
Professor of Economics
Netherlands School of Economics

Boris Pregel, President, American Division
World Academy of Art and Science
Vice-President, American Geographical Society

Conference Coordinator:

John McHale, Director
Center for Integrative Studies
State University of New York
Binghamton, N. Y.

Assistant Coordinators:

Lyle Taylor,
Department of Sociology
University of Colorado

John Colligan
Associate Dean, School of Advanced Technology
State University of New York
Binghamton, N. Y.

5

Aglion, Prof. Raul
Professor of Economics
Beverly Hills, Calif.

Alexandrowicz, Dr. Julian
Director, III Klinika Wewnetrznych
Akademii Medycznei
Krakow, Poland

Baugniet, Prof. Jean
Former Rector
Université Libre de Bruxelles
Brussels, Belgium

Bergier, Jacques
Science Writer
Paris, France

Bergmann, Prof. Ernst
Department of Organic Chemistry
Hebrew University of Jerusalem, Israel

Bjorksten, Dr. Johan
Bjorksten Research Laboratories, Inc.
Madison, Wis.

Block, Eskil
Taby, Sweden

Borgstrom, Prof. Georg
College of Agriculture
Department of Food Science
Michigan State University
East Lansing, Mich.

Boyko, Dr. Elisabeth
Associate Director
International Coordinating Center for Saline Irrigation and Water Purification
Rehovot, Israel

Boyko, Dr. Hugo
President
World Academy of Art and Science
International Coordinating Center for Saline Irrigation and Water Purification
Rehovot, Israel

Braziller, George
Publisher
New York, N. Y.

Bronk, Dr. Detlev W.
President Emeritus
The Rockefeller University
New York, N. Y.

Brown, Lester R.
Senior Fellow
Overseas Development Council
Washington, D. C.

Bushnell, Vivian
American Geographical Society
New York, N. Y.

Caglioti, Prof. Vincenzo
Presidente del Consiglio
Nazionale delle Ricerche
Rome, Italy

Calhoun, Dr. John B.
Unit for Research on Behavioral Systems
Laboratory of Psychology
National Institute of Mental Health
Bethesda, Md.

Calhoun, Mrs. John B.
National Institute of Mental Health
Bethesda, Md.

Cameron, Prof. A. G. W.
Belfer Graduate School of Science
Yeshiva University
New York, N. Y.

Cancro, Dr. Robert
Professor, Department of Psychiatry
The University of Connecticut Health Department
Hartford, Conn.

Cassirer, Dr. Henry R.
Chief
Division of the Use of Mass Media in Out-of-School Education
UNESCO
Paris, France

Catlin, Prof. George E. G.
Vice-President
World Academy of Art and Science
London, England

Christensen, Prof. Harold T.
Department of Sociology
Purdue University
Lafayette, Ind.

Colligan, John
Associate Dean
School of Advanced Technology
State University of New York
Binghamton, N. Y.

Cunetta, Joseph
New York City Water Pollution Board
New York, N. Y.

6

Davis, Dr. William O.
Office of Plans and Programs
Environmental Science Services Administration
U. S. Department of Commerce
Rockville, Md.

de Castro, Dr. Josué
President
Centrē International pour le Développement
Neuilly-sur-Seine, France

Dewey, Dr. Edward R.
President
Foundation for the Study of Cycles
Pittsburgh, Pa.

Dixon, John
Applied Devices Corporation
1660 L Street, N. W.
Washington, D. C.

Dror, Prof. Yehezkel
Hebrew University of Jerusalem
Jerusalem, Israel

Elim, Dr. Raga Sayed
Director General
Universities and the Quest for Peace
Director, International Studies
State University of New York
Binghamton, N. Y.

Eraj, Dr. Yusuf Ali
Family Planning Association
Nairobi, Kenya

Escande, Dr. Léopold
Member of the Institute of France
Director, École Nationale Supérieure
d'Électrotechnique, d'Électronique,
d'Informatique et
d'Hydraulique de Toulouse
Toulouse, France

Fairchild, Dr. Wilma B.
The Geographical Review
New York, N. Y.

Feld, Dr. Jacob
Past President, The New York Academy
of Sciences
Consultant, Civil Engineering
New York, N. Y.

Field, William O.
American Geographical Society
New York, N. Y.

Friedhoff, Dr. Arnold J.
Professor, School of Medicine
New York University Medical Center
New York, N. Y.

Friedrich, Prof. Carl J.
Department of Government
Harvard University
Cambridge, Mass.

Fujimoto, Dr. George l.
Albert Einstein College of Medicine
Bronx, N. Y.

Fuller, Prof. R. Buckminster
Research Professor of Generalized
Design Science Exploration
Southern Illinois University
Carbondale, Ill.

Gardner, Prof. Richard N.
Columbia University Law School
New York, N. Y.

Goitein, Prof. Hugh
Emeritus Professor of Law
Faculty of Social Science
Birmingham, England

Goldstine, Dr. Herman H.
IBM Fellow
Thomas J. Watson Research Center
Yorktown Heights, N. Y.

Hammond, Dr. E. Cuyler
Vice-President
Department of Epidemiology and Statistics
American Cancer Society
New York, N. Y.

Hedén, Dr. Carl-Göran
Research Professor of Microbiological
Engineering
Karolinska Institute
Stockholm, Sweden

Higman, Prof. Howard
Chairman, Department of Sociology
University of Colorado
Boulder, Colo.

Hirschhorn, Dr. Kurt
Division of Medical Genetics
Department of Pediatrics
Mt. Sinai School of Medicine
New York, N. Y.

Hoagland, Dr. Hudson
President Emeritus
Worcester Foundation for Experimental
Biology
Past President, American Academy of
Arts and Sciences
Shrewsbury, Mass.

8

Hollaender, Dr. Alexander
Oak Ridge Laboratory
Oak Ridge, Tenn.

Hotchkiss, Dr. Rollin D.
Professor of Genetics
The Rockefeller University
New York, N. Y.

Isard, Prof. Walter
Wharton School of Finance and Commerce
Regional Science Department
University of Pennsylvania
Philadelphia, Pa.

Jessup, Judge Philip C.
Former Member
International Court of Justice
Norfolk, Conn.

Johnston, Douglas
Professor of Law
University of Toronto
Toronto, Ont., Canada

Kegan, Lawrence R.
Executive Director
Population Crisis Committee
Washington, D. C.

King, Dr. Alexander
Organization for Economic Co-operation
and Development (OECD)
Paris, France

Klineberg, Dr. Otto
International Center for Intergroup Relations
International Social Science Council
Paris, France

Kogbetliantz, Prof. Ervand G.
Professor of Mathematics
Paris, France

Korff, Dr. Serge A.
Professor of Physics
New York University
President, American Geographical Society
New York, N. Y.

Landau, Dr. Erika
Art and Science Youth Centre
Museum Haaretz
Tel Aviv, Israel

Lasswell, Dr. Harold D.
Ford Foundation
Professor of Law and the Social Sciences
Yale University
New Haven, Conn.

Libby, Prof. Willard
Institute of Geophysics
University of California
Los Angeles, Calif.

Lowen, Dr. Walter
Dean, School of Advanced Technology
State University of New York
Binghamton, N. Y.

Lowenthal, Dr. David
American Geographical Society
New York, N. Y.

Lynch, J. Joseph, S.J.
Director, Geophysics Institute
Fordham University
Bronx, N. Y.

Maddux, Mrs. John
The World Bank
Washington, D. C.

Marois, Prof. Maurice
Président du Conseil d'Administration
de l'Institut de la Vie
Paris, France

Marschak, Prof. Robert E.
Department of Physics
University of Rochester
Rochester, N. Y.

McDougal, Prof. Myers S.
Sterling Professor of Law
Yale University
New Haven, Conn.

McHale, Dr. John
Director
Center for Integrative Studies
State University of New York
Binghamton, N. Y.

Mead, Dr. Margaret
Curator Emeritus of Ethnology
American Museum of Natural History
New York, N. Y.

Melman, Prof. Seymour
Columbia University School of Engineering
New York, N. Y.

Moore, Edward C.
Chancellor, Board of Higher Education
Boston, Mass.

Moore, Prof. John Norton
University of Virginia Law School
Charlottesville, Va.

Morain, Lloyd L.
President
American Humanist Association
San Francisco, Calif.

Moss, Dr. N. Henry
Past President, The New York Academy
of Sciences
Associate Clinical Professor of Surgery,
Temple University Health Sciences Center
Albert Einstein Medical Center
Philadelphia, Pa.

Motz, Prof. Lloyd
Department of Astronomy
Columbia University
Past President, The New York Academy
of Sciences
New York, N. Y.

Mudd, Dr. Emily H.
Professor Emeritus
Department of Psychiatry
University of Pennsylvania
Haverford, Pa.

Mudd, Dr. Stuart
Professor Emeritus
University of Pennsylvania
Vice-President, World Academy of Art
and Science
Philadelphia, Pa.

Murphy, Prof. Gardner
Department of Psychology
George Washington University
Washington, D. C.

Ng, Dr. Larry N. K.
Special Research Fellow, National Institute of Mental Health
President, World Man Fund
Bethesda, Md.

Nigrelli, Dr. Ross F.
Director
Osborn Laboratories of Marine Science
New York Aquarium
Brooklyn, N. Y.

Nolte, Richard H.
Executive Director, Institute of Current
World Affairs
The Crane-Rogers Foundation
New York, N. Y.

Nystrom, Dr. J. Warren
Director, Association of American Geographers
Washington, D. C

Owen, Dr. David
Associate Provost
State University of New York
Albany, N. Y.

Ozbekhan, Dr. Hasan
Computer Systems Division
King Resources, Inc.
Los Angeles, Calif.

Parry, Mrs. K. Crouse
Hon. Secretary
The Pierre Teilhard de Chardin Association of Great Britain and Ireland
London, England

Peccei, Dr. Aurelio
Managing Director
Italconsult S.P.A.
Rome, Italy

Perrin, Dr. Francis
High Commissioner for Atomic Energy
Director, Laboratoire de Physique
Atomique et Moleculaire
Collège de France
Paris, France

Platt, Dr. John R.
Mental Health Research Institute
University of Michigan
Ann Arbor, Mich.

Porter, Dr. Arthur
Academic Commissioner
The University of Western Ontario
London, Ont., Canada

Porter, Dr. Richard
General Electric Company
New York, N. Y.

Pregel, Dr. Boris
President, American Division of the
World Academy of Art and Science
Vice-President, American Geographical
Society
Past President, The New York Academy
of Sciences
New York, N. Y.

Price, Prof. Derek J. de Solla
History of Science Program
Yale University
New Haven, Conn.

Reisman, W. Michael
Research Associate and Lecturer
Yale University Law School
New Haven, Conn.

Reiss, Mr. and Mrs. Sidney
New York, N. Y.

Rosenblith, Prof. Walter A.
Associate Provost
Massachusetts Institute of Technology
Cambridge, Mass.

Roze, Dr. J. Arnold
Department of Biology
City College of New York
New York, N. Y.

Sager, Prof. Ruth
Department of Biology
Hunter College
New York, N. Y.

Salk, Dr. Jonas
Salk Institute for Biological Studies
San Diego, Calif.

Salkind, Victor
Engineering Consultant
New York, N. Y.

Schachter, Dr. Oscar
Director of Research
UNITAR, United Nations
New York, N. Y.

Seaborg, Dr. Glenn T.
Chairman, U. S. Atomic Energy Commission
Washington, D. C.

Seitz, Prof. Frederick
President, Rockefeller University
New York, N. Y.

Selikoff, Dr. Irving J.
Past President, The New York Academy of Sciences
Professor of Environmental Medicine
Mount Sinai School of Medicine
New York, N. Y.

Shedlovsky, Prof. Theodore
Department of Physical Chemistry
The Rockefeller University
New York, N. Y.

Simon, Dr. William
Director, Sociology and Anthropology Department
Institute for Juvenile Research
Chicago, Ill.

Taylor, Alice
American Geographical Society
New York, N. Y.

Taylor, George
Executive Secretary
AFL-CIO Standing Committee on Safety and Occupational Health
Washington, D. C.

Tinbergen, Prof. Jan
Netherlands School of Economics
The Hague
The Netherlands

Tiselius, Arne
Head
Institute of Biochemistry
Nobel Institute of the Royal Swedish Academy of Sciences
Uppsala, Sweden

Triffin, Prof. Robert
Department of Economics
Yale University
New Haven, Conn.

Tromp, Dr. Solco W.
Secretary General, International Society of Biometeorology
Director, Biometeorological Research Center
Leiden and International Institute for Interdisciplinary Cycle Research
Leiden, The Netherlands

Tsutsui, Prof. Minoru
Past President, The New York Academy of Sciences
Department of Chemistry
Texas A and M University
College of Science
College Station, Tex.

Tyler, Dr. Ralph W.
Director Emeritus
Center for Advanced Study in the Behavioral Sciences
Stanford University

von Foerster, Prof. Heinz
Department of Electrical Engineering
University of Illinois
Urbana, Ill.

Waddington, Prof. C. H.
Einstein Professor of Biology
State University of New York at Buffalo

Wakefield, Richard P.
Plans and Process Analyst
Center for Studies of Metropolitan Problems
National Institute of Mental Health
Chevy Chase, Md.

Wakefield, Rowan A.
Assistant to the Chancellor
State University of New York
Washington, D. C.

Walmsley, Ambassador Walter N.
Population Crisis Committee
Washington, D. C.

Wellesley-Wesley, James
Secretary
Mankind 2000
London, England

Weston, Burns
Law School, University of Iowa
Iowa City, Iowa

Wolf, Francis
Legal Adviser
International Labor Office
Geneva, Switzerland

Wolf, Dr. Maurice
Former Professor of Internal Medicine
Strasbourg Medical School
Assistant Attending Physician
Mt. Sinai Hospital
New York, N. Y.

Wood, Dr. Walter A.
Chairman of the Board
American Geographical Society
Arctic Institute of North America
Washington, D. C.

Wright, Christopher
Director
Institute for the Study of Science in Human Affairs
Columbia University
New York, N. Y.

Zetlin, Dr. Valentine
Medical Director
Lincoln Institute of Psychotherapy
Counselor of City College of New York
New York, N. Y.

STUDENT OBSERVERS

Martin Blaser
New York, N. Y.

Jorif Brandt
Stockholm, Sweden

Jan Fjellander
Stockholm, Sweden

FOREWORD

Boris Pregel, Harold Lasswell, John McHale

This volume is a selection of papers and discussion from the International Joint Conference on "Environment and Society in Transition" that was held on the premises of The New York Academy of Sciences, April 27–May 2, 1970, under the auspices of the American Geographical Society and the American Division of the World Academy of Art and Science.

In widest perspective, the Conference in New York expresses the cumulative concern of scientists and the public at large for the social consequences and policy implications of scientific knowledge. More narrowly, the undertaking is to be seen as a continuation of the meetings of the World Academy of Art and Science (WAAS, established 1960) in Brussels (1961), in Stockholm (1963), and in Rome (1965). By previous arrangement, it also continues the discussion at the Nobel Symposium 14 (Stockholm, 1969), "The Place of Value in a World of Facts." A number of members of the New York gathering participated in both meetings.

The planning of the New York meeting was begun before the scientific community, the American public, or the public of other countries suddenly became aware of the environmental crisis. Without exaggerating the role of the preparatory work for the New York Conference, there is evidence of the influence of the preliminary memoranda—which were privately circulated among certain international scientific bodies and official agencies—on the content of many meetings and programs.

Such immediate side effects were welcome indicators of the opportune timing of the undertaking. They did not, however, divert the Conference planners from long-range or middle-range objectives. The planners were aware of the perils that usually attend bursts of general enthusiasm on behalf of belated programs of public policy. Insufficiently aware of the complex and delicate nature of ecological balance, programs of superficial plausibility may eventually leave us in a worse state than we are at present.

If the profound policy implications of scientific knowledge for "Environment and Society" are to be grasped, more than an ephemeral and topical approach is essential. The plan therefore was to bring together a Conference whose members would be representative of the best that the contemporary world of science has to offer. There is no question that, as far as several relevant disciplines are concerned, this objective was achieved.

A word about the agenda of the Conference. It began with an authoritative review of the present state of knowledge in the principal fields of knowledge. Each contributor was asked to consider the most significant developments that have occurred in his field of competence as they relate to society and environment in these transitional times. Authors were invited to offer informed speculations about the future evolution of their specialties and, if they chose, to evaluate the social consequences and policy implications.

After this interdisciplinary exposure the members of the Conference divided into multidisciplinary working groups on the policy implications of knowledge for selected sectors of society and environment. Each sector involved

12

a horizontal cut through all fields, since every field must be considered when social consequences and policy implications are at issue. The problems that arise in thinking about "cultivating resources" go beyond the physical setting to the complex interplay of changes in the physical environment with changes in biological and cultural configurations. "Population, health, and family" strategies must go beyond biology and culture to the assessment of meteorological and other features of the resource environment. A similar observation applies to policies designed to affect world "scientific knowledge, education, and communication," "economic and social strategies," and "decision processes." The last was of particular relevance to the aims of the Conference because a major aim was to go beyond the laboratory or the field expedition and to make an impact of policy-forming and-executing processes. In the preparation of its provisional report, each "working group" had the assistance of a legal scholar accustomed to summarizing proposals and justifications for the use of decision makers.

The concluding note to the present volume will indicate how the work of the Conference is to be continued in the immediate and more remote future.

* . * *

It is quite appropriate that we begin by considering the developments of science today. The title of this meeting, "Environment and Society in Transition," can best be understood if we know first where we are. Each of us in his own discipline does indeed "know" to some extent. Perhaps we have a great deal of knowledge; perhaps we know very much and understand very little. I hope we may concern ourselves with understanding more, and understanding more in terms of all the vast amount of knowledge we have accumulated. Here I would like to go back a few thousand years, to 92 B.C., when Lucretius, the popularizer of the theory of the atom, wrote a remarkable poem, "No Single Thing Abides," in which he points out things that are extremely modern. He *understood* things, but, of course, he did not *know* very much; thus he remarks that the systems in the universe are constantly changing: "I see the suns, I see the systems and their forms and even these, the systems and suns, go back to their eternal drift."

Then he makes a notable remark; he says that every atom in existence was, at one time, part of another star, that we could not be where we are or how we are unless we had passed through something else: "This bowl of milk, the pitch on yonder jar, are strange and far bound travelers coming from afar." And finally Lucretius speaks about science, "this voice that stills the pulls of fear and through the conscience thrills."

This is indeed the situation today: if science cannot show us the direction to peace and a better world, all our knowledge, all our understanding will be useless.

Lloyd L. Motz

RECENT ADVANCES IN ASTRONOMY AND SPACE SCIENCES

A. G. W. Cameron

Belfer Graduate School of Science, Yeshiva University
New York, N.Y.

and

NASA Goddard Institute for Space Studies
New York, N.Y.

Introduction

The astronomical sciences have had rather little impact on our day-to-day lives, but from ancient times their philosophical impact has been very great. Primitive man could see that there were many unreachable things in the skies about which he knew nothing; undoubtedly the appearance of the heavens had much to do with the formulation of primitive religions. In recent centuries, as we have learned more and more about the universe and our place in it, man has seemed to occupy an ever-decreasing position of importance in his total universal environment. The skies no longer turn solely about the earth; the other stars no longer turn solely about our sun. The sun has become a normal undistinguished star situated in our galaxy, and our galaxy is but one of billions of similar galaxies strewn throughout space, with no indication that our own position in space has any special significance.

The pace of discovery in astronomy and space science has been particularly hectic during the last decade, and this has led to an intense intellectual ferment in the field. In this article, I will skim quickly over a few of the highlights of the research discoveries that have been made, and at the end I will try to indicate what our present philosophical outlook is toward the universe in general.

The Large-Scale Universe

During the first half of this century, some of the most important developments in astronomy established that spiral nebulae are external galaxies at large distances from our own and demonstrated that there is a correlation between the spectral redshifts of the galaxies and their apparent distance from us. These developments led to the concept of the expanding universe. Modern cosmology has developed from both this observational picture and the underlying theoretical developments of general relativity. For a long time, cosmology has been, basically, a science in search of facts. In the last decade, a number of new facts have been established that have an important bearing on cosmology.

Perhaps the most important of these is the discovery of an isotropic background radiation at microwave frequencies. This radiation appears to be incident on the earth uniformly from all directions, and as far as we have yet been able to determine, it has a blackbody spectrum characterized by a temperature of 2.7° K. The most commonly accepted interpretation of this radiation is that it is leftover light, emitted during the early expansion stage of the universe, when all matter was tightly compressed and the universe was a very hot place, opaque to electromagnetic radiation. Dicke has graphically described this radiation as the remnant of the flash from the primordial fireball. If this interpretation is correct—and much confirmatory experimentation still must be done—then the background radiation is the most primitive observation that can be made in cosmology; it is the earliest light of any kind that it would be possible to detect. We are, in fact, seeing the early universe in its own emitted light, which has been redshifted to the extreme radio wavelengths.

Less primordial, probably less fundamental, but certainly stranger, are the quasars. These are brilliant points of light frequently shining with an intensity 100 times as great as that of the brightest galaxies. Yet this tremendous energy output is generated and emitted from a region less than a light year in diameter. We do not know if quasars are extreme forms of galaxies, although this may be so, but we suspect that they are beacons shining to us from enormous distances in space and enormous intervals back in time. The spectral lines emitted from the quasars have undergone a redshift of as much as three times the original wavelength, and sometimes a little more. Most astronomers interpret these very large wavelength shifts as indications that the emitting objects lie well beyond the faintest galaxy whose redshift has been measured. For about a decade, theoretical astronomers have been completely puzzled about the detailed structure and operating processes of the quasars. Now a new observational statistical puzzle is emerging: there are many quasars up to a maximum redshift, and, suddenly, beyond that redshift, no more are observed. Will this tell us something important about the ancient history of the universe?

Our Galaxy

Perhaps the most exciting discovery that has been made in recent years about the contents of our own galaxy is the pulsars. These clocks, ticking in the sky, shine bursts of radio waves at us every second or so. About a year after their discovery was announced, the tremendous number of observations made by radio astronomers finally seem to provide strong evidence that the pulsars represent the discovery of an entirely new type of star, with an entirely new type of matter, similar to the nuclei of our atoms. This so-called neutron star is an assemblage consisting mostly of neutrons, but it may also have minor components of protons, electrons, muons, and hyperons, all gathered together within a radius of perhaps 13 kilometers and having a mass similar to that of our own sun. Such objects had been predicted by the theoretical physicists during the 1930s, and a few people had continued to study their theoretical properties since that time, but little attention was paid to such theoretical predictions until the discovery of the pulsars. Perhaps there was good historical reason for this: the neutron star was really the first type of major astronomical object to be predicted by theory before it was discovered.

About ten years ago, the new science of x-ray astronomy was born when a rocket ascended above the atmosphere to look for x-rays from the moon and instead found them emanating from a new type of astronomical object, the Scorpius x-ray source. X-ray astronomy can be accomplished only from above the atmosphere, because the atmosphere strongly absorbs x-rays; hence, this was one of the earliest important discoveries resulting from space research. Since then, the x-ray sky has been observed repeatedly by rocket flights, and about 50 x-ray sources have been located. Some of these are very variable in time and rise to a maximum intensity and then fade below the level of detectability. There are two basic types of x-ray sources: a source that has rather high-energy x-rays, typified by the Crab Nebula, which may represent supernova remnants; the other type of source is a blue pointlike object—examples of which have been identified with several of the x-ray sources—whose nature is still unknown. The astronomer has come to expect that when a new region of the electromagnetic spectrum is opened to observation for the first time, completely new types of astronomical objects will be discovered. X-ray astronomy has not let him down in this regard.

Much of infrared radiation is screened from the surface of the earth by absorption in the atmosphere; this absorption is mostly due to water-vapor molecules, but there are a few wavelength windows through which the infrared sky can be observed. It can also be observed from the high atmosphere with balloons or aircraft above the bulk of the water vapor.

Infrared astronomy was a little slow in beginning because of the lack of efficient infrared detectors. In recent years, many important advances have improved infrared detector efficiencies, and, as a result, the science of infrared astronomy has begun.

Here, too, there have been important new discoveries. Certain stars, very faint in visible light, are enormously brilliant in the infrared, probably because surrounding clouds of dust particles largely thermalize the radiation emitted from the star. There has also been at least one distinctly new type of infrared object observed: the infrared nebula. Such an object exists in the vicinity of the Orion Nebula; it has a surface temperature of only about $100°$ K, yet it shines at a rate of about a million times the luminosity of the sun in the far infrared wavelengths. Some astronomers think this may be a location in space where new stars are actively forming.

Some of the most exciting discoveries about our galaxy have been made by the radio astronomers. For years they were rather unhappy that they had only one spectral line to observe, the 21-cm line emitted by neutral hydrogen atoms. Studies with this line were very important for astronomy: they determined the location of large clouds of hydrogen in interstellar space, as well as relative velocities. Now, suddenly, in the last few years, the radio astronomer is deluged with a large number of spectral lines, which started with the discovery of four lines due to OH, the hydroxyl molecule, in interstellar space. The hydroxyl molecule does not behave in a simple fashion; it enhances any or all of its spectral lines in an apparently capricious manner. This enhancement has been attributed to a masering action, stimulated emission resulting from the overpopulation of an excited state of the molecule. One of the truly astonishing accomplishments of the radio astronomers has been the demonstration, with a world-wide radio interferometer, that some of the regions emitting OH lines have dimensions as small as that of the solar system, even though they may exist thousands of light years away from us.

More recently, radio lines due to ammonia (NH_3), water (H_2O), and formaldehyde (CH_2O), have been detected. The ammonia does not appear to have masering action, but the water vapor does. The formaldehyde seems to be a kind of antimaser: it has an overpopulation of the ground state, and hence, an anomalous absorption of radio radiation at its characteristic wavelength. These observations indicate that the interstellar gases have a strange chemistry of their own that we are only beginning to discover.

Our Solar System

Physicists, during the last decade, have been busy with spacecraft measurements determining the presence of energetic particles and magnetic fields in space. They have determined that our planet is surrounded by a shell of energetic particles trapped in the outer reaches of the magnetic field. Beyond this magnetosphere, interplanetary space is filled with an outrushing plasma from the sun, an expansion of the solar corona, which carries its own imbedded magnetic field. Sometimes interplanetary space is also filled with large numbers of energetic cosmic rays accelerated by solar flares. Thus, an apparently uninteresting part of the solar system has turned out to be full of energetic and dynamic phenomena.

Automatic spacecraft have made a preliminary exploration of the nearby planets. Venus has been found to have an extremely thick atmosphere composed predominantly of carbon dioxide and to have a pressure at the base of about 100 atmospheres. The surface of the planet is enormously hot, about $700°K$, and is thermally well shielded from interplanetary space. The atmosphere contains almost no water. Although Venus is a planet with nearly the same mass as the earth and nearly the same radius, its atmosphere is extremely different from that of our own planet. Nevertheless, it seems that the planets are chemically very similar, and many atmospheric physicists believe that the earth would become as Venus is now if it were moved closer to the sun.

Mars also has an atmosphere predominantly composed of carbon dioxide, but this atmosphere has a pressure at the ground of less than one per cent of our own. Indeed, the atmosphere of Mars is in equilibrium with its polar caps, which are also composed of carbon dioxide. The pressure of the carbon dioxide in the Martian atmosphere adjusts itself to equality with the vapor pressure of carbon dioxide over the pole that happens to be experiencing winter at a particular time. There are mere traces of water in the Martian atmosphere, and there is almost zero chance that liquid water can exist on the surface of the planet. Hence it is very unlikely that Mars will prove to be an abode of living beings. Spacecraft pictures of the Martian surface show that it resembles the surface of the moon in many respects, having large numbers of craters. However, the atmosphere makes a difference, and evidently dust on the Martian surface is blown around sufficiently to obliterate the smaller craters in a short period of time.

Probably the most striking results of any of the space programs have been those resulting from the manned exploration of the moon. The Apollo 11 and 12 astronauts brought back many kilograms of lunar rocks and dust that have been subjected to intensive investigations by a wide variety of laboratory techniques. The findings indicate that the moon is largely a chemically differentiated body and that it appears to be deficient in iron as well as in other elements that have relatively low vaporization temperatures for common compounds. In the interior of the moon, where the rocks now on the surface at the Apollo 11 landing site were formed, there was a dramatic absence of water. This resulted in the formation of minerals under very dry and higher temperature conditions that would occur on the earth, with highly significant differences in the mineral content of the rocks.

These tantalizing first glimpses into the bulk chemistry of another planet will provide extremely important clues to the origin of that body and of the solar system as a whole. The moon is markedly unlike the meteorites, the other sources of extraterrestrial material that have been examined in our laboratories. The moon appears to have formed under high-temperature conditions, with a strong absence of the more volatile elements, whereas the meteorites seem to have formed under much cooler conditions, with retention of most of the more volatile elements. Additional landings on the moon will help to extend this information and to render much more certain our preliminary conclusions about that body.

Man's Place in the Universe

These have been a few prominent examples of the many dramatic advances in astronomy and the space sciences that have occurred during the last decade. But these advances are merely the highlights of a tremendous amount of investigation that has been carried out, of both an observational and a theoretical character. The usual pace of research produces a succession of small discoveries, but the intellectual impact of these mounts as the total number of small discoveries grows. Therefore, in this last section, I shall outline some of the principal philosophical points of view that have emerged about man's place in the universe.

The dominant concept resulting from modern research into star formation and the origin of the solar system is that we must expect planetary systems to be extremely common in space. Many stars, perhaps 80 per cent of them, occur in binary pairs. Some investigators view this as merely another manifestation of the tendency for matter to orbit around stars in space; in this case, the stars orbit around each other. However, theories of the origin of the solar system now tend to place the origin of the planets in the general process by which the sun itself formed. Hence, there may be between 10^{10} and 10^{11} planetary systems in our galaxy.

Biochemists studying problems associated with the origin of life tend to conclude that chemical evolution on the earth followed the chemically-most-likely path in every discernable instance. Thus, if a reaction takes place involving complex organic molecules, which may yield many different products, the most common products are the ones that appear to be incorporated in living beings. Therefore, the chemical and biological evolution that has occurred on the surface of the earth appears to be a natural consequence of a favorable environment. It is generally expected that similar chemical and biological evolution will develop anyplace in the universe where conditions are similarly favorable. This means that we should expect the existence of an immense number of planets in our galaxy on which life has originated and evolved.

Will these theoretically prevalent biological systems generate intelligent beings? Obviously, science is not in nearly as good a position to answer this question as it is to answer questions about the likelihood and prevalence of planets favorable to the origin of life. Nevertheless, the biologists tell us that environmental stresses lead toward evolution, that the old doctrine of the survival of the fittest plays a significant role, and that, because intelligence may allow one species to dominate over all others, it is probably only a matter of time before a biological system with ample natural resources will develop intelligent species.

A further question is whether an intelligent species will develop a technology. This may depend on the evolution of a suitable manipulative organ in the intelligent species. On the earth, it appears that the possession of a readily manipulative organ, the hand, played a significant role in the development of intelligence in human beings, and this may also be a general phenomenon. It is a logical consequence of the entire trend in modern interdisciplinary thinking about the prevalence of life and intelligence in the remainder of the galaxy that we should expect enormous numbers of intelligent civilizations to have developed in the vicinity of other stars.

The basic technical requirements for interstellar signaling already exist. The large radio antenna at Arecibo, Puerto Rico, is capable of sending signals to a similar antenna located many hundreds of light years away. It is quite within our technical capabilities to construct even more sensitive antennas capable of communicating across still vaster distances.

Nobody has yet done much listening for extraterrestrial civilizations. But someday, logic tells us, we should find such signals coming from interstellar space. When this happens, astronomy will not be just a science of the immense void that helps shape our philosophy and generate intellectual ferment, but will also help generate some of the most important social consequences that the human race is ever likely to encounter.

A PHYSICIST LOOKS AT GEOGRAPHICAL AND ENVIRONMENTAL PROBLEMS

Serge A. Korff

New York University
University Heights, New York, N.Y. 10453

Introduction

In the last few years, and most especially, very recently, a trend has emerged that seems most desirable from many points of view. This is an increased concern with the quality of our environment. Recently, undergraduates, senators and congressmen, and a wide variety of other persons have all evinced a growing interest and awareness. That this concern is justified admits of no debate. Quite clearly, mankind has managed to spoil so many and such conspicuously large portions of his environment that the problem has reached a crisis stage in many fields. There are, however, some cautionary points to be considered; and I hope that something may be generated that will be a net benefit and will not, because of ill-considered, hasty action, create more problems than it solves.

Trends Already Discernible in the Movement

In the United States we have a great fondness for simple solutions for complicated problems. We love to say, "All you have to do is . . .," and imply that thereby the solution is at hand and the troubles are over. There are all too many illustrations of this characteristic approach. For example, although excessive alcoholism clearly is an evil, the simple solution of making it unlawful drew our country into a situation from which we may not have recovered yet. In the days when we imagined we had only one potential enemy, we promulgated a foreign policy of deterrence based on "massive retaliation." This policy has also led to consequences that plague us long after the policy itself has been abandoned. In the environmental sciences, it is often easy to say, "Oh, we must not do so-and-so," thereby avoiding one set of problems but firmly pushing ourselves into another set.

Another characteristic, which I am afraid will soon manifest itself in the environmental field, is our mercurial tendency to lose interest in an effort, especially after an initial victory has been won. We do not like to stick to a problem, especially when the going gets difficult. Again, all too many illustrations come to mind of the effects of this national characteristic. We are soon going to run into cases in the environmental field where this effect may plague us.

Still another facet of our present operational procedure is that the entire conservation movement has built-in overtones of being against everything. Time and again in the last few years, I have received letters from conservation groups asking me if the organization I represent, or sometimes if I, personally, will not join in an effort to halt something. Perhaps, indeed, the something needs halting. Yet it is astonishingly seldom that any appeal comes to get something positive done; it is always to oppose what someone else wants to do. This posture of opposition to every proposal is not one that bodes well for our future development or for our growth possibilities as a nation or as a world.

It seems to me that we must be positive rather than negative with respect to our manifold problems. The problem is to consider the best way to accomplish something, to consider what the long-term problems are that a given procedure may generate, and then to choose the route that will generate the fewest problems. We will not find solutions that generate no effects. What we must decide is what price is acceptable.

The Basic Choices of Mutual Opposites

A basic reason for our many really difficult problems is the fact that there are two, and often more, mutually exclusive and opposite desiderata. We have thus far been unwilling to take seriously the measures that will limit a rapidly expanding population. We all know what these measures are and we know that technically the job can be done. But politically we have been unwilling to do it. This rapidly expanding population will require increased production of food, of goods, and of services. It will require much more electrical energy. We will have to build many new power plants. The population will want to ride around in cars and will need roads to ride on, while the cars pour out noxious gases. We cannot have it both ways.

If we wish to conserve our environment, it can be done. If we want to limit our population, this too can be done. But the hard fact is: we cannot, at the same time, refuse to limit our population growth and also stop building new automobiles or power plants or extracting more oil or cutting down more forests to build houses for this growing population to use.

It is regrettable that all this is so. A given policy has certain consequences, and they are quite predictable and inescapable. As a nation or as a world, we cannot institute the policy and refuse to accept the consequences. If we have more people, we will do more damage to nature. If Masai tribesmen overrun the African game reserves with their expanding cattle herds, we shall have fewer big animals. If we feel we must feed the world's starving millions, we must put more land into agriculture and use more fertilizer.

Some Specifics

Another one of our great problems arises out of our desire to have certain benefits without paying for them. An excellent illustration of this is the situation in power generation. It is quite clear that we shall have to build new power stations, not only to provide power for the expanding population but also just to take care of the increased demand imposed by the existing groups, which all demand more each year. We want these power stations to be inconspicuous and not to pollute the environment. And yet, what do we do? Some years ago a nuclear plant was suggested for the New York area, but political pressure forced its abandonment. To supply the needed power, the utility companies have had to expand their older steam stations, which are notoriously polluting. Let us take a good look at the problems of power generation.

A steam plant, burning fossil fuels, generates ash and carbon dioxide, the products of combustion. The carbon dioxide is invisible, but who can maintain that the literally millions of tons we are spewing forth do not produce undesirable effects? The effect on the atmosphere's heat retentivity is too well known to need further review. The second type of pollution generated by steam plants is thermal. The condensers must be cooled. This process requires a lot of water, which, in turn, becomes heated and warms the nearby streams when it is discharged. Although a nuclear plant does not produce carbon dioxide, it also generates two problems: one is the disposal of radioactive wastes and the other is, again, thermal pollution.

Much of the pollution can be abated or minimized if we are willing to pay for it. The objection "That would be uneconomic" is the key issue. If we want to keep smoke under control we can do it. We can install smoke precipitators; they cost money. We can recirculate condenser-cooling water; such recirculation also costs more than does the alternative of taking nice, clear, cold water from our reservoirs and discharging it after it is heated. It is possible by chemical treatment to reduce the bulk of nuclear wastes, to concentrate them, and to prepare them for storage for a time long enough for the induced radioactivity to decay. But this costs money. As long as our public service commissions, which fix rates, are busy thinking only of what must be done to keep the rates down, they will not permit the generating companies to charge what is needed to do a really effective job on pollution abatement. We must realize that we have the technical skills *now* to generate vast amounts of power with minimal accompanying pollution, but we have programmed our regulatory agencies not to permit this. If we are sincere about not wanting pollution, we must pay the costs.

Many other illustrations come to mind. Automobiles can be built that will generate less of the fumes that give rise to smog. But the additional devices cost money and many persons are unwilling to face these added costs. Changes in the engines of transport aircraft can greatly reduce the smoke plumes. But in the long run the cost of such changes will show up in higher fares, which many persons and regulatory commissions resist. It is quite clear to anyone who has been stacked up over New York City in an airplane that New York badly needs another major airport. But the chief objection to any proposed location has come from the conservation groups. For some years now, any new airport proposal has been blocked. This hardly seems a good long-range solution. Access to all New York metropolitan airports is by automobile only. Certainly we could prevent the emission of vast amounts of exhaust fumes by constructing subway extensions to the airports from our existing subway system.

Conclusions

We will continue to have problems as long as we continue to insist on following mutually antagonistic policies. Our problem is not a technical one. We know how to do most of what we need to do without generating unacceptable amounts of pollution. But we are quite unwilling to pay the necessary costs, be they in money or in political policies. Science and the technology stand ready to do what needs to be done. If we are sincere in our protestations concerning our environment, we must not just say "no" to every new proposal.

We must, on the other hand, consider how the techniques can be implemented, and if this requires paying more in order to pollute less, we must be willing to pay. If we are unwilling to limit the population explosion, we must face the fact that there will be more people damaging our environment, and more industries to serve those people tomorrow. If we insist on electric power and industrial output as cheap as possible, we will find that these same industries produce more wastes. What we need to do is to apply our knowledge, with the desired end in view, and with the further stricture that if the end we desire is not the most economic one, we are willing to pay for the differences. With such an approach we can still make progress. At the moment I do not see any acceptable alternative.

* * *

Motz: I should like to address myself to another aspect of the problem of saturation. It appears that the problem facing us is, essentially, saturation of population, of contaminants in the atmosphere; I should like to propose that possibly we are also reaching an intellectual saturation, that is, the striving of people toward new intellectual ideas seems to be reaching a kind of leveling off, and it is difficult to see where new things are going to come from. And I particularly should like to consider the problem of physics.

If we look at the laws of physics, we discover that the modern period is governed by two great theories, the theory of relativity and the quantum theory, and these have proved adequate to explain many things. As an example, I might indicate the development of stars. Today we know pretty well how stars evolve, how they develop, and this, in a sense, has given us a solution to the important problems that were presented to scientists, astronomers particularly, at the beginning of this century. We know now that a star is a natural concomitant of the evolution of the universe itself. It is part of an overall cosmological problem, and if we apply the laws of relativity and quantum mechanics, we can understand quite well what goes on.

We know today that the atoms all around us, the atoms in our solar system, evolved or were formed through the evolution of stars, through the evolution of certain types of stars—the older stars, which then, in some manner or other, expelled vast quantities of material, and from this material solar systems like ours were formed.

We can also investigate various other problems, such as those presented by Dr. Cameron, and in each case we apply the two theories that we understand, the theory of relativity and the quantum mechanics.

Is it possible that we have, in a sense, come to the end of the formulation of new, great theories? I know this is a shocking idea; many people in scientific research will argue that if such is the case, it is hardly worth living.

But let us examine the situation. These two theories, the theory of relativity and the quantum theory, which in a sense introduced two very basic constants, the speed of light and the quantum of action, were developed at about the same time: in 1900 the quantum theory was introduced, and in 1905 the theory of relativity.

Then there was a period of tremendous activity with the discovery of what we call the quantum mechanics, that part in nature of wave characteristics. In the period from 1900 to 1925, during which all of this developed—a kind of golden age of physics—we find two basic principles were discovered, the basic principle that is the theory of relativity and the basic principle that led to the quantum mechanics, the notion of a quantum of action. From 1925 to the present, some 45 years, with the greatest concentration of scientific minds working in this area, there has been no basic discovery, no new discovery or development of a basic theory. Even such developments as the laser and the discovery of antiparticles, the positron, all stem from the basic theories. We would not dream of trying to understand these new technological discoveries in terms of theories other than the theory of relativity and quantum mechanics. The laser, for example, which is a representation of one of the most advanced technological developments, was discussed in a paper by Einstein in 1917. It is not a new basic principle. The pulsars, which are important discoveries in astronomy and are, indeed, newly observed, must be understood in terms of relativity and quantum mechanics.

Consider the discovery of antiparticles: the positron was discovered in 1934 by Carl Anderson, but the positron as antielectron was predicted in 1928 by Dirac on the theory of relativity and quantum mechanics. This was a purely mathematical extension of the theory of relativity.

Because we now find that the rate of discovery of basic principles has been practically nil since 1925, we must ask ourselves where is science to go? In what direction is science to proceed? It is true we shall discover many, many more facts, we shall discover many new technologies, but will we have new theories to work on? Will the excitement of science be the same in the future as it was from 1900 to 1925, as least in the physical sciences? When it is argued that there is a great deal to discover, I must answer that indeed there is: facts. Let us consider, for example, the field of particle physics, which is a kind of force that is not at all understood. Vast numbers of facts. But again, they must be treated within the realm of the two basic theories.

It is true that new mathematical techniques are introduced, new approaches and new skills are developed, but there is no departure from the basic theories. Therefore, the question I wish to propose, and that I hope may stimulate some discussion is this: Are there new basic theories to be discovered or are we simply looking for new forces? It is true that whenever we find a new force, when we go from a study of the structure of the atom itself—the electron circling the nucleus—to the nucleus, we run into new forces. We speak of nuclear forces, but we study the nucleus again in terms of quantum mechanics and relativity theory.

This is a topic worth considering because we find more and more young people turning away from science. Here I am limiting myself to the physical sciences; I think there is still excitement in the biological and life sciences, but in the physical sciences these excitements have perhaps disappeared or, at least, do not present the same kind of challenge as they did to men like Einstein.

There appears to be a kind of saturation, and if this is indeed the case, then perhaps our approach to the education of youngsters—what we tell them about the intellectual rewards science has to offer—ought to change. We ought not to hold out promises of the same kind that were promised the earlier explorer in this field, and ought perhaps to indicate that science today, at least in the physical sciences, is a matter of very hard, tough work with no great brilliant ideas that will burst upon a person all at once.

If we are going to talk about environment and the relationship of science to environment, I think we ought also to talk about the intellectual environment and consider whether science can offer, or scientific training and learning can offer, the same excitement to young people today as it did in the past.

THE BIG VIEW

Richard W. Porter

General Electric Company
New York, N.Y.

Astronomy is the most ancient of the sciences. Beginning, perhaps, with a sense of awe and wonder at the regular movement of the life-giving sun, the beautiful moon and the stars across the heavens, and the terror induced by an occasional eclipse, it progressed in the minds of the more curious and intelligent to a study of these motions and eventually an ability to predict them with some degree of accuracy. Because of the significance of these celestial objects in early religions, this primitive scientific knowledge was carefully accumulated and passed on at first by word of mouth and then in written records within the priestly hierarchy. There was much nonsense, of course, from our 20th century viewpoint, but the essence of science was there: observation, accumulation, and categorization of information; attempt at rational explanation, prediction, and application.

It would appear that the beginnings of astronomy were probably in the cradle of civilization, Mesopotamia, begun perhaps by the Sumerians and continued even more strongly by the Babylonians, spreading gradually into China, India, and Egypt. The noted astronomer Fred Hoyle [a] suggests that in Mesopotamia the work was largely numerical and empirical, perhaps a kind of code-cracking exercise. Meanwhile, a parallel development occurred in early Greece, where the subject was of more philosophical than religious value, and it seems that the Greeks were the first to try to think of astronomical problems in terms of geometrical models. That these early scientific studies had practical significance can hardly be doubted. Architects laid out temples and tombs precisely in accordance with celestial references; navigators steered their primitive ships over the open seas using only the sun, moon, and stars to guide them; farmers timed the seasonal planting and harvesting of crops by observed positions of the heavenly bodies.

Various technological developments, in particular the invention of accurate clocks and chronometers, the telescope, the photographic plate, the spectrograph, the sensitive photometer and photoamplifier, have greatly extended and refined the kind of observations that can be made; the burgeoning field of radio astronomy has added a second spectral "window" through which to look at the universe; and space technology has eliminated completely the observational limitations imposed by the earth's atmosphere and made possible the new fields of ultraviolet and x-ray astronomy. Furthermore, the development of even more sophisticated mathematical systems, the availability of powerful computational aids, and the many significant inputs from physical and chemical laboratory experiments have contributed a basis for understanding the observational results of modern astronomy and extending the underlying and unifying theoretical structure that makes astronomy a science.

As in all branches of science, the pace has been accelerating rapidly in recent years. It seems as if each scientific question answered gives rise to

[a] Fred Hoyle, *Astronomy*, MIT Press, 1964.

25

several new and generally more difficult questions, and each new advance in observational capability opens whole new areas of scientific controversy. Not surprisingly, there has been considerable splintering or specialization within the overall science of astronomy. The International Astronomical Union now has a total of 38 different commissions. However, it might be possible to speak informally about three divisions by type of object to be studied, such as solar astronomy, lunar and planetary astronomy, and stellar and galactic astronomy; and of four divisions according to technique, for instance, optical and infrared astronomy, radio and radar astronomy, high-energy (ultraviolet, x-ray, and energetic particles) astronomy, and celestial mechanics.

Solar astronomy is of direct importance in studies of the terrestrial environment, because the sun supplies all the energy that makes life possible on earth and seems to influence many commonplace phenomena in ways that are not yet well understood. Each of the four technique areas is concerned with and contributes to solar astronomy. Because it is the nearest star, the sun is also of profound interest to stellar astronomers, and as the progenitor of the solar system, it is important in the study of the moon and planets.

The brightness of sunlight, as we see it here on earth with eyes that are sensitive only to wavelengths from about 0.3 to about 0.8 micrometers, is relatively constant, with a small seasonal variation caused by the ellipticity of the earth's orbit and an irregular variation that seldom exceeds ±2 per cent. However, in the ultraviolet and x-ray regions of the spectrum, the radiation we receive from the sun is highly variable. During an intense solar flare, the brightness of the sun increases by as much as 100 per cent in the ultraviolet and by an order of magnitude in the "hard" x-ray region. The flare also produces several distinctive kinds of radio noise bursts and, sometimes, high-energy protons (solar cosmic rays) in sufficient quantity to be a potential radiation hazard to space travelers. There is always some evaporation and escape of less energetic electrons and protons from the very hot (about 1,000,000,000° K) solar corona and these charged particles flow more or less radially outward from the sun, constituting what is generally called the solar wind. When a flare occurs, strong pulses of very hot plasma are ejected through the corona and travel outward through the solar system at a higher speed than the normal solar wind velocity, creating magnetohydrodynamic shock waves of vast proportions. It is the interaction of these shock waves and the plasma concentrations behind them with the earth's magnetosphere that is primarily responsible for the phenomenon known as a magnetic storm.

It would be highly desirable, for many reasons, to be able to forecast these solar-flare events reliably, and therefore much attention is being given to developing the kind of scientific understanding that would make such forecasting possible. It is now well known that the flare is associated with a complex magnetic field configuration surrounding groups of sunspots on the surface of the sun. One major accomplishment of modern solar astronomy has been the development of a capability to measure magnetic fields on the surface of the sun with good resolution both in geometry and in time from telescopes on earth.

Our knowledge of the high-energy electromagnetic radiation and particles emitted by the sun is largely dependent on measurements from artificial earth satellites and deep spacecraft. The Orbiting Solar Observatory satellites, of which six have been successfully launched by NASA, must certainly be considered as a major new observational capability for solar astronomy. An unmanned spacecraft called HELIOS, to be built in Germany and launched by the United States during the mid-'70s, will carry 10 or 11 rather sophisticated experiments to within 0.3 AU (approximately 30,000,000,000 miles) of the sun. The improved detail with which specific areas of the sun can be observed and photographed above the turbulence of the earth's atmosphere and the importance of making observations at wavelengths that do not penetrate the atmosphere have led to the design of a solar telescope to be carried and operated on the first manned orbital workshop of the Apollo Applications Program.

Lunar and planetary astronomy had become an almost deserted field until rather recently. Improved optical telescopes were, of course, being designed, like the 200-inch Hale telescope at our Mt. Palomar, but these big new telescopes were important primarily for their increased light-gathering capability and their consequent ability to "see" farther out into the remote regions of the universe. The ability to make clear, high-resolution photographic images of objects like the moon and planets from the surface of the earth is limited by atmospheric turbulence to such an extent that telescopes of larger diameter than 30 to 60 inches are not useful. In fact, some of the best lunar photographs prior to the space age were made with the 36-inch refractor at the Lick Observatory. Consequently, when most of the details that could be seen through a terrestrial telescope had been mapped for the moon and Mars and the orbits of these and other visible objects in the solar system had been determined with reasonable precision, many astronomers lost interest and moved on to other fields.

It is perhaps interesting to note, as an aside, that during the first half of the 18th century, astronomers were eagerly seeking ways to solve the longitude-determination problem of celestial navigation, which is essentially one of an accurate determination of universal time on a ship at sea. Almost a hundred years earlier, Galileo had thought about using the moons of Jupiter for this purpose, but the idea had proved to be unfeasible. Newton had a simpler idea: the observation of our own moon against the background of the stars. However, it was not until Tobias Mayer, with the aid of Euler's new mathematical techniques, carried out the calculations necessary to predict the apparent motion of the moon, hour by hour, and the first British Astronomer Royal, John Flamsteed, had determined the positions of nearly 3,000 stars to within an accuracy of 10 seconds of arc, that this idea became practical. Oddly enough, it was just about this time that the inventor John Harrison produced a chronometer not regulated by a pendulum, which proved itself capable of keeping accurate time over long periods at sea. It turned out to be at least as accurate for most purposes as navigation by the moon and was considerably easier to use.

In recent years, with the advent of infrared, radio, and radar astronomical techniques and the urgent need to know as much as possible about the moon and planets in preparation for space missions, there has been a reawakening of interest in these bodies by astronomers. It is, indeed, remarkable how much can be learned about a planet by looking at it from earth in all available regions of the spectrum. Radio, radar, and infrared mapping of the moon have provided the gross thermal and electrical characteristics of the lunar surface to a depth of at least several meters, and even though these data can now be obtained at a few points by spacecraft that land on the moon, the astronomical data continue to be a valuable source of information about large-scale inhomogeneities in subsurface properties. Laser retroflective measurements from earth using the optical corner reflectors placed on the moon by Apollo missions are increasing by several orders of magnitude the accuracy with which one can measure the moon's orbit and motion around its own center of gravity. A precision of ± 30 centimeters has already been achieved, and improvement to better than ± 15 centimeters seems probable. Measurements of this kind are useful in trying to unlock some of the secrets of gravity, and may also be valuable in obtaining a better understanding of the way in which the continents seem to be drifting over the face of the earth.

Modern astronomical observations of Venus have shown that this planet is, to say the least, peculiar. Almost a twin of the earth in diameter and mass, it is perpetually shrouded in clouds and its surface is extremely hot, on the order of $600°$ K ($620°$ F). It has a very dense atmosphere (about 60 times more dense than that of the earth), which consists almost entirely of carbon dioxide. It is now known by radar measurements that Venus rotates very slowly in the "wrong" direction, compared with earth, Mars, and the other planets, and that its rotation is synchronized with its own and the earth's orbital periodicity around the sun, so that it always shows the same face toward earth when the two planets are close together. It is also known that some areas on Venus are relatively more reflective than others at radio frequencies, and by a method known as the Doppler-delay technique it has been possible to construct a crude map of the radar reflectivity of the "near side" of the planet. Previous astronomical information has generally been confirmed and extended by the successful U.S.A. and Soviet spacecraft missions to Venus.

Mars also has been extensively studied by astronomers and, more recently, at close range by unmanned scientific spacecraft of the U.S.A. Its atmosphere is surprisingly thin, just as that of Venus is surprisingly thick, and it has no significant cloud cover. Furthermore, the side facing the earth at nearest approach is fully sunlit; because the planet rotates at about the same speed as the earth, its entire surface is readily available to all the techniques of optical, as well as radio and radar, astronomy. Much has been learned by these techniques, including a great deal about atmospheric composition, tempperature, wind velocities, and surface material. However, the image resolution of the best terrestrial optical telescopes is not quite good enough to show what the surface looks like. Consequently, it was something of a surprise to everyone when the first crude telemetered pictures from Mariner IV showed Mars to be heavily cratered, much more like the moon than the earth. Both the astronomical observations from earth and the Mariner results agree that there can be no liquid water on the surface of Mars, except perhaps in the form of highly saline solutions deep in crevasses or otherwise protected in such a way as to form a local "microclimate." Furthermore, it appears from the limited available pictorial evidence that there may never have been extensive oceans, lakes, or rivers on Mars, at least during the last billion years or so. Some scientists are, therefore, tending to lose hope that Mars may contain or ever have contained indigenous living organisms, similar to but having evolved separately from those on earth. Because other conditions all seem so favorable, however, it would seem important to carry out a careful study of the extent to which indigenous chemical and biological evolution may have taken place on Mars and, to this end, to avoid contamination of the planet by living organisms from earth or by large quantities of easily dispersible terrestrial organic materials until after such a study has been completed.

Most interesting from the viewpoint of planetary astronomy at present is Jupiter, the largest planet of all. More than 11 times bigger in diameter than earth, and 300 times heavier, Jupiter contains more than twice as much matter as all of the other planets together, yet its density is remarkably low, only about one and one-third greater than that of water, or less than one-quarter the average density of the earth. It spins around its axis in less than 10 hours, faster than any other planet, and it has 12 known moons, one of which is as large as the planet Mercury. It is almost certainly composed mostly of hydrogen, with some helium, ammonia, methane, and enough iron, silicon, and other heavy elements to make a central core about the size of the earth. At the very high pressures existing deep within Jupiter, it seems likely that hydrogen can take on a form more dense than ordinary "frozen" hydrogen, in which the electrons would not be bound to the nuclei but would be free to move through the solid material, as in a metal. Thus, this form of hydrogen would be a good conductor of electricity and heat, and if it does comprise a significant part of the central body of Jupiter, it might explain some of the interesting electric and magnetic properties of this giant planet.

Infrared measurements at different wavelengths indicate that Jupiter is intrinsically warm, i.e., that it radiates more energy than it receives from the sun. Radio measurements at centimeter wavelengths give additional thermal information, presumably at greater depth in the atmosphere. At decimeter wavelengths, Jupiter produces strong synchrotron radiation, from which it can be determined that the planet has a strong magnetic field, tilted about 10° from the rotation axis, and that it is surrounded by belts of trapped radiation. At longer wavelengths, Jupiter emits incredibly powerful bursts of radio energy that last from a few thousandths of a second to several seconds and follow one another in rapid succession during Jovian radio storms. These waves are now believed to be generated by the "dumping" of energetic particles from the magnetosphere into the atmosphere. Exactly what triggers this dumping process is not well understood, although the phenomenon must be similar to that which causes brilliant auroral displays on earth. The radio storms on Jupiter appear to be correlated in some way with the position of the planet's nearest big satellite, Io. Presumably, this satellite, or its magnetic wake, disturbs the Jovian magnetosphere in such a way as to enhance the dumping process.

Jupiter's heavily clouded atmosphere perpetually rages with storms, undoubtedly producing snow and hail of frozen ammonia, perhaps at a lower level even water droplets and ice, and almost certainly violent electrical discharges. It contains all the ingredients thought to be necessary for chemical and biological evolution, except perhaps the placid shallow seas of earth. Whether, in fact, Jupiter even has a true surface, or whether there is just a gradual transition from atmosphere to frothy liquid to solid, is still a matter of scientific conjecture. If life has developed "on" Jupiter, it seems likely that it would have taken a form capable of floating at some appropriate level in the dense atmosphere. The most unusual atmospheric feature is the Great Red Spot, a giant oval about 30,000 miles long and 8,000 miles wide, which has been observed for at least 100 years. It always remains in the southern tropical zone, although it seems to drift about slowly within the zone, and changes color from bright red to dull gray on an irregular basis. Most scientists now believe that it is a large-scale atmospheric phenomenon associated in some way with a topographic irregularity deep within the planet.

Saturn, with its beautiful and distinctive rings, Uranus, Neptune, little

Pluto, the thousands of asteroids, the comets, and the interplanetary dust that pervades our solar system, all seem to hold secrets that can now be discovered by modern astronomy and space research and that will help us to understand how and why we happen to be here, whether or not we are alone, and what our ultimate fate may be.

But most scientists who call themselves astronomers look farther out and concern themselves with the birth and death of stars, with galaxies, the interstellar medium, and the nature of the universe itself. Until about 30 or 40 years ago, because astronomical observations were restricted to the small part of the electromagnetic spectrum that could penetrate the atmosphere on a clear night and be recorded on photographic film, a large part of astronomical research was devoted to collecting data that could not be well understood. Very hot or very cold objects or those that emitted their radiation by nonthermal processes were generally beyond adequate comprehension because so many important data were lacking. Then, one by one, the other regions of the spectrum were opened up. Cosmic rays provided some clues to high-energy processes in the universe. Karl Jansky's observations of radio waves from the Milky Way, although of little apparent interest to astronomers at first, eventually led the way to the giant radio-telescopes that have revealed both high-energy and nonthermal phenomena on a grand scale. Finally, the availability of high-altitude balloons, aircraft, and space technology has opened up the far infrared, ultraviolet, x-ray, and gamma-ray portions of the spectrum. It is now possible to devise and carry out observations to test almost any hypothesis in astrophysics. Perhaps it is why both astronomers and physicists are beginning to regard astrophysics as one of the most interesting and exciting branches of physics.

Among the most interesting subjects of modern astronomy, surely the following must be listed: [b]

1—The strange properties of quasars and Seyfert Galaxies, which appear to involve some of the most energetic processes known, and which may be the oldest and most distant objects ever recognized by man.

2—The mysterious pulsars, which emit powerful radio pulses, so accurately timed that at first radio astronomers thought they must be receiving signals from a distant civilization.

3—The possibility that dust clouds surrounding some stars may indicate the process of formation of systems of planets like our own.

4—The observation of intense maser beams of hydroxyl (OH) radiation from intersteller space, brighter than could possibly be produced by any normal thermal source, even with a temperature of trillions of degrees.

5—The discovery of radiations indicating the presence of water, ammonia, and organic molecules in cold interstellar space, which suggests that perhaps the simplest steps of chemical and biological evolution did not occur only after the earth was formed but were part of the same process that formed the stars.

6—The million-degree temperatures in the solar corona, surrounding the much cooler surface of the sun, when every schoolboy knows that heat flows from hot to cold places and not the other way around!

7—The microwave background of space; is it the reverberation of a primordial Big Bang?

[b] These are adapted from "A Long-Range Program in Space Astronomy," Astronomy Missions Board, NASA, 1969, NASA SP-213.

In its position paper of July, 1969, the NASA Astronomy Missions Board points to the Crab Nebula as a subject for study by all the disciplines of astronomy. This remarkable object, known to be the remnant of a supernova "explosion" that occurred, appropriately enough, on the Fourth of July in 1054 A. D., presumably represents the termination of the life of a star—with a bang, not a whimper. It is believed that, as a star begins to run out of nuclear fuel, its own gravity causes it first to collapse rapidly, then to explode violently, releasing more energy during one year than it had given off in its entire previous life. Theorists have predicted that the residue might include a neutron star—a star so compressed that its atoms would lose their individuality and become, in effect, continuous nuclear matter. Such a star would be only 10 km or so in diameter, but each cubic centimeter would weigh a billion tons or more.

When looking at the Crab Nebula, optical astronomers see a faintly glowing nebulosity, interlaced by a network of delicate reddish filaments that are expanding at 1,000 km/sec. In the center is a faint star, unlike any known stellar types. The light from the nebulosity is strongly polarized, and when this information is combined with that resulting from the strong radiation seen by radio astronomers, it becomes evident that the nebulosity is not just a glowing hot mass of gas but a plasma containing electrons with energy exceeding 100 billion electron volts (velocities near the speed of light), permeated by a relatively weak magnetic field that extends over a region light-years in extent. With the advent of space technology, it became known that the Crab was also a potent emitter of x-ray energy. And only a year after the discovery of the first pulsar, it was discovered that the peculiar star within the Crab Nebula was indeed a pulsar, flashing 30 times a second—much faster than the typical pulsar—in the visible and x-ray portions as well as in the radio part of the spectrum. The total rate of energy radiation by the Crab Pulsar appears to be more than 100 times that of our Sun.

The x-ray, radio, and optical observations all support the model of the Crab Nebula as being energized by a rotating neutron star, rotating at 30 times per second, which resulted from the collapse of the original star and conservation of its angular momentum. The collapse also would have compressed the normal magnetic field previously present in the star to a value on the order of a trillion gauss. According to the model, the electric field induced by this rapidly rotating magnetic field would accelerate particles that would drag the magnetic field with them, emitting synchrotron radiation as they go. This process would extract energy and angular momentum from the star at a calculable rate, which seems to agree with the recent observation that the Crab Pulsar period is slowly increasing.

It will be of great interest to learn whether all supernovae produce a pulsar or neutron star to mark the final stage of stellar evolution or whether some may involve, as has been postulated, an even more fantastic collapse, resulting in a lump of matter so highly condensed that no radiation of any kind, and hence no information, can escape its gravitational pull. Such a gravitational singularity would indeed be a black hole in the universe—into which things and information could enter, but from which nothing would ever return.

Just as it seems impossible today to discuss astronomy without mentioning the contributions of space technology to this venerable science, so it may soon be in other scientific areas. We have learned much about geodesy by carefully measuring the small perturbations in the orbits of artificial earth satellites and about geology from large-scale photographs and images at various wavelengths of the earth from satellites. Physical, chemical, and mineralogical studies of material returned from the moon are already providing a wealth of new information about our satellite, and it is not hard to predict that they will ultimately tell us something worthwhile about the earth, too. Less than 15 years ago, our understanding of magnetic storms and aurorae on earth was essentially limited to the fact that they were in some way related to disturbances on the sun. Now, because of what has been learned by means of instruments on satellites and space probes, we have a working knowledge of the earth's upper atmosphere and magnetic environment and can intelligently construct and test hypothetical models of solar-terrestrial interaction. We even begin, now, to have some ability to predict what will happen to living organisms when they are transplanted into a space environment.

Trivial as it may seem, after talk of planets, stars, and galaxies, the ability to observe the weather quantitatively at hundreds of thousands of points each day and to track thousands of balloons and recover data from hundreds of ocean buoys may be the key that will enable meteorological science to develop a practical capability to forecast accurately terrestrial weather for as long as two weeks in advance. Some of the same techniques will also be much in demand for tracking the migration of animals, including some sea-going species, and even of large birds.

As the use of space technology matures, I think we shall not speak so much of space science. After all, space is not a science, but a place, even if a rather big place. Instead we shall include space research, that is, research accomplished with the help of space technology, as a part of astronomy, physics, chemistry, geology, geodesy, biology, and so on. In considering the consequences of astronomy and space research for mankind, we are therefore really asking about the consequences of science as a whole, for astronomy is one of the purest of the sciences, and space research is a part of many sciences.

"The pursuit of science," according to Prof. I. I. Rabi, "satisfies a basic desire of aspiration just to know, to find out, or perhaps make order out of the otherwise chaotic jumble of immediate experience. It is an aspiration shared with all mankind, but more with youth and childhood than with adults." [e] Members of the scientific community, continues Professor Rabi, "possess an inner solidity which comes from a sense of achievement and an inner conviction that the advance of science is important and worthy of their greatest effort. This solidity comes in a context of fierce competition, strongly held conviction, and differing assessments as to the value of one achievement or another. Over and above all this too human confusion is the assurance that with further study will come order and beauty and a deeper understanding."

There is little question but that the joy of the pursuit and the quiet exultation that is felt when something new and exciting is revealed are reason enough for a scientist to be a scientist, for an astronomer to be an astronomer. Is it reason enough, however, to ask for public support of science, especially when the tools of science are becoming exceedingly expensive, and when many important and diverse claims are being made on the public purse? It is necessary to examine the consequences of science for society as a whole and to ask how big a price tag can be considered reasonable.

There are, of course, practical consequences. It is surely no accident that down through the course of history the societies that made their mark

ᵉ I. I. Rabi, "Science and the Satisfaction of Human Aspirations," Proc. National Academy of Sciences, December 1963.

were those that encouraged scientific activity. In one particular instance, history provides a clear picture of the way in which a nation, suddenly emerging, was able to expand vigorously into the upper ranks of the developed nations. The example is Japan, which just 100 years ago, broke with its own centuries-old tradition of isolation and invited the introduction of science, along with other aspects of Western civilization. After an incubation period of 10 or 15 years, while foreign scientists were imported and a first generation of Japanese scientists was trained, scientific activity in Japan took off exponentially, at a doubling period of about 10 years. The results in terms of industrial development, international trade, standard of living, and original technological achievement are well known. Japan now ranks third in the industrial world, being exceeded only by the U.S.A. and the U.S.S.R., and is the fourth nation to have developed and orbited an artificial earth satellite, using only its own financial and technological resources. Had it chosen to ignore science and import only technology, it does not seem likely that Japan would yet, if ever, have broken out of the status of a clever copier of Western products.

It is rarely possible to trace the individual progress of ideas from the first scientific revelation to the final useful product. However, consider the tremendous consequences that followed when astronomers asked themselves what is the source of energy of the stars, and set to work to find out. Perhaps within a hundred years, the future of mankind will depend on an understanding and ability to control this source of energy. Now astronomers are asking how it can be that certain galaxies emit prodigious quantities of radio energy—more even than can be expected from nuclear transformations. What will be the practical consequences of studying this problem?

We have seen the development from a study of the movements of the moon and planets to the creation of artificial moons and planets, capable of carrying man-made instruments, machines, and even men themselves. In 13 short years, this capability has expanded our communication capability in ways and to an extent that could not have been imagined less than a generation ago. We do not yet know what the sociological consequences of this new communication capability will be, but it could contribute to the elimination of much of the starvation in India, to a revulsion against war as an instrument of national policy, to a common language in this tower of Babel we call the world. We have already seen the development of means to observe and warn against deadly hurricanes and have envisaged a World Weather Watch which will constantly pour the necessary information into giant computers that can tell us what the weather will be at any point on the globe one or two weeks in advance. With such knowledge and understanding in hand, it seems certain that some measure of beneficial control will follow.

We have measured the size and shape of the earth to an unprecedented accuracy. We have brought back from space pictures of remote areas of the earth, which are useful in designing highways and prospecting for mineral resources, and we hope to expand this activity to enable us to make far better use of our most precious resource, fresh water, and to trace the development and thereby aid in the elimination of diseases of crops and forests. We have developed the capability to photograph military and industrial installations from space, so that very little can be done in secret anymore. We hope that this capability will lead at least to a stabilization of strategic armaments and thereby help us avoid the mutual nuclear catastrophe that haunts our lives.

We can even predict that someday, in orbital workshops free from the ubiquitous pull of gravity, it will be possible to produce materials that no terrestrial factory could ever hope to achieve—special alloys, special glasses, high-purity vaccines, to mention only a few. And these, in turn, will have consequences that we are not yet able to imagine.

But beyond these practical consequences, there is an area of spiritual consequence that is most difficult to define. The urge of science, as well as of religion, is to comprehend the visible and invisible universe and to find man's place within it. In these matters, religion has historically taken the lead. Questions about man's place in the universe have demanded answers in each generation, and the ancient Hebrews could not wait for the discovery of the neutron, or the development of a theory of stellar evolution, or for biology to explain the variety of life and the origins of man. By means of dramatic imagery, the great religions have provided explanations that have given immediate satisfaction to man's yearning for order in the world and for guidance in his life.

The more prosaic, plodding progress of science, while it may in some cases destroy beautiful edifices of thought that are enshrined in history, at the same time may be preparing the ground for new and even more beautiful edifices to come. In any case, no true humanitarian or religious leader can fail to welcome the advances of science, for they enable him to advance with greater certainty and deeper understanding.

Society today is indeed "in transition." It always has been; it always will be. So long as man is alive, he will continue to ask, "Who am I? Where am I? Why am I here? What is this around me? How can I make it better for myself, my children, and my children's children?" It is the business of science and, in some special ways, of astronomy and space research to provide the knowledge and understanding that alone can answer these questions.

* * *

NIGRELLI: I am going to discuss our aquatic environment and its potential to support and sustain an increasing population, which, it is estimated, will be about six billion in the year 2000. Many estimates of this potential have been optimistic, but most predictions are less optimistic, and rightly so, because no sure methods have yet been developed to estimate the effects of fishing on fish population.

It is true, of course, that with the increase in fishing efforts there has been a gradual increase in the annual harvest of marine fishes in recent years. According to statistics published in 1969 by the Food and Agriculture Organization, the amount of fish caught between 1958 and 1968 increased from 33 million metric tons to 64 million metric tons. The optimists believe that 200 million tons of fish per year can be harvested from the seas on a sustained basis. This represents the total amount of essential animal protein required for six billion people.

Most of the exploitation for food from the sea occurs on the continental shelves of the various nations. It is in such environments that the herring and herringlike fishes, cod and flatfishes, are the major inhabitants, and the species most intensively exploited. These fishes, together with the basses, mackerels, jacks, and tunas, make up a group referred to as the *cardinal species* and comprise about 80 percent of the world catch, mainly by the fishing efforts of about 19 nations, with Japan and the U.S.S.R. the leaders. The remaining 20 percent of the catch is of unrelated species and is usually taken by countries that lack sophisticated facilities and technical abilities for large-scale oceanic fisheries.

Many scientists believe that the continental shelves are being overexploited. Several fishing areas now produce less than they did ten years ago. For example, the cod population in the Barents Sea is now about ten percent of its former level. In addition, the increase in fishing pressure on the stocks are shown by the smaller average size of the fish and the greater effort needed to catch the same amount of fish.

International commissions are now active in the conservation of the cod and other valuable stocks that are beginning to show a decline in populations. In addition to being overexploited, the available stocks on the shelves are being subjected to increasing contamination by toxic pollutants, some of which may reduce the reproductive potential or make the fish unfit for human consumption. Extending ocean fisheries beyond the continental shelves will no doubt increase the annual production but will require more sophisticated methods in locating and harvesting fishes in the open seas and at a much greater cost per ton.

The tropical regions of the oceans are the least exploited. The catch from Oceania, for instance, doubled in quantity from 1958 to 1968, but the maximum was only 210,000 metric tons. There may be a reason for this, because eating fish from such areas, especially in areas delineated by latitudes 35° N and 34° S, is fraught with danger since many species are poisonous. Fish poisoning is synonymous with icthyotoxism or icthyosarcotoxism; but it does not include botulism or poisoning caused by microbial or radioactive contamination or by absorption of pesticides and such industrial pollutants as cyanide or mercury— as has recently occurred in the United States' Great Lakes.

There are several large groups of fishes whose flesh is poisonous when eaten. The most powerful poison is found in the flesh and organs of the puffer family, a group of fish widely eaten by Orientals. The purified poison is a highly lethal neurotoxin with a pharmacological activity 160,000 times more potent than cocaine in blocking nerve conduction. In Japan, this type of poisoning is called *fugu* and is responsible for about 42 percent of the total annual food intoxication. The Japanese Ministry of Public Health and Welfare reported that in the 10-year period from 1954 to 1963 almost 2,000 cases of *fugu* poisoning and 82 deaths have occurred.

The most widely distributed form of icthyosarcotoxism is called ciguatera. More than 400 species from 11 orders and 55 families of bony fishes have been reported at one time or another as being poisonous, with such desirable species as snappers, jacks, basses, and porgies most frequently affected, especially in the tropical and subtropical waters. About 5,000 cases of intoxication and many deaths have been reported. The development of this poison in otherwise normal fish appears to be associated with the food chain cycle. Carnivorous fish, such as those just mentioned, are most toxic and apparently derive the poison from herbivores that eat toxic blue-green algae, a species of plant life that is more apt to appear as the environment becomes more polluted.

It is apparent, then, that the seas in tropical regions may not prove to be a panacea for protein-starved populations. The reduction of trash fish to fish-protein concentrate, although an important endeavor, must be a highly selective operation. A small number of ciguatoxic fishes accidently reduced to fish concentrate along with normal fish may render the whole batch toxic.

Mariculture, or fish farming as a means of increasing the production of fish and shellfish, has caught the imagination of fishery scientists and the business community. Although methods can be developed to increase the production of such highly priced items as oysters, clams, mussels, shrimps, mullet, pompano, flounders, and yellowtails, to mention a few, the total tonnage that can be produced on a sustained annual basis will make very little dent in the amount of animal protein that will be required to feed six billion persons, especially the low-income group. It will take many years before fish farming will be economically feasible. It took trout-hatchery scientists more than 50 years to develop their technology to reach the state of production at which we are in 1970.

Many of the same, and even more difficult, problems of management will be encountered before any appreciable numbers of marine organisms can be cultivated and marketed. Fish and shellfish are subject to diseases caused by viruses and other infectious agents, and to diseases caused by nutritional deficiencies, genetic and hormonal imbalances, cancer, and to physiological abnormalities caused by changes in environmental factors. Many problems related to each of these items will have to be solved and at present we know very little about diseases of marine and brackish animals, some of which occur naturally and are known to be transmissible to man.

However, fish comprise only a small fraction of the community of life in the sea. The sea is alive with literally thousands of species of plant and invertebrate life. Planktonic forms are the base of the food web, and the variety of links in the food chain, except for the edible fish species, have been generally ignored. Because of economic, esthetic, and ethnological reasons, most of these living forms, except for certain mollusks and some crustaceans, are not exploited as human food source.

But there is a constant need for fresh sources of drugs, especially new antibiotics, to combat resistant strains of pathogenic bacteria. The unique life mode of many marine organisms suggests interesting chemical and biochemical constituents, and this has in recent years given added impetus to the field of natural-products chemistry. Thus, such animals as sponges, jellyfish, etc., marine worms, and plants of the sea are being studied by biologists, chemists, and pharmacologists as a source of drugs. The number of potentially useful substances that have been isolated from such marine animals and plants has highlighted the need for more intensive biochemical studies. Experimentally, substances from marine organisms have shown a wide variety of pharmacological properties, including antiviral, antimicrobial, cancer-inhibiting, anticoagulant, nerve-blocking, and hypotensive activities. Many have already made their way into the Materia Medica. Eledosin, extracted from a species of Mediterranean octopus, has proved to be an extremely potent hypotensive agent, actually 10,000 times more powerful than quinidine sulfate, the standard drug used as a cardiac depressant.

The living resources of the sea can be exploited for food and drugs to a greater degree, with conservation measures to be imposed on a world basis for some of the more economically important species that are being overexploited. One can no longer consider any area of the total planet in isolation. The increased population has its effects not only in terms of the actual resource it removes from the environment but also by how it deleteriously affects those resources while they inhabit their environment.

THE ENVIRONMENTAL CONSEQUENCES OF MAN'S QUEST FOR FOOD

Lester R. Brown

*Overseas Development Council
Washington, D.C.*

Man may have first appeared on earth as early as two million years ago. During perhaps 99 percent of his existence he hunted and gathered wild food, living as a predatory animal. While he was dependent on hunting and gathering for his food, he probably never exceeded ten million in number—the estimated human population that the earth could support under these conditions.

Then, about ten thousand years ago, man learned to domesticate animals and plants, thus initiating the great transition from hunter to tiller. Initially, only a handful of men were attracted to the new agrarian way of life, but through the millenia it has become the preferred way of life. Today only a small fraction of one percent of the human race lives by hunting and gathering. The transition from hunter to tiller is virtually complete. Man has substituted the vicissitudes of weather for the uncertainty of the hunt.

The capacity of man the hunter for intervening in his environment was exceedingly limited, but man the tiller developed a seemingly unlimited capacity for altering his environment, shaping it to his ends. Initially quite simple and limited in scope, his interventions became successively more complex and widespread. Eventually some of the consequences of those interventions exceeded man's understanding of them, creating some worrisome problems. Some of these problems are largely local, but the more serious are global in scale; no respecters of national boundaries, they affect rich and poor alike and are irrespective of ideology. The threat to the environment, and to many forms of life, posed by the steadily growing use of chemicals in farming underlines the common predicament of all mankind as few things do.

Scholars agree that, though still a mystery just being unraveled, the earliest beginnings of agriculture probably occurred in western Asia, in the hills and grassy northern plains surrounding the Fertile Crescent. Wheat and barley grew wild there, as did sheep, goats, pigs, cattle, horses, and deer. To this day, wild barley and two kinds of wild wheat (emmer and eikorn) flourish in the region.

In simplest terms, agriculture is an effort by man to shape his environment so that it better suits his needs. Man selected certain species of animals and plants, which in nature were quite useful to him as a source of food or clothing, and began favoring these species above all others. For example, wheat, a cereal growing naturally in certain areas of the Middle East, is now planted on nearly 600 million acres of land once covered by grass or forest.

The natural cover of some three billion acres, or 10 percent of the earth's land surface, has been removed and planted to crops that man has domesticated. Expressed in these terms, the area cleared by man seems small, but in terms of the area supporting vegetation, it is quite large; in terms of the area potentially capable of supporting crops useful to man, it is still larger, perhaps even predominant. Two-thirds of the cropped area is planted to cereals: rice, wheat, corn, rye, oats, and barley. As a result of man's favoritism to certain species of animals, the relative importance of various species has been greatly altered. Hereford cattle roam the Great Plains, once the home of an estimated 30 to 40 million buffalo. In Australia, the kangaroo has been largely displaced by cattle of European origin, and the domesticated water buffalo now inhabits all the major rice-growing river valleys of Asia.

Knowledge of farming spread from the Middle East throughout the world. It moved westward across Asia, southward into the Tigris-Euphrates Valley, then along the Nile into Africa and northwestward into Europe, through the Danube Valley and along the Mediterranean coast. Agriculture apparently had independent origins in the Americas and perhaps even in the Far East as well. But these developments came later than the Neolithic achievements in Southwestern Asia.

The great Neolithic achievements of agriculture and husbandry gave man a more abundant and secure food supply, allowing him to increase his numbers and establishing the base for civilization. Grain fields fed growing urban populations. But the problem of obtaining enough food remained; it has plagued man since his beginnings. Technological advances, such as the discovery of irrigation, the use of animals for draft purposes, and the exchange of crops between Old World and New, plus many other technological breakthroughs in more recent times, have greatly expanded the earth's food producing capacity. Spurts in food production have permitted man's numbers to increase, and these increases have in turn exerted pressure on the food supply, forcing man to innovate and devise still more effective means of producing food. In a finite biosphere, we must assume that at some point this reinforcing cycle must be broken.

The history of agriculture is filled with technological advances of varying importance, each of which increased the earth's population-sustaining capacity. A few stand out. The development of irrigation agriculture some 6,000 years ago in the Middle East, probably beginning on the Tigris-Euphrates flood plain, greatly enhanced man's food-producing capacity. It also required a high degree of social organization and closely coincided with the emergence of the early civilizations.

Another of the early farming breakthroughs was the discovery by man at least 3,000 years B.C. that, to augment his own limited muscle power, he could harness certain animals much stronger than himself. Once man learned to use draft animals, he was able to convert roughage into a usable form of energy. Early animal-drawn implements were crude, little more than pointed sticks, and hitches were inefficient implements, often attached directly to the horns of animals. But harnessing this new-found source of energy enabled man to greatly expand his food supply over time.

Among the strongest supporting evidence for the independent origins of agriculture in the Old World and the New World is the quite different crops domesticated in the two regions. Thus, when Columbus established the link between the Old World and the New, he set in motion an exchange of crops between the two worlds that continues to the present. Interestingly, some of the crops were better suited to the world in which they were introduced than to the one in which they originated. As this exchange of crops progressed, the earth's population-sustaining capacity increased steadily.

Perhaps the classic example of this was the introduction of the potato into northern Europe, where it greatly augmented the food supply, permitting marked increases in population. This was most dramatically illustrated in Ireland, where the Irish population increased rapidly for several decades on the strength of the expanded food supply made available by the potato. Only when potato blight (*Phytophthora infestans*) devastated the potato crop was population checked in Ireland.

Interestingly, the principal source of vegetable oil in the United States and its principal farm export is the soybean, a crop introduced from China several decades ago. Grain sorghum, now the second-ranking feed grain in the United States, after corn, crossed the Atlantic from Africa in the form of food stores on the early slave ships.

The principal source of vegetable oil in the Soviet Union is the sunflower, a plant that originated in the southern Great Plains of the United States. Corn, the only cereal crop indigenous to the New World, is now produced on every continent. The movement of crops between the Old World and the New, and among continents and countries, continues as man alters the nature of the crops themselves through genetic manipulation, as he alters the environment, and as these physical factors interact with the changing demands of the marketplace.

The same factors have influenced the movement of livestock from their areas of origin to other parts of the world. The New World is particularly indebted to the Old for all of its livestock and, with the exception of the turkey, all of its poultry. During the 16th and 17th centuries, the principal dynamics in world agriculture was related to the exchange of crops between the Old World and the New.

Over the past two centuries, scientific advances in the harnessing of mechanical power, in soil chemistry, and in plant genetics have opened new horizons in man's quest for food, permitting him to intervene in his environment more extensively than ever before. Development of the steam engine in the latter half of the 18th century, though it did not profoundly affect agriculture, set the stage for mechanization and the later development of the internal-combustion engine. The internal-combustion engine in the form of the farm tractor introduced a new era in agriculture. Man suddenly had at his command a vast new source of energy, petroleum, which he could use to produce more food. It was now possible to substitute petroleum for oats and hay, to substitute the products of eons-old photosynthesis for that on present-day farms. The substitution of fossil fuels for oats, corn, hay and other feedstuffs increased the potential energy supply per person working the land severalfold. The displacement of horses by tractors in the United States during the first half of the 20th century also released some 70 million acres, once used to feed the nation's horses, for other purposes

During the early 19th century the foundations for another major technological advance in agriculture were established by von Liebeig of Germany, the father of modern soil chemistry. He identified the importance of the nutrients nitrogen, phosphorus, and potassium in plant growth. He demonstrated that one could restore or enhance the soil's natural fertility by adding these nutrients in the proper proportions and, on the basis of this, recommended the use of mineral fertilizers by farmers. Eventually, farmers in their continuing efforts to expand the food supply were to learn to substitute fertilizer for land as the frontiers disappeared.

At the time of von Liebeig's findings, there was still ample opportunity in much of the world for expanding the area under cultivation. Thus, it was not until the 20th century that the use of chemical fertilizers became widespread. As of 1970, the world's farmers are using more than 40 million metric tons of nitrogen fertilizer alone on three billion acres of cropland, or nearly 30 pounds per acre. Usage varies widely among countries, with intensive use in the densely populated, industrial countries and light use in the poor countries.

In high-rainfall countries practicing intensive agriculture, such as the Netherlands and Japan, a major share of the food supply is attributable to the use of chemical fertilizers, applied in both countries at more than 200 pounds per acre yearly. Stated otherwise, soil fertility would decline rapidly in the absence of chemical fertilizers, dropping food production by half or perhaps as much as two-thirds. On a global basis, perhaps one-fourth of man's total food supply is attributable to the use of chemical fertilizers.

Along with the development of mechanical power and soil chemistry, the third major breakthrough that has contributed so much to man's food-producing capacity was the breakthrough in plant genetics, made possible by the work of the Austrian monk Mendel. From the work of Mendel and his more modern successors, such as Burbank, plant breeders have learned to select, classify, and combine various sources of germ plasm in order to alter plant characteristics. Plant breeders can make plants more tolerant to cold, more resistant to drought, less susceptible to disease, more responsive to fertilizer, and higher in protein content and can alter literally hundreds of other characteristics. To cite some specific examples, they have extended the northern limit of commercial corn production 500 miles within the United States, developed wheats that yield well close to the equator, and increased the oil content of the soybean.

Although the scientific advances that made possible farm mechanization, widespread use of chemical fertilizers, and the dramatic improvement in plants and animals through breeding occurred largely during the 19th century, it was not until well into the 20th century that this knowledge was applied on a widespread commercial basis.

From the time agriculture was first invented until roughly the beginning of the 20th century, the world's food supply increased much more as a result of expanding the area under cultivation than of raising the productivity of land already being farmed. Throughout most of history, man increased the food supply by moving from valley to valley, from country to country, and from continent to continent. Only during the 20th century, as the frontiers gradually disappeared, was man forced to concentrate his energies on raising the yields of existing cultivated areas. The first yield takeoff—the transition from a condition of nearly static yields to that of rapid sustained increases—apparently occurred in Japan at about the turn of the century. Japan's rice yields, which began their sustained rise in the early years of this century, have risen steadily for seven decades except for the interruptions of war. Yield takeoffs apparently occurred in some of the countries of western Europe, such as the Netherlands, Denmark, and possibly Sweden, at about the same time.

But the ability to generate yield-per-acre advances was confined to a relatively small number of countries with advanced farm technologies. The yield takeoff did not occur in North America until about 1940. Between the Civil War and the beginning of World War II, per-acre yields of corn, the principal cereal in the United States, were essentially unchanged.

Although the poor countries of the world did not have advanced farm technologies capable of supporting a rapid increase in output per acre, they were beginning to receive from the industrial countries new medical technologies that greatly reduced death rates, particularly among infants. As a result, their populations began to accelerate rapidly after World War II, causing the demand for food to outstrip their food-production capacity. Asia, Africa, and Latin America, which had all been net cereal exporters prior to World War II, exporting largely to Europe, suddenly found that their increased populations were consuming what had once been their exportable surpluses.. By 1950, the less-developed world was importing several million tons of grain yearly from the developed countries, principally North America. By 1966, this flow of food had increased to 32 million tons, a large part as food aid shipments from the United States. The population explosion in the poor countries thus dramatically altered the pattern of world grain trade.

Cereals are a convenient indicator of food supply because they occupy more than 70 percent of the world's cropland and provide the major share of man's food-energy intake. Wheat and rice each supply one-fifth of man's food-energy supply. With corn and other cereals, they provide more than 53 percent of directly consumed calories and a sizable part of the remainder consumed indirectly in the form of meat, milk, and eggs.

The other important starchy foods are potatoes and other tubers and roots, which provide ten percent of man's caloric intake. Animal fats and vegetable oils, including butter, lard, and oils from peanuts, soybeans, and sunflower, account for nine percent, sugar seven percent, fruits and vegetables ten percent, livestock products 11 percent, and fish one percent. Within any particular locality, diets can vary greatly from this overall global pattern.

Although man's capacity to increase food production has increased several hundredfold since the discovery of agriculture, the majority of mankind is still not well nourished. On the plus side, man has been able to assure the food supply of perhaps one-third of the human race and to eliminate the threat of famine for the remaining two-thirds. For that one-third of mankind living in North America, Western Europe, Eastern Europe, Australia, Japan, and a handful of the smaller countries elsewhere in Asia, Africa, and Latin America, food supplies are now assured. Since the industrial revolution, the productivity and income of those living in the above regions and countries has risen to the point at which, except for a small segment with marginal incomes, diets are now more than adequate. Many people are in fact overfed.

Throughout most of history the threat of famine has been present, usually as a result of droughts, floods, or crop diseases. Occasionally famines are man-made, caused by war or man's inhumanity to man, as in the Soviet Union in 1935–37 and in Biafra in 1969–70; but in the past few decades, the world has been spared from massive famine, except for the man-induced famine in Biafra in 1969–70. Shortly after World War II, the United States decided it had the agricultural know-how and the food production capacity to avoid famine anywhere in the world. This policy was adhered to even when, as in 1966 and 1967, one-fifth of the U.S. wheat crop was required to stave off famine in India. Sixty million Indians were fed entirely by food shipments from the United States for a period of two years.

Since World War II, the global production and distribution system, both among countries and within countries, has evolved to the point where large quantities of food can be moved from one country to another on a sufficient scale to avoid famine caused by any readily envisaged natural catastrophe, such as the monsoon failures on the Indian subcontinent in the mid-60s. Although the specter of famine of the kind that occurred in West Bengal in 1943, costing some two to four million lives, has been eliminated for at least the time being, human nutrition on a global scale is still in a sorry state. Malnutrition exacts an enormous toll in both the physical and mental development of a great majority of the youngsters in the world today.

The effects of nutrition on physical development is dramatically evident on the streets of Tokyo today. Japanese teen-agers, well nourished from infancy as a result of the enormous rise in income and proper nutrition in Japan, tower inches above their elders. Much of the difference is attributable to a much greater protein intake during early years of life. The effect of good nutrition on physical development is also evident in other ways. In the summer of 1968, India held its Olympic tryouts in New Delhi, with the purpose of selecting a team to go to the Olympics in Mexico City in the fall. Unfortunately, not a single Indian athlete, male or female, met the minimum qualifications for participating in the Olympic competition in any of the 32 track and field events. This, of course, was only partly due to nutrition: dated training techniques and lack of support for athletics were also factors.

Recent studies of malnutrition and mental development indicate a relationship at least as striking as that for physical development. A clinical study of a group of youngsters, severely malnourished before the age of five, conducted in Mexico over a period of several years, indicated an average I.Q. 13 points below that of a carefully selected control group. Studies conducted elsewhere bear out these findings. Unfortunately, these setbacks to development of the brain and central nervous system in early life are irreparable. No amount of feeding or educating in later life can offset it. Protein shortages in the poor countries today are depreciating their human resources for at least a generation to come.

Although the prospects for feeding the projected increases of numbers in the poor countries over the next decade or two appeared bleak in 1966 and 1967 at the time of the Indian food crisis, they have improved measurably ovei the last few years. The principal reason is the successful introduction and rapid spread in several of the larger poor countries, such as Mexico, the Philippines, Indonesia, Pakistan, India, and Turkey, of the new high-yielding varieties of wheat and rice. This phenomenon, popularly referred to as the Green Revolution, is buying additional time with which to stabilize population growth.

The Green Revolution, based on the much greater responsiveness of the new wheats and rices to fertilizer, which enables them to roughly double the yields of the traditional varieties they replace, is not a universal phenomenon in the poor countries. Cropwise, it is limited largely to wheat and to rice, and geographically, it is limited to those areas with an adequate, controlled water supply. In terms of total cereal acreage, it still accounts for only a small share of the total in the developing countries. Even with these constraints, however, some unprecedented gains in cereal production have been realized over the years. Mexico, which imported one-third of its wheat a generation ago, is today exporting wheat, corn, and rice. Cereal production per person increased each year during the sixties, climbing nearly 40 percent during the decade. The Philippines has ended half a century of dependence on rice imports. Annual increases in wheat yields in Pakistan and India, following the introduction of Mexican dwarf wheats, are double the increases in corn yields in the United States following the introduction of hybrid corn a generation ago. Food aid imports into these two countries are at the lowest level in a decade. The introduction of the new seeds and their associated technologies is clearly not a solution to the food-population problem, but it is buying time with which to stabilize population growth.

In considering future food requirements we must keep in mind that there are two sources of additional demand for food: the first, population growth, is well understood; the second, rising incomes, is not so commonly understood. Present projections indicate world population nearly doubling over the remaining three decades of this century. Assuming that the end-of-century population is no better or worse fed than at present, this would require a doubling of the world's food production capacity.

Rising incomes, an objective of virtually every national government in the world today, will greatly increase claims on the world's agricultural resources. This is perhaps best illustrated by looking at the grain requirements of high- and low-income people. The two billion or so living in the poor countries have available an average of 400 pounds of grain per year. With perhaps 10 percent used for seed, about 360 pounds, or a pound a day, remains for actual consumption. With only a pound of grain available, virtually all must be consumed directly simply to meet minimum energy requirements and to keep body and soul together. As incomes rise, the amount of grain per person that is consumed directly declines, leveling off at about 150 pounds, but total grain used rises steadily, eventually reaching a ton per year as in North America at present. The great bulk of this is consumed indirectly in the form of meat, milk, and eggs. Thus, the claims exerted against the earth's food-producing capacity by those living in the richest countries, say the United States or Canada, are perhaps five times as great as those living in the poor countries.

Advances in food technology are beginning to offset some of the increases in grain requirements in the industrial countries by substituting vegetable products for animal products. This has been most pronounced in the substitution of vegetable oils and shortenings for butter and for lard. In 1940, the average American consumed 14 pounds of butter and 2 pounds of oleomargarine. In 1970, the figures were 6 pounds of butter and 10 pounds of oleomargarine. Lard has been almost entirely replaced with vegetable shortenings. Some 65 percent of whipped toppings purchased in the supermarket are of nondairy origin, as are 35 percent of the coffee whiteners.

Food technologists are now concentrating their efforts on developing acceptable vegetable protein substitutes for various meats. The first major breakthrough has come with a commercially successful imitation bacon derived largely from soybeans. Livestock are rather inefficient in their conversion of vegetable materials into animal protein and fat products, returning only 10 per cent of their energy intake to man in the form of edible food products. Their advantage is that they can consume roughage otherwise of very little use to man. The other 90 percent they use to keep warm and to reproduce and for other purposes. The inherent inefficiency of livestock and advancing food technology have convinced food technologists that they can largely replace many of the livestock products consumed today with products of vegetable origin that are comparable in nutritional quality.

Despite these gains in food technology and the growing acceptance of vegetable-derived livestock substitutes, the overall prospect for the next few decades is for substantially increased claims on the world grain supply as a result of rising incomes. If population projections materialize and economic growth targets are realized, the combined increase in demand could nearly triple world food requirements by the end of the century.

The net effect of the projected doubling in man's numbers and hoped-for future gains in income, particularly among the world's poor, is growing pressure on the earth's agricultural ecosystem. This pressure takes two essential forms: one is extensive, essentially expanding the land under the plow; the other is intensive, the intensification of cultivation on the existing agricultural land area through the increased use of agricultural chemicals, more productive seeds, and the greater use of water.

The price of the extension of cultivation into marginal areas is often the loss of that thin mantle of topsoil, measured on most of the earth's surface in inches, on which man depends for his food supply. Increasing human population invariably results in the clearing of more and more land, either for the production of more food or as a result of the demand for fuel. In the more densely populated poor countries, fuel demands have long since exceeded the replacement capacity of local forests. The result is that the forested area has declined until in many parts of the world there is little left. In these circumstances, as in the Indo-Pakistan subcontinent, people are reduced to using some other source of fuel, such as cow dung, for cooking purposes. The number of people in the world today who rely on cow dung for fuel probably far exceeds those using some of the more modern fuels, such as natural gas.

As man's numbers increase, cattle numbers also increase, denuding the countryside of its natural grass cover. Prof. Robert Brooks, of Williams College, who lived in India for several years, provides a graphic example:

> A classic illustration of large-scale destruction is afforded by the spectacle of wind erosion in Rajasthan. Overgrazing by goats destroys the desert plants which might otherwise hold the soil in place. Goatherds equipped with sickles attached to 20-foot poles strip the leaves of trees to float downward into the waiting mouths of famished goats and sheep. The trees die and the soil blows away two hundred miles to New Delhi where it comes to rest in the lungs of its inhabitants and on the shiny cars of foreign diplomats.

The progressive deforestation that is a result of the population expansion of the last few decades is totally denuding the countryside, creating conditions for rapid soil and water erosion. Millions of acres of cropland in Asia, the Middle East, North Africa, and Central America are being abandoned each year because severe erosion by both wind and water have rendered them unproductive or at least incapable of sustaining the local inhabitants with existing technologies.

It takes centuries to form an inch of topsoil through natural processes, but man is managing to destroy it in some areas of the world in only a fraction of that time. But the problem associated with the loss of topsoil does not end with the abandonment of the severely eroded land. Much of the topsoil finds its way into streams, rivers, and irrigation canals, and eventually into irrigation reservoirs. A dramatic example of this indirect cost of soil erosion is provided by the Mangla Reservoir, recently constructed in the foothills of the Himalayas in West Pakistan as part of the Indus River system. The feasibility studies undertaken in the late 1950s, justifying the eventual investment of $600,000,000 in this irrigation reservoir, were based on a life expectancy of at least 100 years. As the rate of population growth in the watershed feeding the Mangla Reservoir has increased, so has the rate of denuding and soil erosion. The result is that this reservoir is expected to be completely filled with silt within 50 years or less, rendering it useless for water storage.

History provides many examples of man's abuse of the soil that sustains him. North Africa, once fertile, highly productive, and the granary of the Roman Empire, has lost the natural fertility it once had and is today heavily dependent on food imports under the U.S. food aid program.

The United States has also learned of the costs of overplowing and abusing its rich agricultural inheritance. As a result of overplowing and overgrazing of the Southern Great Plains during the early decades of this century, wind erosion gradually worsened, culminating in the dust-bowl era of the 1930s. The United States responded to this situation by fallowing 30,000,000 acres and by constructing literally thousands of miles of windbreaks in the form of rows of trees across the Great Plains. Had not the U.S. responded in this fashion, much of the Southern Great Plains, like the once-fertile fields of North Africa, would have been abandoned by now.

The alternative to bringing more land, much of it increasingly marginal, under the plow is to shift to more intensive use of existing farmland with agricultural chemicals and other inputs. This brings another set of ecological problems. It is now clear that the use of DDT and other chlorinated hydrocarbons for agricultural and such other purposes as malaria eradication is beginning to threaten at least some species of animal life. DDT is today found in the tissue of many forms of life, ranging from penguins in Antarctica to children in the villages in Thailand. DDT tends to concentrate in animal tissues, particularly fats; among the various species, the highest level of concentration occurs in the more predatory ones.

There is growing evidence that the use of DDT threatens the survival of some of the more predatory species of birds, such as the peregrine falcon, the bald eagle, and others because it affects their calcium formation capacity. The result is thin-shelled eggs, which often do not survive the nesting period. It is unfortunate that the adverse effect of DDT on certain species, particularly some of the predatory birds, is such that some may become extinct before the growing accumulation of DDT in the environment can be arrested. It is ironic that a Swiss scientist was awarded the Nobel Prize in 1947 for developing DDT and that less than a generation later its use is being banned in a number of countries.

Man is perhaps more likely to accumulate higher concentrations of DDT in his tissues than many species because he is an omnivore. The average American, for example, consumes more than 100 pounds of beef each year from animals, which, by virtue of their grazing capacity, are concentrators of DDT. Concentrations of DDT in mothers' milk in the United States now exceed the tolerance levels established for foodstuffs by the Food and Drug Administration.

In his efforts to increase his food supply, man has lately been faced not only with a scarcity of land but an increasingly acute scarcity of water as well. To acquire the amount of water needed, he has been gradually bringing the rivers under control through large irrigation systems. These irrigation systems, such as the Aswan in the United Arab Republic, the Volta River Project in Ghana, the Colorado River system in the United States, and some of the even more massive irrigation systems in Asia, are altering the environment in undeterminable ways. In addition to reducing the annual deposition of fertile silt on the river flood plains and displacing population in the vicinity of the reservoirs, they are creating ideal conditions for the spread of schistosomiasis, a debilitating parasitic affliction that is transmitted to man via snails. Like emphysema in the rich countries, schistosomiasis is an important and growing health problem today because of alterations man has wrought in his habitat in the poor countries. Also known as snail fever, this disease today afflicts an estimated 250,-000,000 people, more than any other disease, now that the incidence of malaria has declined.

As the demand for food in the future multiplies as a result of both population growth and rising incomes, particularly among two-thirds of mankind now poor and malnourished, man will be forced to intervene more and more in his environment, often with little understanding of the consequences of so doing. The demand for water for agricultural purposes will probably far exceed that available from conventional sources. Man will then be forced to desalt sea water for irrigation purposes or, what may prove more feasible, alter the world's climatic patterns to shift some of the rain now falling over the oceans over some of the earth's land masses.

This latter requires essentially two technological capabilities. The first is an enormous capacity for gathering information on temperatures, humidity, air mass movements, precipitation, ocean currents, and many other factors. Enough information must be collected to permit construction of a model of the earth's climatic system. Once this is in hand, a computational facility capable of simulating the earth's climatic system, indicating the effect of man-induced changes in climate, is required. Earth-orbiting satellites provide the information-gathering capacity and the latest generation of computers—the computational capability. As is often the case, technology is far ahead of institutions needed to use it effectively. Until a supranational institution is created to analyze, coordinate, and regulate activities in this field, individual countries and firms must exercise restraint in various activities designed specifically to alter climate, particularly those entailing alteration of rainfall patterns.

Forces already in motion, namely the projected increases in world population and planned increases in income, can be counted on to greatly expand the demand for food, requiring man to seriously consider even more far-reaching interventions in his environment well before the end of this century. With a limited amount of new land that can be economically brought under cultivation, future increases in food production must come largely from intensifying cultivation of land already under cultivation. This, in turn, implies that the use of chemical fertilizer must climb sharply, perhaps quadrupling 1970 levels by the end of the century. The use of pesticides must also increase greatly as crop stands become more specialized and more lush and as year-round cropping in the tropics spreads.

At this point in history, it is very clear that there is a need for a new supranational institution to manage the environment, including weather, on a global scale, much as a generation ago the International Monetary Fund was required to preserve and strengthen the international monetary system. Such an institution is needed to provide a framework for global cooperation. It would measure and analyze the impact of man on his environment. It would need to set standards and tolerance levels and to acquire the influence and power to enforce adherence to the standards it set.

Where man's supply of food is concerned, the question is no longer whether we can produce enough food but, rather, what are the environmental consequences of doing so. As we confront the latter question, there is growing doubt as to whether the earth's agricultural ecosystem can both sustain an end-of-century population of 6.5 billion and accommodate the universal desire on the part of the world's hungry for a much better diet.

* * *

MOTZ: That ends the formal part of our session, and the floor is now open to discussion and questions. I hope that enough ideas have been thrown out to stimulate many questions.

DAVIS: One of the things that strikes me is that we are moving not to the end of science, but perhaps to a new era in science. Science has become more and more specialized, and perhaps we have reached the point where the mere pursuit of increasing specialization will yield very little in the way of results. In the case of the environment this is particularly clear. We are reaching a stage in our work with the environment where we cannot any longer deal merely with one branch; you cannot talk about meteorology, you cannot talk about two-week weather forecasts, without knowing the state of the ocean to a depth of 1,500 meters. You cannot really study the long-term behavior of the atmosphere without knowing the solar cycle and its variations, without knowing the perturbations of the earth on its axis and the interactions this may have with earthquakes.

We are reaching a stage, fortunately, at which, at about the time we are acquiring a universal problem we are also acquiring a universal tool; specifically I refer to the ability to observe the earth's spectrum. We will soon be operational with an entirely new satellite system; in addition to day and night photographs, it will also begin to observe the sun and ionosphere, and within a year or two will sound the atmosphere for both temperature and moisture. By the end of the decade we expect to be able to observe the dynamic behavior of the solid earth, of a motion larger than, say, 10 centimeters. We expect soon to be able to observe the general circulation of the oceans, at least on the surface. We expect soon to be able to observe the temperature of the oceans on a global basis.

All of this represents a universal input, and I think that we are being forced away from specialization into a marriage of disciplines. I hope that perhaps this is the most fruitful direction for science to take. Certainly the fact that we are all sitting here today representing, probably, several dozen disciplines, not only in science but in many other areas of life, is encouraging, and perhaps this is the earliest stage of a new era. In a way, we have arrived at the point in the history of the race where, if we do not start to include our brains in the ecology as well as the rest of us, we are going to be in trouble. And, at this stage, we are beginning to have the tools to do so; I can foresee the time, perhaps 15 years from now, when with a little help we can probably literally computerize the ecology.

But this has some grave consequences for the social scientists, and I hope that this will be discussed sometimes during the meeting because individual freedom has become an issue. Already you no longer have the right to burn trash on Saturday afternoons, and very soon you will lose other freedoms, and nations will lose freedoms. You will not have the freedom to reverse rivers, or stop snow over the Great Lakes if such action is going to have consequences elsewhere in the world.

TISELIUS: Dr. Korff, you have brought out, very carefully, the conflicts between political and technological possibilities; in the pollution area I wonder whether one should perhaps consider waste more as a resource than as a pure nuisance. There are so many examples of this. I mean the possibility of deriving sulfur ferrin from sulfur dioxide. For another example, there is a lot of talk about aquaculture, and there have been suggestions about using low-cost electricity to pump up the rather nutritious deep water from the sea in order to fertilize coastal waters.

What I want to ask, really, is this: Isn't our problem that we are emotionally fragmented in our politicalizations and we don't really integrate to the extent we should? If we look at our problems in much bigger pieces, we will probably come up, from the environmental point of view, with much more acceptable solutions.

KORFF: I don't know whether I am the expert to answer this question, but I think there is no doubt at all that if we regard our waste as a resource, we can do a great deal with it. There are abundant illustration of this. Modern sewage disposal plants turn out useful products from waste matter. It is equally true that a great many other wastes can be processed and used usefully. Ash from burning coal can be used for landfill. There are products from the fishing industry that used to be thrown out; certainly they can be processed and used.

So I think there is no doubt at all that, by the application of the technology that we already have, we know how to do these things. We could do a great deal to benefit ourselves. Our problem is, to some extent, educational. There are many communities in the United States that do not have modern sewage disposal plants, and if you suggest to them that they should install one, they say, "Oh, my, that would cost a lot of money." Therefore we have an educating problem to do.

WADDINGTON: Speaking as a biologist, it has always seemed to me a fantastic situation that half of the energy produced by mankind is thrown away as waste heat; only about half of what we produce we actually use. The majority of standard physicochemical processes cannot make much use of low-temperature heat, but of course biological and enzyme-based processes can. Add an extra five degrees to the temperature of water and you can greatly increase the production of algae. This seems to be a fantastic challenge to biology, which we have not yet properly taken up. These are things we don't yet know the technology for to utilize waste heat, but it is obviously sitting there to be developed. It may be used for cultivating fish or other organisms, which has been investigated in a very small and totally inadequately financed way, or maybe something like solid-state enzymology.

But we have got and we do know of systems that can use low-temperature energy, and there is quantity of energy equal to that which we use now regarded as waste, but waiting to be utilized.

PERRIN: It was stated earlier by Dr. Motz that during the first 25 years of this century two new theories were developed which introduced a fundamental constant: the quantum theory in 1900; and the relativity theory in 1905, with the fundamental constant the velocity of light.

The theory of general relativity was also mentioned, and I would say that this is also really associated with a new fundamental constant, the expansion of the universe. This is not as accurately known, but it may be of basic importance in cosmology, and it is completely new when compared with the relativity theory.

During the past 45 years, essentially no new fundamental basic principles have been discovered; I agree on this point; however, I would note that that period was the end of fundamental physics, and we have since discovered in physics new, surprising facts, that seem to lead to new, fundamental principles for understanding them. I think that all the effort in high-energy physics, which required a great deal of money, as you know, was well worth it, for example, in seismic research, in which, in the next decade, new fundamental principles will be discovered, similar in importance for understanding the universe to those that were discovered during the first part of this century. I think this is really the justification for all the money spent in this direction. It is essentially for the intellectual benefit of mankind; I don't think that any practical application can come from this new discovery, even if it *is* quite fundamental. It is not because of possible practical application that so much money should be spent on such research; it is for the profit that accrues to the spiritual development of mankind, which may be a luxury, but is quite fundamental to all our modern society.

A few words more about what could be discovered in fundamental science in branches other than high-energy physics: I think that astronomy, with all the marvelous discoveries that have been made, is a field of research in which new fundamental laws in cosmology will perhaps be discovered in the near future; or we may just miss completely, but a large effort in this direction in manpower and money is necessary. And also, in this case, with no practical application in view, it is essential for the better understanding of the universe in which we live that all this work and financial expenditure be carried out.

Of course, in the field of biology it is possible that a new fundamental law of nature may be discovered. But, anyway, the understanding of extremely complex systems, even if there is no new fundamental law of biology involved, is certainly something that, if not basic, is certainly most important for mankind.

In this case, as differing from the other two fields in which I consider that basic research should be developed, we may expect extremely important applications for mankind from the continuing progress of fundamental biology.

So, both for progress in our intellectual understanding and for progress in practical application to medicine, agriculture, etc., we should concentrate on developments in fundamental biology.

DE CASTRO: I believe most of you would agree that the problem of pollution and environmental contamination is primarily a political problem. And, having suggested this, I also suggest that politics is a science also. When people asked Einstein why man, who had discovered atomic energy, had not utilized this energy for the benefit of humanity but to kill people *en masse*, he answered that it was, perhaps, because politics is a much more complicated science than physics. I believe he was right.

It seems to me that pollution is some kind of social disease specialized in by the well-developed areas of the world. The underdeveloped areas, two-thirds of the world, have not yet been entirely contaminated by progress, by industry, and by pollution. So perhaps one point is to avoid spread of this disease by contamination of the other areas that have not yet reached this very bad condition. That is a point that we should perhaps emphasize in a discussion about contamination.

Why has this contamination and pollution developed in such large scale in the overdeveloped areas? Because industrialization and its special technology have no ethics. Until today it has been considered an instrument of interest to increase productivity and its concomitant benefits, but man has been completely forgotten as a piece of the instrument of productivity. But we have never considered that we must protect man from contamination.

I would like to ask: Don't you think we must be careful not to advise the underdeveloped countries to imitate the well-developed countries? Should they undergo the same kinds of development? In other words, that which I call the Utopia of exportation would not be exported because the price would be the increase of bad seeds, and unfortunately we—all the world—will be contaminated.

It seems to me that pollution is something like underdevelopment; it is mainly a political problem. The decade of the development of the undeveloped nations has been transformed into the decade of deception, of frustration, because we have not had any kind of political policy or plan for development. Everything has been improvised, everything has been without a plan.

I think what is essential is that these countries must create their own techniques, so that current techniques will not be encrusted. We must see to it that the ecosystems of these developing areas are not disturbed by the creation of pollution.

BJORKSTEN: Regarding seasons, we have seen that the reproductive response of certain organisms is extremely affected by 15 minutes' variation per day of daylight. And certainly the differences in foods, the greater supplies of vitamins, and different kinds and quantities of proteins that go with various seasons are bound to greatly influence development.

PREGEL: I would like to note that in discussions of contamination and pollution we can't speak about the whole globe, because the contamination is not equal everywhere, so all these problems are not quite the same everywhere. But we certainly cannot dream of reducing the production of energy, as some have suggested. We can find safe ways to produce energy, but nobody can stop technical developments. We have to see how we can use the technology more effectively and safely, and how we can help countries in the developing stages.

ON RE-DOING MAN *

Kurt Hirschhorn

Mount Sinai School of Medicine
New York, N.Y.

The past 20 years and, more particularly, the past five years have seen an exponential growth of scientific technology. The chemical structure of the hereditary material and its language have essentially been resolved. Cells can be routinely grown in test tubes by tissue-culture techniques. The exact biochemical mechanisms of many hereditary disorders have been clarified. Computer programs for genetic analysis are in common use. All these advances and many others have inevitably led to discussions and suggestions for the modification of human heredity, both in individuals and in populations: genetic engineering.

One of the principal concerns of the pioneers in the field is the problem of the human genetic load, that is, the frequency of disadvantageous genes in the population. Each of us carries between three and eight genes that, if present in double dose in the offspring of two carriers of identical genes, would lead to severe genetic abnormality, or even to death of the affected individual before or after birth. In view of the rapid medical advances in the treatment of such diseases, it is likely that affected individuals will be able to reproduce more frequently than in the past. Therefore, instead of a loss of genes due to death or sterility of the abnormal, the mutant gene will be transmitted to future generations in a slowly but steadily increasing frequency. This is leading the pessimists to predict that we will become a race of genetic cripples requiring a host of therapeutic crutches. The optimists, on the other hand, have a great faith that the forces of natural evolution will continue to select favorably those individuals who are best adapted to the then current environment. It is important to remember in this context that the "natural" environment necessarily includes man-made medical, technical, and social factors.

Because it appears that at least some of the aspects of evolution and a great deal of genetic planning will be in human and, specifically, scientific hands, it is crucial at this relatively early stage to consider the ethical implications of these proposed maneuvers. Few scientists today doubt the feasibility of genetic engineering, and there is considerable danger that common use of this practice will be upon us before its ethical applications are defined.

A number of different methods have been proposed for the control and modification of human hereditary material. Some of these methods are meant to work on the population level, some on the family level, and others directly on the affected individual. Interest in the alteration of the genetic pool of human populations originated shortly after the time of Mendel and Darwin, in the latter part of the 19th century. The leaders were the English group of

* Based on a paper originally given at a symposium jointly sponsored by Marymount College, N.Y., and Commonweal Magazine. Reprinted by permission from Commonweal **88**: 257–261 (1968), 232 Madison Ave., New York, N.Y.

eugenicists headed by Galton. Eugenics is nothing more than planned breeding. This technique, of course, has been successfully used in the development of hybrid breeds of cattle, corn, and other food products.

Human eugenics can be positive or negative. Positive eugenics is the preferential breeding of so-called superior individuals in order to improve the genetic stock of the human race. The most famous of the many proponents of positive eugenics was the late Nobel Prize winner Herman J. Muller. He suggested that sperm banks be established for a relatively small number of donors, chosen by some appropriate panel, and that this frozen sperm remain in storage until some future panel had decided that the chosen donors truly represented desirable genetic studs. If the decision is favorable, a relatively large number of women would be inseminated with these samples of sperm; proponents of this method hope that a better world would result. The qualifications for such a donor would include high intellectual achievement and a socially desirable personality, qualities assumed to be affected by the genetic make-up of the individual, as well as an absence of obvious genetically determined physical anomalies.

A much more common effort is in the application of negative eugenics. This is defined as the discouragement or the legal prohibition of reproduction by individuals carrying genes leading to disease or disability. This can be achieved by genetic counseling or by sterilization, either voluntary or enforced. There are, however, quite divergent opinions as to which genetic traits are to be considered sufficiently disadvantageous to warrant the application of negative eugenics.

A diametrically opposite solution is that of euthenics, which is a modification of the environment in such a way as to allow the genetically abnormal individual to develop normally and to live a relatively normal life. Euthenics can be applied both medically and socially. The prescription of glasses for nearsighted individuals is an example of medical euthenics. Special schools for the deaf, a great proportion of whom are genetically abnormal, is an example of social euthenics. The humanitarianism of such efforts is obvious, but it is exactly these types of activities that have led to the concern of the pessimists, who assume that evolution has selected for the best of possible variations in man and that further accumulations of genes considered abnormal can only lead to decline.

One of the most talked-about advances for the future is the possibility of altering an individual's genetic complement. Since we are well on the way to understanding the genetic code, as well as to deciphering it, it is suggested that we can alter it. This code is written in a language of 64 letters, each one determined by a special arrangement of three out of four possible nucleotide bases. A chain of these bases is called deoxyribonucleic acid, or DNA, and makes up the genetic material of the chromosomes. If the altered letter responsible for an abnormal gene can be located and the appropriate nucleotide base substituted, the corrected message would again produce its normal product, which would be either a structurally or enzymologically functional protein.

Another method of providing a proper gene, or code word, to an individual having a defect has been suggested from an analysis of viral behavior in bacteria. It has long been known that certain types of viruses can carry genetic information from one bacterium to another or instruct a bacterium carrying it to produce what is essentially a viral product. Viruses are functional only when they live in a host cell. They use the host's genetic machinery to translate their own genetic codes. Viruses living parasitically in human cells can cause such diseases as poliomyelitis and have been implicated in the causation of tumors. Other viruses have been shown to live in cells and to be reproduced along with the cells without causing damage either to the cell or to the organism. If such a harmless virus either produces a protein that will serve the function of one lacking in an affected individual, or if it can be made to carry the genetic material required for such functions into the cells of the affected individual, it could permanently cure the disease without additional therapy. If carried on to the next generation, it could even prevent the inheritance of the disease.

Transplanting Nuclei

An even more radical approach has been outlined by Lederberg. It has become possible to transplant whole nuclei, the structures that carry the DNA, from one cell to another. It has become easy to grow cells from various tissues of any individual in tissue culture. Such tissue cultures can be examined for a variety of genetic markers and thereby screened for evidence of new mutations. Lederberg suggests that it would be possible to use nuclei from such cells derived from known human individuals, again with favorable genetic traits, for the asexual human reproduction of replicas of the individuals whose nuclei are being used. For example, a nucleus from a cell of the chosen individual could be transplanted into a human egg whose own nucleus has been removed. This egg, implanted in a womb, could then divide just like a normal fertilized egg, to produce an individual genetically identical to the one whose nucleus was used. One of the proposed advantages of such a method would be that, as in positive eugenics, one could choose the traits that appear to be favorable, and do so with greater efficiency by eliminating the somewhat randomly chosen female parent necessary for the sperm bank approach. Another advantage is that one can mimic what has developed in plants as a system for the preservation of genetic stability over limited periods of time. Many plants reproduce intermittently by such vegetative or parthenogenetic mechanisms, always followed by periods of sexual reproduction for the purpose of elimination of disadvantageous mutants and increase in variability.

Another possibility derives from two other technological advances. Tissue typing, similar to blood typing, has made it possible to transplant cells, tissues, and organs from one individual to another with reasonably long-term success. During the past few years, scientists have also succeeded in producing cell hybrids containing some of the genetic material from each of two cell types, either from two different species or from two different individuals of the same species. Very recently, Weiss and Green at New York University have succeeded in hybridizing normal human culture cells with cells from a long-established mouse tissue-culture line. Different products from such fusions contain varying numbers of human chromosomes and, therefore, varying amounts of human genes. If such hybrids can be produced which carry primarily that genetic information which is lacking or abnormal in an affected individual, transplantation of these cultured cells into the individual may produce a correction of his defect.

These are the proposed methods. It is now fair to consider the question of feasibility. Feasibility must be considered not only from a technical point of view; of equal importance is the effect of each of these methods on the evolution of the human population and the effect of evolution on the efficacy of the method. In general, it can be stated that most of the proposed methods either now or in the not too distant future will be technically possible. We are, therefore, not dealing with hypothesis in science fiction but with scientific reality. Let us consider each of the propositions independently.

Positive eugenics by means of artificial insemination from sperm banks has been practiced successfully in cattle for years. Artificial insemination in man is an everyday occurrence. But what are some of its effects? There is now ample evidence in many species, including man, of the advantages for the population of individual genetic variation, mainly in terms of flexibility of adaptation to a changing environment. Changes in environment can produce drastic effects on some individuals, but a population that contains many genetic variations of that set of genes affected by the particular environmental change will contain numerous individuals who can adapt. There is also good evidence that individuals who carry two different forms of the same gene, that is, are heterozygous, appear to have an advantage. This is true even if the gene in double dose—that is, in the homozygous state—produces a severe disease. For example, individuals homozygous for the gene coding for sickle cell hemoglobin invariably develop sickle cell anemia, which is generally fatal before the reproductive years. Heterozygotes for the gene are, however, protected more than normals from the effects of the most malignant form of malaria. It has been shown that women who carry the gene in single dose have a higher fertility in malarial areas than do normals. This effect is well known to agricultural geneticists and is referred to as hybrid vigor. Fertilization of many women by sperm from few men will have an adverse effect on both of these advantages of genetic variability because the population will tend to be more and more alike in its genetic characteristics. Also, selection for a few genetically advantageous factors will carry with it selection for a host of other genes present in the same individuals, genes whose effects are unknown when present in high numbers in the population. Therefore, the interaction between positive eugenics and evolution makes this method not feasible on its own.

Abnormal Offspring

Negative eugenics is, of course, currently practiced by most human geneticists. It is possible to detect carriers of many genes that, when inherited from both parents, will produce abnormal offspring. Parents, both of whom carry such a gene, can be told that they have a one-in-four chance of producing an abnormal child. Individuals who carry chromosomal translocations are informed that they have a high risk of producing offspring with congenital malformations and mental retardation. But how far can one carry out such a program? Some states have laws prescribing the sterilization of individuals mentally retarded to a certain degree. These laws are frequently based on false information regarding the heredity of the conditions. The marriage of people with reduced intelligence is forbidden in some localities, again without adequate genetic information. While the effects of negative eugenics may be quite desirable in individual families with a high risk of known hereditary diseases, it is important to examine its effects on the general population.

These effects must be looked at individually for conditions determined by genes that express themselves in a single dose (dominant) or in double dose (recessive) and those which are due to an interaction of many genes (polygenic inheritance). With a few exceptions, dominant diseases are rare and interfere severely with reproductive ability. They are generally maintained in the population by new mutations. Therefore, there is either no need or essentially no need for discouraging these individuals from reproduction. Any discouragement, if applicable, will be useful only within that family but not have any significance for the general population. One possible exception is the severe neurological disorder, Huntington's chorea, which does not express itself until most of the patient's children are already born. In such a situation it may be useful to advise the child of an affected individual that he has a 50% chance of developing the disease and a 25% chance of any of his children's being affected. Negative eugenics in such a case would at least keep the gene frequency at the level usually maintained by new mutations.

The story is quite different for recessive conditions. Although detection of the clinically normal carriers of these genes is currently possible only for a few diseases, the techniques are rapidly developing whereby many of these conditions can be diagnosed even if the gene is present only in single dose and will not cause the disease. Again, with any particular married couple it would be possible to advise them that they are both carriers of the gene and that any child of theirs would have a 25% chance of being affected. However, any attempt to decrease the gene frequency of these common genetic disorders in the population by prevention of fertility of all carriers would be doomed to failure. First, we all carry between three and eight of these genes in single doses. Secondly, for many of these conditions, the frequency in the population of carriers is about one in 50 or even greater. Prevention of fertility for even one of these disorders would stop a sizable proportion of the population from reproducing and for all of these disorders would prevent the entire human population from having any children. Reduction in fertility of a sizable proportion of the population would also prevent the passing on to future generations of a great number of favorable genes and would, therefore, interfere with the selective aspects of evolution, which can function only to improve the population within a changing environment by selecting from a gene pool containing enormous variability. It has now been shown that in fact no two individuals, with the exception of identical twins, are likely to be genetically and biochemically identical, thereby allowing the greatest possible adaptation to changing environment and the most efficient selection of the fittest.

The most complex problem is that of negative eugenics for traits determined by polygenic inheritance. Characteristics inherited in this manner include many measurements that are distributed over a wide range throughout the population, such as height, birth weight, and intelligence. The last of these can serve as a good example of the problems encountered. Severe mental retardation in a child is not infrequently associated with perfectly normal intelligence or in some cases even superior intelligence in the parents. These cases can, a priori, be assumed to be due to the homozygous state in the child of a gene leading to mental retardation, the parents representing heterozygous carriers. On the other hand, borderline mental retardation shows a high association with subnormal intelligence in other family members. This type of deficiency can be assumed to be due to polygenic factors, more of the pertinent genes in these families being of the variety that tends to lower intelligence. However, among the offspring of these families there is also a high proportion of individuals with normal intelligence and a sprinkling of individuals with superior intelligence.

All of these comments are made with the realization that our current measurements of intelligence are very crude and cannot be compared between different population groups. It is estimated that, on the whole, people with superior intelligence have fewer offspring than do those of average or somewhat below average intelligence. If people of normal intelligence were restricted to producing only two offspring and people of reduced intelligence were by negative eugenics prevented from having any offspring at all, the result, as has been calculated by the British geneticist Lionel Penrose, would be a gradual shift downward in the mean intelligence level of the population. This is due to the lack of replacement of intellectually superior individuals from offspring of the majority of the population, that is, those not superior in intellect.

Current Possibilities

It can be seen, therefore, that neither positive nor negative eugenics can ever significantly improve the gene pool of the population and simultaneously allow for adequate evolutionary improvement of the human race. The only useful aspect of negative eugenics is in individual counseling of specific families in order to prevent some of the births of abnormal individuals. One recent advance in this sphere has important implications from both a genetic and a social point of view. It is now possible to diagnose genetic and chromosomal abnormalities in an unborn child by obtaining cells from the amniotic fluid in which the child lives in the mother. Although the future may bring further advances, allowing one to start treatment on the unborn child and to produce a functionally normal infant, the only currently possible solution is restricted to termination of particular pregnancies by therapeutic abortion. This is, of course, applied negative eugenics in its most extreme form.

Euthenics, the alteration of the environment to allow aberrant individuals to develop normally and to lead a normal life, is currently being employed. Medical examples include special diets for children with a variety of inborn errors of metabolism who would, in the absence of such diets, either die or grow up mentally retarded. Such action, of course, requires very early diagnosis of these diseases, and programs are currently in effect to routinely examine newborns for such defects. Other examples include the treatment of diabetics with insulin and the provision of special devices for children with skeletal deformities. Social measures are of extreme importance in this regard. As has many times been pointed out by Dobzhansky, it is useless to plan for any type of genetic improvement if we do not provide an environment within which an individual can best use his strong qualities and obtain support for his weak qualities. One need only mention the availability of an environment conducive to artistic endeavor for Toulouse-Lautrec, who was deformed by an inherited disease.

The feasibility of alteration of an individual's genes by direct chemical change of his DNA is technically an enormously difficult task. Even if it became possible to do this, the chance of error would be enormous. Such an error, of course, would have the diametrically opposite effect of that desired; in other words, the individual would become even more abnormal. The introduction of corrective genetic material by viruses or transplantation or appropriately hybridized cells is technically predictable and, since it would be performed only in a single affected individual, would have no direct effect on the population. If it became widespread, it could, like euthenics, increase the frequency in the population of so-called abnormal genes, but if this treatment became a routine phenomenon, it would not develop into an evolutionarily disadvantageous situation. It must also be constantly kept in mind that medical advances are occurring at a much more rapid rate than any conceivable deterioration of the genetic endowment of man. It is, therefore, very likely that such corrective procedures will become commonplace long before there is a noticeable increase in the load of disadvantageous genes in the population.

The growing of human beings from cultured cells, while again possibly feasible, would interfere with the action of evolutionary forces. There would be an increase, just as with positive eugenics, of a number of individuals who would be alike in their genetic complement, with no opportunity for the high degree of genetic recombination that occurs during the formation of sperm and eggs and which is evident in the resultant progeny. This would diminish the adaptability of the population to changes in the environment and, if these genetic replicas were later permitted to return to sexual reproduction, would lead to a marked increase in homozygosity for a number of genes with the disadvantages pointed out before.

Who Will Be the Judges?

We see, therefore, that many of the proposed techniques are feasible although not necessarily practical in producing their desired results. We may now ask the question, which of these are ethical from a humanistic point of view? Both positive and negative eugenics as applied to populations presume a judgment of what is genetically good and what is bad. Who will be the judges, and where will the line be between good and bad? We have had at least one example of a sad experience with eugenics in Nazi Germany. This alone can serve as a lesson on the impossibility of separating science and politics. The most difficult decisions will come in defining the borderline cases. Will we breed against tallness because space requirements become more critical? Will we breed against nearsightedness because people with glasses may not make good astronauts? Will we forbid intellectually inferior individuals from procreating despite their proved ability to produce a number of superior individuals? Or should we, rather, provide an adequate environment for the offspring of such individuals to realize their full genetic potential?

C. C. Li, in his presidential address to the American Society of Human Genetics in 1960, pointed out the real fallacy in eugenic arguments. Man has continuously improved his environment to allow so-called inferior individuals to survive and reproduce. The movement into the cave and the putting on of clothes protected the individual unable to survive the stress of the elements. Should we then consider that we have reached the peak of man's progress, largely determined by environmental improvements designed to increase fertility and longevity, and that any future improvements designed to permit anyone to live a normal life will only lead to deterioration? Nineteenth-century scientists, including such eminent biologists as Galton, firmly believed that this peak was reached in their time. This obviously fallacious reasoning must not allow a lapse in ethical considerations on the part of the individual and by humanity as a whole, just to placate the genetic pessimists.

The tired axiom of democracy that all men are created equal must not be considered from the geneticist's point of view, since genetically all men are created unequal. Equality must be defined purely and simply as equality of opportunity to do what one is best equipped to do. When we achieve this, the forces of natural evolution will choose those individuals best adapted to this egalitarian environment. No matter how we change the genetic make-up of individuals, we cannot do away with natural selection. We must always remember that natural selection is determined by a combination of truly natural events and the artificial modifications that we are introducing into our environment in an exponentially increasing number.

With these points in mind, we can try to decide what, in all of these methods, is both feasible and ethical. I believe that the only logical conclusion is that all maneuvers of genetic engineering must be judged for each individual and in each case must take into primary consideration the rights of the individual. This is by definition impossible in any attempt at positive eugenics. Negative eugenics in the form of intelligent genetic coun-- seling is the answer for some. Our currently unreasonable attitude toward practicing negative eugenics by means of intelligent selection for therapeutic abortion must be changed. Basic to such a change is a more accurate definition of a living human being. Such restricted uses of negative eugenics will prevent individual tragedies. Correction of unprevented genetic disease, or that due to new mutation, by introduction of new genetic material may be one answer for the future; but until such a new world becomes universally feasible, we must on the whole restrict ourselves to environmental manipulations from the points of view both of allowing affected individuals to live normally and of permitting each individual to realize his full genetic potential. There is no question that genetic engineering will come about. But both the scientists directly involved and, perhaps more important, the political and social leaders of our civilization must exercise utmost caution in order to prevent genetic, evolutionary, and social tragedies.

* * *

WADDINGTON: I wanted to make some general remarks about biology, but first I would like to make one or two comments on Dr. Hirschhorn's statement. I think the possibilities of biological engineering, particularly of the kind he talked about, sound very much like science fiction. But you are not necessarily very far away from practical solution. Let us say they are not far away if anyone is prepared to spend the money to develop them. It seems that nuclear transportation is extremely easy, and we can clone frogs, we can transplant nuclei from one egg into another, and we can do much more radical things than merely cloning individuals. It is quite possible to make hybrid cells by somatic cell fusion. Grow cells in culture, particularly in the presence of a certain killed virus, and they very often stick together though they are somewhat unstable in their behavior at the division.

And you can select out from such a culture nuclei that have different combinations of chromosomes from different species. Such nuclei have been made. There are hybrid cells, hybrid between man and mouse. Nobody has ever yet injected such a nucleus into an egg capable of developing. The thing has not yet been done in frogs, but it could be done tomorrow, making transplantation between hybrid and toad chromosomes into a frog. It cannot yet be done in mammals, but people are working on developing methods of transplanting nuclei into the mammal—which is, of course, much more difficult to handle. But it is perfectly possible to do.

But to make it a practical proposition that could be relied on would cost quite a lot. It would not cost as much as developing the atom bomb, but it is not the sort of thing that can be developed in secret without everybody's knowing about it. It will be developed only if the taxpayer agrees to foot the bill. But I think it is a definite question of whether such types of work will be developed, and for what purpose.

I think it is most likely that in the next ten or 20 years such hybridization will be developed in order to produce better domestic animals for, say, tropical Africa. Plant breeders have, for decades, put disease resistance from a wild species of potato into cultivated potatoes by arranging to transfer the chromosome. Animal breeders have not been able to do this in the past, but they will be able to in about ten or 15 years if they want to. And I think it is quite likely they will want to for breeding really suitable animals for domestic life in the infested parts of Africa. Once this has been done in cattle, it can be done in other species of animals, and people will have to think whether they want it done with hominoids and themselves.

Having said that in relation to Dr. Hirschhorn's paper, I would like to make some remarks on my own. I would like to make a theoretical, almost philosophical, point.

Civilization as we know it in recent years has been based on and dominated by physical-chemical methods. We are essentially a mechanically based civilization. I think the major problems in the world today and the immediate future are really much more biological than they are physical and chemical.

It is physically impossible for the whole world to become developed in the sense of the highly industrialized developed nations of the West today. There simply are not enough raw materials of various kinds to go around. Each American is supported, I believe, by the order of a thousand tons of steel and several tons of aluminum, titanium, all sorts of minerals. It has been calculated that, to provide everyone in the world with a four-sheet daily newspaper and four sheets of toilet paper would require production of all the forests in the world to be multiplied by four times. I think there is no practical possibility of the whole world's developing into our present kind of quick-turnover, usually wasteful type of civilization.

If the world is going to develop, it must develop into something based on long lifetimes of artifacts that either last for a very long time or, when used up, are recycled. That is much more the kind of thing we find in biology. In physical systems, one is usually analyzing systems with a relatively small number of components, and one is usually interested in maximizing the output or turnover rate of one of the components. Biological systems normally involve very large numbers of components, and the only thing one thinks of in the way of maximizing is maximizing from extremely complex function of all the components. The only thing biology really maximizes is this peculiar index called evolutionary fitness, and this is such a complex concept that nobody, in my opinion, ever really satisfactorily defines what it means. It is a component something like the value of a hand of cards that become bridge. It is a capacity to cope with whatever may turn up over the next few years. It is highly complex, but it is the only thing that goes in for maximizing.

Everything else goes into optimizing, and I think this is a totally different type of outlook, one much more appropriate to the kinds of work we are going into. We have to be able to think up systems in which all the components are optimized in relation to some global, overall, end product that is, as I say, the function of a vast number of variables. Take, for instance, the agricultural revolution. A whole of civilization has been based on artificial landscapes, artificial landscapes that are really types of agriculture. They may be the European open-field systems, or the development of enclosed fields with rotation of crops, or terraced rice fields in the East, etc., but they are all artificial, man-made, functioning ecosies. But the basis of civilization was laid in systems that were relatively self-stabilizing.

Now we have based agriculture on monocultures, maximizing some things, such as the output of wheat—maximizing it by putting in vast amounts of fertilizers, playing with the risk of a new pest from which the crop is completely unprotected—and we are gambling on some unstable non-self-stabilizing ecosies.

I think we are going to have to try to think of designing self-stabilizing ecological systems, and I think this is one of the great challenges to biology; what do you do with the great masses of bush in Africa? We simply do not know how to turn those things into productive, self-stabilizing, new forms of landscape suitable for civilized life. Europe has done it, China has done it, India perhaps did it for a bit, but there are vast areas of the world where this re-creation of true nature into a man-adapted nature has not been done. And I believe this is really much more challenging and ultimately much more immediate than the problem of environmental pollution that we, in the West, are so excited about at the moment. In point of fact, the major thing we have done to our environment in the last 50 years is to clean it up. We have cleaned it up biologically. We have got rid of yellow fever, the plague, most of the intestinal parasites, the nematodes, what have you. Our environment is much cleaner than it was a hundred years ago or than it is now in India and Africa.

The particular types of pollution we are putting into it now are, of course, very sinister and certainly have to be controlled. But on a world scale, this is somewhat second-order, and if you walk through the cities of Bombay or Calcutta you are not really likely to think that New York is a lot dirtier. I'm afraid you will find they are a lot dirtier than New York. But they are dirty not only with industrial pollutants but also with biological pollutants of disease, etc. And I feel the great problem for the world as a whole and for 80 percent of the developing world is first to clean it up biologically and second to design new, self-stabilizing functional ecosystems that produce the agriculturally useful crops we have.

POTENTIAL OF APPLIED MICROBIOLOGY

Carl-Göran Hedén

Karolinska Institute
Stockholm, Sweden

Before man learned to use fire, he probably found that seed grains left soaking in rain water would turn into a stimulating brew. Small chisel blows of trial and error slowly shaped a range of primitive fermentations that preserved perishable foods and improved their taste, texture, and nutritive value. We now know that those processes are based on activities linked to a microcosm of unicellular organisms that always surround us, fighting a constant battle with each other and with the higher plants and animals. The importance of this invisible world is indicated by the fact that the upper few centimeters of any well-cultivated soil contains some 300–3,000 kilograms of living microorganisms per hectare. Many of these microorganisms we are able to classify only in general terms; some we cannot even cultivate. Lacking automatic screening devices, we must resort to guesswork about such important matters as the ecological effects of herbicides, pesticides, etc., but this will certainly change as a consequence of recent advances in numerical taxonomy. We will then have tools that will be very powerful for turning microbial ecology to our benefit.

Living beings not only have developed defense mechanisms adapted to their normal environment but also have often learned to derive benefits from other organisms that fit into the same ecological niche. Man, who has developed highly advanced protective mechanisms, has not only been able to expand his niche by countering the defenses of other organisms; he has also learned to exploit other living systems in a conscious and systematic fashion. Microbiology concerns itself with both aspects: the neutralization of pathogens (by vaccines, chemotherapeutic agents, interferon inducers, sera, etc.) and the use of selected strains to yield valuable products (amino acids or vitamins) or perform desirable conversions (steroid synthesis, mineral leaching, etc.). Sometimes—as with most antibiotics—the product we seek is part of the armory a microorganism has developed to guard against competition in a mixed population. We select the organism on the basis of certain desired features, then grow it without imposing the stresses of its normal environment (semistarvation and a temperature that is just as suboptimal as the dissolved-oxygen concentrations it has to put up with in nature). This means that we may well prevent derepression of potentially important genetic sites, but our technique also excludes from consideration products that require cooperation between several different types of microorganisms. Because symbiosis is so often observed in nature, it would be logical to think that cooperation in building barriers around common ecological niches would be fairly common.

Obviously then there is much to be said for systematic studies of "synthetic" mixed populations, especially now that continuous-culture methods have given us such a versatile tool. It is now possible to study the competitive forces regulating the carbon and nitrogen cycles in nature, as well as to begin development of an automated optimization of microbiological production processes.

Until now, applied microbial ecology (wine, beer, cheese, sauerkraut, vinegar, and Oriental fermented foods) has gone its own empirical way, unaffected by the remarkable developments in microbial physiology and genetics. This will certainly change, and we will probably devote much attention to symbiotic fermentations. One interesting example has been provided by M. Tveit and co-workers, who grew a fodder yeast (*Torula*) on waste starch partially decomposed by a mold (*Endomycopsis*). And V. R. Srinivasan demonstrated the decomposition of cellulose by one bacterium (*Cellulomonas*), assisted by another (*Alcaligenes faecalis*) that has the capacity to eliminate a primary breakdown product (cellobiose) that would otherwise arrest the breakdown.

Such efforts are important, because we are now starting to touch uncomfortably many of the walls around our ecological niche. One wall is in outer space, where the astronauts may be exposed to a new size range of infectious particles floating about them, and where they may also risk loss of some of the protection afforded by the normal flora they carry along from earth. Such a mixed crowd of "stowaways" could include pathogens, and these might well play havoc with the crew of an orbiting laboratory or a planetary base. The orbiting personnel might previously have been shielded from the subclinical minor infections that stimulate our defensive mechanisms here on Earth. Obviously, microbial ecology could become a critically important feature of health surveillance in space, but in a more distant future, the well-being of certain microorganisms could well become the second most important consideration. This would, for example, be true if it were decided to use microbiological gas exchangers (algae or *Hydrogenomonas*) for controlling the atmospheres in confined spaces.

Space research offers a dramatic reminder that man depends on his microbiological environment, but our ignorance is disturbingly great. For instance, the intestinal flora with which we live in balance provides us with valuable vitamins (K and the B group), but the details of what it does to our food are largely unknown. However, we know that aromatization of certain organic compounds by the intestinal flora may yield harmful products and that many anaerobes produce potent toxins. One cannot help wondering what inoculation at birth with a cocktail of proved beneficial strains could mean to health and aging. In particular, one wonders if it would not be possible to establish some strains with a potent nitrogen-fixing ability or with defective regulatory genes in order to yield essential amino acids that might supplement a deficient dietary protein later in life. Maintenance of the "artificial" flora would probably require occasional reinoculation and, possibly, some adjustment of food habits, but it would be worth the trouble. Under all circumstances, it might become an established practice to deposit one's normal flora, as well as germ cells and bone marrow, in deep-freeze storage. This might be considered reasonable preparation for later exposures to broad-spectrum antibiotics, radiation, or leukemia (virus?).

Space research may remind us of ecological walls, but it is, of course, here on earth that the walls are closing in on us. The planet is getting crowded and full of biologically significant stresses, increasing the probability of infectious disease here, no less than in the confined space cabin. Again, the normal flora is jeopardized: by indiscriminate use of antibiotics in medicine, by chemicals in industry and agriculture, by widespread cleaning and disinfecting compounds in the household. The consequences, for instance in the form of disturbed mineralization processes, are now so common that we take them for granted and forget that applied microbiology can offer many remedies.

Let us first consider the microbiological implications of increased population density, which go far beyond influences on the epidemiology of infectious disease. There is, for instance, a need to increase food production drastically if predictions of serious famines are not to come true. Here, the first microbiological priority should be concerned with the reduction of waste (fermentative preservation techniques and control of toxigenic fungi and bacteria). Second in importance is an expansion of the traditional animal protein resources (fortification of animal fodder by yeast grown on hydrocarbons or sulfite liquor or by algae grown on sewage; extended conversion of cellulose waste to meat by use of microorganisms that operate inside the herbivores; veterinary vaccines against rinderpest, foot-and-mouth disease, and Newcastle virus; aerosol vaccination, etc.) and of the normal agricultural crops. An important aspect of the latter has been referred to as "the green revolution"—expanding monocultures of high-yielding plant varieties grown under a chemical umbrella of insecticides. The microbiologist should be aware that the latter may become increasingly unattractive as environmental concern spreads from the industrialized nations to the developing countries of the world. Because breeding of insect-resistant plant varieties is a time-consuming affair, microbiology should focus attention on biological control by insect pathogens. An example of the current interest in this area of study is the recent establishment by a number of academies of an entomological field station in Nairobi. It is fairly safe to predict that biological control will become an important activity and that it will be greatly accelerated by recent advances in aerobiology. The relation between particle size and penetration of a jungle canopy has for instance, been established, and microencapsulation has provided new possibilities to protect bacteria and viruses suspended in the air. In addition, insect tissue-cell culture may lead to novel production techniques.

The "green revolution's" requirements for water and for nitrogen fertilizers (as much as 100 kilograms per hectare) give the applied microbiologist additional food for thought. Nitrogen fertilizers already ensure the existence of one-sixth of the globe's population; by the turn of the century we must obviously produce some 50 million tons per year, which may not appear to be very much when compared with the 100 million tons of nitrogen fixed by soil microorganisms, but it is still a lot to transport. Nitrogen is the major nutrient whose lack often limits plant development in many parts of the world, and India now uses one-fifth of its foreign exchange to buy fertilizers. Consequently, an extended use of blue-green algae, or bacteria that can utilize the nitrogen in the air, or plants (alfalfa, clover, soy) that can use microorganisms for the same purpose, seems essential. But the proper organisms for use as soil or seed inoculants must be found, and this still requires extensive investigation. Genetic research aimed at giving nitrogen-fixing ability to such common grains as wheat might also be undertaken, and many sectors of soil microbiology should be stimulated. There is every reason to believe, for example, that we can increase the economic return per kilogram of nitrogen fertilizer used in the field. This may actually be quite low in some areas because denitrifying bacteria cause volatilization of the nitrogen that is applied to the soil or formed during ammonium oxidation. In other cases, the assimilation processes are so slow that inorganić nitrogen is washed out of the soil by percolating water before it can be used by plants. Knowledge about such processes, about the depression of biological nitrogen fixation by mineral fertilizers, and about the microbial production of substances that cement dust together to form aggregates that improve aeration and root development, more information about all these phenomena will certainly indicate solutions to many current problems. Our example of this is the mineralization process in waste treatment, which becomes more and more important as the recirculation concept becomes the central theme in pollution control. Waste will come to be regarded as a resource rather than as a nuisance. As pointed out by J. W. M. La Rivière: "Molasses is an industrial waste in a country where there is no industry to use it, but it is a valuable raw material to a fermentation industry elsewhere. The designation 'waste' merely signifies a corresponding lack in our technological potential as expressed in a low price, which is negative when treatment is required."

All waste can be converted into fertilizer by anaerobic digestion, and since this process will handle cattle, bullock, and pig manure, chopped straw, sugar-cane trash, beet leaves, coffee pulp, and all other sorts of vegetable materials, it early attracted attention in agricultural communities. There, the main interest was in the combustible methane formed during the process, because the gas can be used for grain drying, for hothouse heating, and even as a tractor fuel. However, little attention has been given to methane's heat value for temperature control in auxiliary fermentations or to the carbon dioxide that is generated when it is burned. This gas could be used not only for increasing the productivity in hothouses but also for algal cultures (for instance *Spirulina platensis,* which has been used for food by the natives of Chad for generations). However, where the illumination available for photosynthesis is inadequate, it might be more economical to convert the methane directly into fodder (or possibly even food) by using microorganisms. As indicated by studies in my own laboratory, this is quite feasible, although we are currently using methanol as the carbon source. By novel catalytic techniques, this alcohol can be produced very cheaply from methane.

It is highly probable that such gas conversions will become most important if we are ever forced to turn waste into human food. After all, we can easily clean and pipe the gases to a separate installation, provided with its own salt supply and nitrogen-fixing facility. The use of microbial cells grown directly in sewage is not so attractive because of the toxic metal salts and other industrial pollutants that are difficult to eliminate.

Most waste materials, as well as hydrocarbons, can thus be converted into a homogeneous organic material with the aid of microorganisms. We first think of this conversion in terms of food and fodder, amino acids, and vitamins, but it might also constitute the starting point for an integrated biochemical industry. Either the functional characteristics of the original material (enzymatic activity or physicochemical properties) will attract attention or we will use it as a starting point for subsequent conversions that can yield specialized molecules. It is reasonable to assume that specialized media will be produced for submerged growth of plant tissue cells or perhaps even for the cultivation of human diploid cells. The latter could serve not only as a substrate for virus vaccine production but also for making substances compatible with our body (growth hormone, fibrinolytic enzymes, interferon, or even genetic repair materials). However, even such foreign materials as bacterial enzymes will also assume great medical importance. They now provide tools for laboratory diagnosis and serve as digestive aids, but the work of Chang and his co-workers in Canada indicates a growing therapeutic significance. Urease and an ion exchanger were microencapsulated in a material that is permeable to urea but not to antibodies. The resulting powder was then enclosed in a small tube that could be attached to an extracorporeal shunt in uremic animals, which responded very favorably to the treatment.

The use of enzyme technology based on solid carriers will also influence that branch of bioengineering concerned with fermentation. It will, for instance, be feasible to increase the permeability of cells at one stage in a process by adding substances that can be easily removed in later stages. Add to this the potential of microbial genetics, for example, permitting the automatic selection of resistant strains that make "protected fermentations" possible, and it is no wonder that bioengineering is an active field at most universities. However, the advances in molecular biology and basic genetic research come faster than the applied sciences can keep up with, so special efforts are needed.

Fortunately, we have the facilities and the personnel available; obviously there is also some statesmanship required. I am thinking of President Nixon's unilateral renouncement of biological warfare. This constructive initiative has set the stage for a system of cooperating laboratories, and if there is a reasonable amount of reciprocity an International Organization for Applied Microbiology might well be created. It would certainly be an important step toward improving the human environment.

* * *

STUART MUDD: As multicellular organisms, we live in an environment that literally swarms with unicellular forms of life. We are surrounded and we are immensely outnumbered, as metazoan organisms, by the microorganisms that are in our earth and water and air and in our total environment.

Dr. Waddington has indicated that our relation to microorganisms is twofold. In one respect, they are enormously important as allies, and this possibility for aid to us has been sketched by Professor Hedén, who is a world authority. It seems appropriate for me, if I may, to sketch the inimical aspects of microorganisms to human beings, that is to say, to sketch a little bit of host and parasite interaction in the infectious diseases and in the allied rejection of foreign tissue, as we see it in the present wave of organ transplants, skin grafts, and other aspects of rejection immunology.

The beginning of adaptive immunity has been traced by Dr. Robert A. Good and Sir Macfarlane Burnet to the beginnings of vertebrate life. Up to this time the invertebrates got along quite well in the midst of their microbiological environment by mechanisms that are not well understood but did not involve what we now know as adaptive immunity.

In the last part of the 19th century, there was great debate as to whether the mechanisms of this adaptive resistance to the microorganisms that are a threat to metazoan organisms was vested in the humors—that is the blood, the serum, the lymph—or in the cells. We had the great proponent of humoral immunity, Buchner of Germany, and the great proponent of cellular immunity, Metchnikoff of the Institut Pasteur. Like many controversies, it turned out that both sides had a great deal of truth in their positions, and it fell to Denys and Leclef in 1895 to show that, in the special case of streptococcal infection, the resistance was twofold: the infection gave rise to circulating antibodies, and these neutralized the toxins or attached themselves to the surface of the streptococci and caused them to be capable of being ingested and destroyed by the sentinel cells of the body. So this, in a sense, resolved the basic controversy. It resulted, however, in a concept that was incomplete with regard to the total mechanisms of interaction of host and parasite.

The great American microbiologist Theobald Smith has outlined the requisites of a successful parasite—that is, a parasite that is able to coexist in the presence of its host. He pointed out that the process of evolution in the host-parasite relationship was toward mutual tolerance. This is a very important point, I think, philosophically as well as biologically, because the succssful evolution either of host or parasite is not in the destruction of the enemy but in an accommodation that makes it possible for the host and parasite to coexist in the same ecosystem. If the host is destroyed, that is the end of the road for the host, and if the parasite is destroyed, that is the end of the road for the parasite, but as they coexist in some of the older diseases, such as syphilis and tuberculosis, both can survive and multiply.

This subject has been studied in brilliant detail in regard to myxomatosis. This virus was introduced into Australia as a means of ridding the country of rabbits, which ate most of the grass needed for sheep pasturage. Fenner and Burnet introduced the Brazilian myxomatosis virus into Australian rabbits of European descent, and it has been found that Theobald Smith's prediction was actually realized. The initial virulent disease has given rise, in the course of time substantiated by individual investigators' studies, to a state where there is a reciprocal tolerance, and neither are the rabbits exterminated by the parasite nor vice versa, but by and large the two species survive in a state of tolerated coexistence.

The first 75 years, I would say, after the Denys and Leclef study, tended to indicate that the pattern that they developed, namely of circulating humoral components or antibodies that neutralized toxins or attached to the surface of the parasite and caused these parasites to be phagocytosed and killed, has stood up. But we have just had the privilege of submitting papers in which the situation has been amplified in regard to at least one pathogenic species, namely, *Staphylococcus aureus*.

It is found that in the special case of ordinary staphylococci, the action of the components of human serum in promoting phagocytosis and intracellular killing is not significantly improved by circulating antibodies. However, the bulwark of general bodily defense known as the reticuloendothelial system can be significantly augmented in its functioning by induction of hypersensitivity and elicitation of the activated state of defensive cells known as macrophages. This is cell-mediated resistance. Conditions for evoking this nonspecific cellular resistance may be rendered particularly favorable by the circumstance that most individuals are already hypersensitive to staphylococci, and elicitation of cell mediated resistance may be safely practiced with staphylococcal antigens. I believe it is not quixotic to hope that by such means we may, in a measure, get the better of such infections as those that cause, for instance, the upper-respiratory infections, which exact such a heavy burden of suffering and reduced efficiency.

HOTCHKISS: When the higher organisms began to climb out of the ocean, they were taking some readings on the environment as it was then, and they developed means, with various pumps and pipes and strainers, to carry the atmosphere as they experienced it, so that they could pump the atmosphere and the various saline solutions they needed through their bodies. We can take, as a first approximation, what man needs even now by seeing what his various strainers are trying to filter out of the atmosphere and the very dilute saline that he takes in. Unfortunately, the niche that he was reaching for at that time was 20 percent of the planet's surface, and it seemed big enough. Now, of course, we are coming to the second approximation; the things that man filters out are becoming part of the environment themselves.

I think that man is a part of the environment and that he has taken his last evolutionary reading, at a time when the whole world was his and he could reach out for it all.

Dr. Hirschhorn has brought us very sharply up to date and reminded us that there are movements afoot that might change man's actual heredity.

I have the name for being involved simply because we take genes from one kind of bacteria and put them into another. This sort of genetic engineering is like robbing Peter to pay Paul, taking something already highly developed in one organism and transporting it into another. I do not actually consider this engineering, because the word engineering connotes, to me, some kind of concerted action operating on a more or less broad plane and actually this is one of the most intimate of all the things that man is trying to do.

The Shope papilloma virus, which is a dangerous enough thing for rabbits, forms the kind of arginase that seems to be characteristic of the virus. There are some questions about the details, but for a while, this was a cure looking for a disease. If anybody had too much arginine, this virus might be good for him. A couple of months ago, Stanfield Rogers told me that he discovered two siblings with a metabolic disease that indicated high arginine. The older child was in a very bad state; the younger child was going through the same premonitory symptoms and was simply at an earlier stage. These children have been given some of the virus to see if it will help. And this may be illustrative. I do not call this engineering, but genetic intervention. What sounds like science fiction is genetic manipulation, willful changing of properties that seem to be hereditary and, in higher organisms, are largely predetermined at a quite early age. One more of Dr. Hirschhorn's points is that what is ethical is anything that takes into account man's freedom of choice. I think it is important for all of us to realize that the very freedom of choice that has worked against eugenics was that we did not want to interfere with a person's right to choose another life partner because that was what one chose in genetics. The genes were carried by whole organisms and by whole nuclei from whole organisms.

Now we have the possibility before us, a very real one, that we can intervene with bits of gene fragments taken from anywhere. They are no longer stolen from Peter because they can be cultivated *in vitro,* and they were in a sense borrowed from Peter in order to enrich Paul. They are becoming more available all the time, and the freedom of choices we would be faced with is just the freedom between the right to take a small improvement and your right to reject it or bypass the opportunity. So I think the very sentiment that made eugenics rather distasteful to most people is exactly the sentiment that makes these new things very tempting in the future.

UNIDENTIFIED DISCUSSANT: I would like to comment on Dr. Hirschhorn's definition of the ethical aspects of this, which I thought very brave of him, because we all know it is a very controversial issue. If we define the ethical aspects exclusively in terms of the free choice of people, what are we to say of generations in the future that are affected by our decisions now? I realize this is a very complicated question, but I merely throw it up for some discussion at some point in this conference, and I do not restrict that merely to genetic manipulation or engineering.

It seems to me that the people who will live after the year 2000 will have much less control over their presence than we today have over their presence. That is, the decisions we either make or do not make today will affect their lives in a greater dimension than they themselves will be able to affect their ilves. And if that is true, then the ethical aspect of that becomes exceptionally complicated, because we have no way of asking them what kind of world they want to live in, and most of us will not be around to live in it. But I do believe that this is an ethical question that goes beyond the concept of free choice of the people who are to participate in the consequences.

BRONK: This reminds me of a few years ago, when President Kennedy asked me to organize, from the National Academy, a study on our natural resources. After about eight or nine or ten months he asked how we were getting on, whether the 120-odd people who were studying the various aspects of the problems had come to any definitive position that he could use in any of his addresses. I told him it was too soon to have any very definite opinions. Then I suddenly thought and I said, "With one exception, Mr. President; I find that everybody is concerned about their own conditions and environment and availability for what they wish and need during their lifetime; almost everybody is concerned about the welfare of their children, practically everybody is concerned, to a certain degree, with the welfare of their grand-children. But, Mr. President, I find in my discussions thus far that nobody gives a good Goddamn about their great-grandchildren's future." And I think this is one of our social problems, to be able to abrogate our own special private interests as of the moment, the immediate future, and look forward to the world and the conditions we are leaving for the others, including our biological environment.

UNIDENTIFIED DISCUSSANT: I think that the point that ethics is connected with the number of choices available is extremely important, and I think the point raised here should really be pursued. The interesting thing is that there is an answer to your question. The answer to the question of what decisions we have to make so that later generations may have their freedom of decision must be such that the number of decisions in later generations will be larger.

SAGER: I would like to focus not so much on what is possible in the future but on what is desirable, and, in fact, I have been anticipated, happily, by the last comments, which I welcome. We are all aware of the explosion of new knowledge in biological science and of the fact that this knowledge gives us a great range of possibilities for the future. And our problem is not so much to try to predict what will, in fact, happen, but rather to focus upon what we consider desirable, because it is only in that sense, I think, that we can have some influence on the future.

The comments that I am going to make stem from my experience in my own daily life. I am a research scientist, and, as such, I am impressed with the excitement and the fun of research in a laboratory as a way of life and also as a teacher. But I have discovered how difficult it is to communicate this excitement to my students.

First, I would stress what I consider to be the state of the art in biology at the moment, because I think this really sets the stage for our considerations, both realistic and idealistic. And I would like to stress in my view the outstanding achievement of biological research of the last 20 years—the establishment of molecular biology. Molecular biology tells us how cells work, and all future advances in basic and applied biology will rest on this solid framework. And I think one should not underestimate the power of what has been learned.

Already we see that the sudden burst of development of understanding in embryology and differentiation—to some extent in neurobiology and, hopefully, in the very complicated area of ecology—which is just beginning, will rest on this solid framework. The problems that we face, therefore, are not the problems of not having a solid framework on which to build research. Our problems are political problems.

Our problem is, first, will we live in a world affluent enough to support this basic research which will lead to a greater understanding of the natural world? Will we even live in a world able to afford the kinds of applied research that we will need so that we can solve the problems in order to do the basic research? And, perhaps most important, will there be any young people who want to do this research? Finally, as academic people and on the elderly side of this generation gap, where can we intercede in this problem?

In other words, what we have in our hands are powerful possibilities for good and powerful possibilities for evil. What I am going to propose is that we formulate a statement of a reordering of priorities in biological research. This reordering of priorities does not call for less research; it calls for more research, because a firm and continuing basic understanding of the fundamentals on which all of these applications are to be based is really just at the beginning. It would be a great error were one to say, we can stop doing basic research and apply some of this, because the applications have built into them the need for more knowledge. What we need is more, but we need a reordering of those priorities.

I would also like to propose an explicit ethical basis in the choice of research problems, in the choice of what one does and does not do. And finally, I would like to propose, in the panel discussion, a very preliminary and tentative biological bill of rights for man. Essentially this bill of rights deals with the rights of every individual to the kind of healthy and normal development that we would like to see for ourselves and for our children and for our friends.

And finally, I would like to stress that, as the health sciences, the biological sciences provide a new focus, a focus that I think we should take advantage of, because it is science that really focuses on the individual. All other areas of natural science can be thought of in very gross terms as dealing with statistics, people by the millions. As the planet becomes more crowded and as the problems become exacerbated, the importance of the individual and the dignity of the individual are bound to decline. One of the great contributions of emphasis on biology, I think, will be a reordering of our priorities to recognize and remember the importance of the health and dignity of the individual.

BRONK: I would like to endorse most heartily what Professor Sager has just said. Our problem now is not to find more money, not to find more appropriations, but to sensibly use what we have. And for that reason, our allocation of the primary priorities along specific narrow lines, whose effort can come to the most, will be the foremost thing that could be done.

SAGER: May I just say I am for more money for everybody.

BRONK: But I believe that before asking for more money we should make the best possible use of what we have. Relative to that, let me say that those who have been deeply involved in Washington for many, many years, had a very strong group of friends in Congress, people like Senator Lester Hill, Leverett Saltonstall, Warren Magnusen, Albert Thomas, and others. But times change, people die, some retire, many get defeated for re-election, and one of the great difficulties we have at the present time is that most of the people who are urging funds for support of research are engaged in this, and we are naturally suspect as having special interests. In this country—and I know England well enough to know the same applies there, and I know what goes on in Sweden—we do not have an adequate clientele willing to speak for what we think is important. Legislators and foundation executives have heard us say what we say many, many times. In industries that are dependent on the continued development of science, they speak very seldom on our behalf.

I remember several years ago when the National Science Foundation suddenly came up against a particular crisis and, in order to fulfill our obligations to universities, we wrote to about 20 university presidents and said, "Look, this is your problem as well as ours. We are trying to help you, you have a senator, you have two senators from your state, get busy." And within a couple of weeks the whole picture was changed.

So that we have, as Dr. Sager has said, to start with beginning education, develop a greater awareness of the role and the benefits as well as the handicaps and disadvantages of certain types of technology. We have to spread this more widely unless we are going to have a very different type of government than most of our countries have.

UNIDENTIFIED DISCUSSANT: I am just an observer from Sweden, but I would like to give a point of information on this. I have come from a conference on the future in Kyoto, Japan, and there was a most interesting paper from the University of Heidelberg. They had done opinion polls about research politics, and it was found, more or less irrespective of educational background, that they had a common framework about what was important, and that was in the biological sector and was, to a great extent, at odds with the actual politics of the German Government, which stressed nuclear research and military research.

HIRSCHHORN: In thinking about some of the comments that have been made about the question of ethics for future generations, I think we must be careful here to divide those aspects of individual freedom of choice which do not, in fact, affect the current individuals and those that do. For example, I can very well agree with the comments that if we continue to allow complete individual freedom of choice of how many children every family is going to have, this may well have a strong detrimental affect on the biological environment of future generations.

On the other hand, because we have no concept at this time, in view of the rapidly changing environment—which, over the years becomes exponentially even greater—of what future adaptive values will be, we cannot sensibly plan a biology to adapt to an unknown environment. Nevertheless, we must strongly consider at this point that any of the genetic engineering type of changes not only will affect those future generations, but will, in fact, very strongly affect the individual on whom such manipulation is practiced.

UNIDENTIFIED DISCUSSANT: I work in environmental medicine, somewhat on the firing line because I am often called about people who have specific problems about environmental effects and human disease. And I was very unhappy to hear what Dr. Bronk said, because if it is true—and it may very well be—and if Dr. Sager's help is not realized for the future, I think we are in for a great deal of trouble.

I do not know whether even the scientific community is thoroughly aware of the very scant resources that our society in the United States and elsewhere is now applying to solutions of environmental problems. In this country, for example, this year's budget of the National Institute of Environmental Health Sciences, which is charged with all the basic research for the study of the biological effect of environmental changes in-house, extramural support, etc., is $18 million, and that is all. The other agency in our Government which has any relation to these problems, and which in the past had the major responsibility, is the Bureau of Occupational Safety and Health, and their budget this year for noise and accidental death and asbestos and carbon monoxide, etc., is $10 million for all research—their own, extramural, their administrative costs, etc. There is, for practical purposes, very little in the way of actual support, of putting our money where our discussions are, in this country. I don't think the situation is very much better in Great Britain, where there are some 14 research associations now looking at problems of water pollution, eight looking at problems of air pollution, and some at noise as well, and their total finances this year are two and-a-half million pounds, one and-a-half million for airport noise.

I would like to compound this pessimism by pointing out that the excellent discussion that we heard from Dr. Waddington and others really is almost like fighting the last war. The problems of the question of environmental changes are more complex than the microbiological approaches, which are basically a one-to-one relationship in many of these effects, whereas in environmental changes they are long-term, low-level, multiple-factor, and highly complex. The human effects are even more complex than this: they involve lifetimes, because most disease occurs in adults and their study requires detailed and long-term observation. Therefore, if we do not reorder our priorities in the sense, not of exchanging the funds that are now available for basic research for the long-term environmental research, but rather of a great extension of support available for environmental-factor research, we really are in for a great deal of trouble.

WOLF: I am the legal adviser of the International Labor Office. I was very much interested in what Dr. Ruth Sager had to say about the Bill of Rights, and I am very happy she said it. Indeed, we have a Bill of Rights. We have a Bill of Rights on an international level, but what we need is more implementation of what exists. And it is possible.

In my organization, for instance, we started to protect the workers and indeed, all those who work, which is the whole world, against certain risks due to the environment. The predecessors of the ILO, just before the First World War, elaborated an international treaty against the use of white phosphorus. We have always continued in such fields, framing international treaties to protect the working man from all the dangers to which he is exposed. We were the first to lay down international regulations for the protection of ionizing radiation. And now we are preparing an international instrument on protection against benzine.

I think that through regulations on definite problems, a lot can be done to implement those rights you describe, Dr. Sager. The trouble is that we lawyers—and I think we are on equal footing with scientists and with politicians—despite all our efforts, lack imagination because we cannot foresee all the dangers. I will give you an example outside the sphere of the law of international regulation.

I live in Geneva, near the lake. Ten or 15 years ago, nobody thought that lake would be completely polluted, neither the lawyers nor the politicians nor the scientists. In a way I think the only human beings who had sufficient imagination to foresee this probably were the poets, because they are probably more sensitive to things than we are. But today that beautiful lake is completely polluted. I think a great effort has to be made by all of us to increase our imagination and to try to implement what we all believe.

WADDINGTON: I want to make one remark in relation to the last two speakers, and this is the question of some of the fundamental studies of environmental pollutants, and the low-level, long-term, toxic effects. This was quite a fashionable notion in the early days of nuclear bomb testing, when the public got very excited about the genetic hazards from ionizing radiation and fallout, and there was a strong movement to set up one large international laboratory under the auspices of the WHO to operate the facilities simply to test radiation damage alone. Of course, this is only one of the sort of low-level, long-term pollutants that emit thousands of substances about which we have no notion of what their long-term effects are going to be except that they are going to have long-term effects. It seems to me this is really too big a job probably even for the United States to take care of, particularly in its present situation. It is an eminently suitable job for international organizations because, unfortunately, much of this work boils down to pretty boring routine, and you are really going to have to test long-extended experiments at low levels. And it is a question of statistics' continuing to go up from the national level of one or two effects, and so on.

On the other hand, it could provide very good training grounds for countries whose scientific efforts are just getting off the ground and who are perhaps turning out more people capable of undertaking scientific research than they actually have facilities of their own to employ. I believe it is an extremely suitable subject to be organized by the international network of laboratories for testing environmental effects. And I believe that to work towards this is one of the things that an international meeting of this kind might endorse as a suitable policy.

THE WORLD WE WILL LIVE IN *

Jonas Salk

Salk Institute for Biological Studies
San Diego, Calif.

Every sentient human being is aware of the dangers as well as of the opportunities that now confront man. He is also aware of the pressure of forces within himself and the pressure of forces around him. Caught between his inner world and the outer world, he feels dissonant more often than consonant. He feels stress, longing, and frustration. His soul feels weary, and he yearns to find, and to express, the meaning of his life. Some among the most "successful," and many in the midst of "great riches," appear to hope for something else.

Science and technology are indicted by some as the cause of this malaise, and some even recommend antiscience and antitechnology remedies. However, in the absence of a true understanding of the current syndrome, it is unwise to ascribe the causative role to something that may not be the cause or to prescribe as a remedy the elimination of something that could even help in the development of remedies. Although science and technology are unlikely as a cause they are, nevertheless, a force with which man must cope.

The views just expressed are similar to those contained in the preface of a book called Living Tomorrow Today! which says:

> As we are brought closer in touch with the world in which we live, it is apparent that *the physical and spiritual health of a single individual, and mankind as a whole,* should be the final yardstick in measuring both progress and problems. As tomorrow comes closer to becoming today, we are reminded that, to a degree far greater than in any time in history, *the future will depend on whether our ethics and values can keep pace with our scientific and technical knowledge.*

And in the prologue of this book, in reference to scientific and technical growth, Isaac Asimov says the following:

> But all these advances, coming upon us faster and faster from every direction must be correctly handled, and that isn't easy. Living tomorrow today involves confronting *the enormous problems which must be worked out in order that our limitless opportunities may be properly used to the benefit of all people.* All of society is altering widely under the impact of advancing science, and *all mankind is being carried furiously forward in a mad race toward a not completely visible goal. We must stay in control so that tomorrow's life will be worth living and that today's hope will beget greater hope.*

---

* A version under the title The Two Epochs was originally presented before the American Institute of Aeronautics and Astronautics in October, 1969 and the present version was delivered before the National Association of Manufacturers in December, 1969.

If the progress and the problems of man are to be measured in terms of effects on the single individual and on mankind as a whole, we need to examine the nature of the relationship between the individual and mankind, and if the future of man depends on ethics and values, let us also examine ethics and values in relation to the individual and to mankind. Because ethics, values, individuals, and mankind are all bound together in the effects expressed in the physical and spiritual health of man and the quality of human life, as is the survival of mankind, not only do we need to look into the past but we must also try, through our imaginations, to see as far as we can into the future.

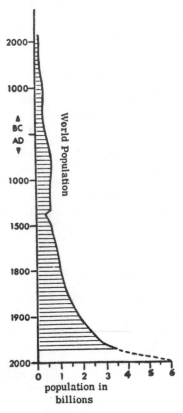

FIGURE 1. World population estimates, 2000 B.C. to A.D. 2000.

We must not overlook the long evolutionary history before the beginning of man's own history. Man's population history is shown in Figures 1 and 2. Here we see the long gradual increase in world population until recent time, when the rise became almost exponential. The only interruption in this progression was a dip due to the effects of the bubonic plague, in the 14th century when the world population was reduced by about one-fourth. It soon returned to the level the resources then available could support and man's biological and ecological condition would permit. As time advanced, as territories opened, as resources were developed, and with the improvement in hygienic conditions and food supply and hence in man's physical health, there was a prolongation of life expectancy, and the annual excess of births over deaths resulted in the enormous increase in numbers now present on the face of the earth.

Projections based on current trends reveal, in FIGURES 1 and 2, an unfinished picture, through the year 2000. A world undergoing such rapid change will affect both the individual and mankind—with profound implications not only for ethics and values and the quality of life but for the survival of man.

Shifting our attention to another species, we see what happens to a fly

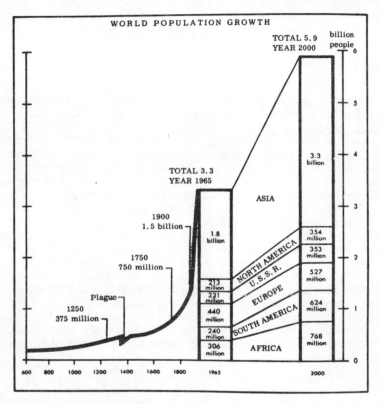

FIGURE 2. Population by world areas, A.D. 600–2000. By permission of John McHale and George Braziller, Inc., from The Future of the Future, © 1969.

population in a closed system—a system in which space and resources are finite. In FIGURE 3, the curve of growth of a fruit-fly population is shown. This curve, the first part of which resembles the world population curve, reveals a deflection at the mid-point, along the steep slope, because of a change from acceleration to deceleration and then a plateau. This deflection is emphasized in FIGURE 4 by interruption of the curve at the point of change from progressive increase to progressive decrease; in FIGURE 5 the difference is still further emphasized by displacement with the creation of curve A and curve B.

It is of particular interest, in the present context, to note that curves such as

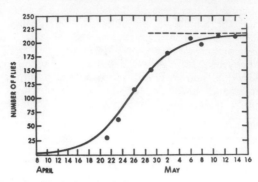

FIGURE 3. Growth of fruit-fly population.

these resemble curves that describe other self-controlled or self-regulated phenomena in living systems that are referred to as homeostasis, a term coined by
physiologists. Homeostasis refers to the maintenance of "optimal levels" through
control and regulatory mechanisms, and is sometimes metaphorically regarded
as evidencing the "wisdom" of the body—or, more broadly, the "wisdom" of
nature—the seeming "purpose" of which is "survival" of the individual and
the species. Such innately programmed phenomena, in which deceleration of
production occurs when "enough" is produced, are seen, for example, in the
formation of a specific hormone, or an antibody, or blood cells after blood
loss, or tissue cells in repair after injury.

"Optimal levels" are the cumulative and integrated result of the natural
selection of forms that best fit prevailing circumstances. The biological
"wisdom" so revealed expresses itself through a complex constellation of
physiological systems that sense and signal "not yet enough," or "enough," and
then respond appropriately. Such systems for sensing and responding vary in
degree of sensitivity and in degree of responsiveness—some react to small
stimuli, others to large; some have a short lag, others a long one; some have
short-term effects, others long; some are reversible, others irreversible; but all
are part of an order that we recognize functionally as health and that evolved
by trial and error.

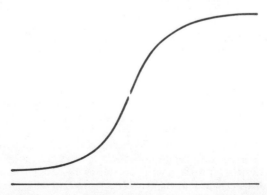

FIGURE 4. Schema of population growth and decrease in acceleration.

As I look at these curves, and feel myself in them, my imagination becomes playful and begins to ascribe human attributes to the fly population. I imagine that if flies possessed the human attributes of "foresight" and "insight," the *outlook* of each succeeding generation would differ, and furthermore, the *outlook* in what I think of as Epoch B, would be qualitatively different from that in Epoch A.

I even imagine, in each generation of this fly population, a sense of "responsibility" for the *future* and a need for a mechanism for discharging this "responsibility." If control is at least partly "voluntary" rather than entirely "autonomic' or "physiologic,' (as is, in fact, true of the fruit-fly population) the flies would have to "evaluate" and "decide" before "taking action." This would mean that for guiding effective action the "system of values" in Epoch B would have to be different from that of Epoch A. These are the thoughts that occur to me as I imagine myself a part of these curves.

When I then look at the curves of population growth of man, I sense the existence within us of an innate "sense of order" similar to the "sense of

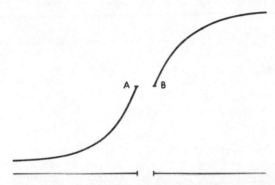

FIGURE 5. Illustration of break between population growth and decrease in acceleration.

order" that exists in such physiologically regulated biological systems, as systems of molecules, or a population of flies in a closed system; and that through this innate "sense of order" we become aware when we begin to have "enough." I also sense that we are now becoming aware of a need for a fundamental change in our value system—a change appropriate to the difference between Epoch A and Epoch B.

In historical terms, I conceive Epoch B as starting with the generation now maturing. From all parts of the world there is widespread evidence suggesting that Epoch B is already well underway and that a *new ethic* for the *new epoch* has already begun to evolve.

In the new epoch, we need to look at ourselves from a biological viewpoint, rather than from such other viewpoints as have been dominant in Epoch A, to be able to shape a *new value system* for our time.

The idea that the biological viewpoint might provide a point of departure for developing a workable and livable ethic for the new epoch requires further elaboration. The biological way of thought is here used in an attempt to develop an understanding of what appears to be an epochal change in man's view of himself at a point in time when his survival is threatened by products of his own creation and when the need for self-discipline to avert disaster is required more than ever before.

It would seem, from the present state of man and the world, that an uncompromising ethic based upon something as real as "survival" is needed.

Evolution is a game of life and death in which there can be no loser and no winner for evolution itself to survive. Thus, life and death must coexist for the survival of life in evolution. In a similar way, life's problems and solutions are inextricably related, because solutions almost always lead to the creation of other problems, as we see, for example, in problems created by the solutions produced through science and technology. In the same way, species and the individual are inseparable. It is not possible to have a species without individuals or individuals without a species. It follows from this that any ethic for man requires that whatever is regarded as of value for the individual must also be of value for the species, and, conversely, what is judged to be of value for the species must also be of value for the individual. Such an ethic would be based upon the idea that *man and species are one*—and, ultimately, that *nation and mankind are one*.

Species that have survived possess well-developed, gene-controlled, self-regulating mechanisms that respect the environment and are "species-oriented" rather than "individual-oriented." As far as I know, there is no other species that is "individual-oriented" to the extent seen in man. In an "individual-oriented" society is to be maintained, then an ethic and a system of values must be developed that will make this of value for the species as well as for the individual. Such a culture would require an equal sense of responsibility for others and for oneself. Greed and self-service would be of negative value in such a culture and would have to be regulated and controlled.

Thus, an ethic and a value system are supplementary to biologically operated mechanisms of regulation and control and can also be inhibitors of the natural mechanisms for regulation and control. Changes in ethics and values, and changes in moral and religious codes, should be looked on as part of the evolution of effective human control and regulatory systems toward the goal of an ethic in which *man and species are one*—when *nation and mankind can then be one*.

For the foregoing ethic to become operable so that we may then be able to use our limitless opportunities for the benefit of all, it will be necessary to understand self-regulation and self-control, both within individuals and within groups. Man seems to know this intuitively and seems to desire this. Moreover, he seems to be moving in this direction. At times, in utopian euphoria, he goes too far one way—and then to opposite extremes, as he continues to acquire the experience necessary to support the faith and confidence that, given the opportunity, man can more often be generous, kind, and loving.

This is the sense that moves young people to reject the prevailing ethic and to look at earlier metaphysical philosophies for guidance in their search for meaning in their lives. The need for a philosophy could be satisfied by a realistic idealism, such as can be developed from the biological point of view. This will require an enormous amount of imagination and thought. Simple in the abstract, its accomplishment is perhaps the most difficult and crucial challenge with which man will ever be faced. What is involved is the ultimate question of man's survival and the nature of the quality of his life in the future.

A self-consciousness of evolution could be of great value. Will man supplement his less experienced wisdom with the more experienced wisdom of nature? Will the ethic he espouses and the value system he develops follow patterns of proved value in nature?

Life is an error-making and an error-correcting process, and nature in marking man's papers will grade him for wisdom according to the judgments he makes as measured both by survival and then by the quality of life of those who survive. Although the lives of individuals have been prolonged and made more secure, and man has been able to go to the Moon, the survival of the species here on Earth is threatened by uncontrolled population growth as well as by an atomic holocaust—which one individual could touch off. All these are the accomplishments of science and technology.

The dinosaur, in its day, was jeopardized by environmental change and by predators, both of which led to its extermination. Man is now the cause of major environmental and ecologic change and makes predators out of his fellow men through an ethic and a value system that permits and encourages us to so conduct our lives. Man, therefore, is pitted against his species.

It is necessary for man to see himself as he really is and examine the paradoxes of his life if he is not to destroy his species by a progressive elimination of options. Will man see himself as he is, in relation to his future, and act appropriately on behalf of his own survival?

Those in positions of leadership, who are able to influence the course of events, will soon reveal whether they possess within themselves and whether mankind possesses within itself the capacity for long-term survival—with meaning and satisfaction in life—or merely for a chronic, agonizing, short-term survival. It will not be long before we know. The dangers are great. Will we see that the opportunities are greater—and that the choice is ours to make?

Will we help the evolution of an ethic and a value system appropriate for our time, and will we contribute toward the development of a trajectory for man that will lead to the destination he seems to want to reach? The price for this is high in terms of his personal ego or image and his national ego or image, but the reward is great in terms of his true and natural self. Will man have the wisdom, the courage, and the strength to take the course to long-term survival, or will he—in ignorance, fear, or weakness—accept the world we now live in as the best we can create?

WHY FERTILITY CONTROL?

Hudson Hoagland

President Emeritus
Worcester Foundation for Experimental Biology
Worcester, Mass.

Continuation of the present rate of population growth is a threat to all mankind. The application of techniques of modern medicine and public health since World War II to underdeveloped countries has markedly reduced death rates, leaving birth rates, always large, either unchanged or increased as the result of improvements in health and longevity. In many of these developing countries, increases of 20 years in life expectancy have occurred in the course of one or two decades—increases that took roughly a century to bring about in the United States and in western European countries. Population growth today depends only on the difference between birth rates and death rates, because there are no more vacant lands for colonization by people from crowded countries.

World population today stands at some 3.6 billion and will double in 35 years at its current growth rate of two percent, so that by the turn of the century we will have some 7 billion people on earth, and in an additional 35 years, at our present growth rate, there will be 14 billion of us. Thus, a child born today and living on into his 70s would know a world of this number of people, and his grandson would live to see 60 billion people. Even the most optimistic can hardly think that our planet can support such a burden of humanity.

Most concern with the population explosion has been expressed for the underdeveloped countries, which are being held back by the production of excessive numbers of children who are consumers, but not producers, who grow up hungry, and often with less education and job opportunities than were available to their parents. Capital for industrialization, for agricultural machinery, and for fertilizer plants is very hard to get, because produce for exports to obtain capital must be consumed locally. Competent economists and demographers predict mass starvation on a scale hitherto undreamed of in countries of Asia, Africa, and the Near East in from ten to 20 years. The so-called green revolution, the production of improved rice and wheat crops, is no answer. At best it can fend off the day of reckoning by only a decade or two.

But, quite aside from food, there are other very serious considerations, because direct correlations have been made between occurrence of wars and revolutions, poverty, and rates of population growth, so that the likelihood of drastic population reductions by a third World War, with nuclear, chemical, and biological weapons, is increasing as human misery expands with numbers of hungry, ignorant, and frustrated people.

Studies of animal societies have demonstrated conclusively the pathology of crowded living—of "pathological togetherness," as John Calhoun has called it. Animal societies are organized in terms of dominance hierarchies, and their structure makes possible effective social living with a high biological survival value. Under crowded conditions, even with ample food, the social systems of gregarious animals collapse because of the stress of crowding, which operates via the hypothalamic-pituitary-adrenocortical axis. This has been demonstrated for many species, both in the natural state, in zoos, and under experimental laboratory conditions. The social collapse results in much fighting, highly abnormal behavior, loss of sex drives, failure to make nests, and cannibalism of the young. As a consequence, the population density falls to levels in which social organization can be re-established effectively, then the population builds up again and the cycle repeats. There is evidence that crowding of people in concentration camps also has resulted in pathological behavior, and many of the problems of the slums and ghettos may very well be a byproduct of the stresses of social overcrowding, although this has not been adequately investigated.

We often overlook the tragic consequences of our increases in population in the developed countries. A one percent growth rate, which is an average for some major industrialized countries, e.g. the United States, the U.S.S.R., and Japan, will double populations in 70 years. The United States, with our present growth rate, will have some 300 million people by the turn of the century, and it is the rich countries that are doing most to destroy our biosphere by pollution. So serious is this question of pollution of land, air, and water and the consumption of irreplaceable metals, fuels, and natural resources by the industrially advanced countries that competent scientists believe the world cannot indefinitely support even the 3.6 billion people we have on earth today, let alone the horrendous numbers anticipated for the relatively near future.

We are assaulted by noise and poisoned by carbon monoxide and other noxious fumes from a hundred million motor vehicles. Thousands of tons of pollutants, many of them highly toxic, are being pumped into the air from industry, jet planes, and automobiles, producing widespread smog, which is increasingly visible over relatively deserted areas far from industrial centers. Already a thousand million acres of arable land have been lost to erosion and salt through human activity, and two-thirds of our forest lands have been lost. A hundred and fifty species of birds and mammals have vanished altogether in the past two centuries, primarily as the result of human activities.

In the United States, annual wastes and pollutants amount to 142 million tons of smoke and noxious fumes, 7 million junked automobiles, 200 million tons of paper, 48 billion cans, 20 billion bottles and jars, three billion tons of waste rock and mill tailings, and 50 trillion gallons of hot water. These figures have been furnished by the staff of the Secretary General of the United Nations.

Reports from West Germany note that millions of dead fish were found floating down the Rhine, poisoned by the accidental spilling of some 300 pounds of an insecticide into the river. The danger of contamination was so serious that Amsterdam, Rotterdam, and the Hague switched to emergency supplies of drinking water. Prior to this accident the Rhine was referred to as an 800-mile-long sewer. Pesticides in 1960 wiped out vast numbers of fish in the lower Mississippi, and industrial wastes long ago exterminated shad in the Delaware and in the Connecticut rivers, shellfish in the Merrimack, and all edible fish in the Hudson nearly to Albany. Various hawks, including ospreys and eagles, at the end of food chains that concentrate DDT, are facing extinction. The United Nations report links the dangerous deterioration in world resources to the simultaneous world growth in population. The problem is indeed a global one. The fat of seals on the Antarctic icecap has been found contaminated with DDT and other pollutants—and, of course, our burgeoning upsurge in numbers of people makes the extensive use of pesticides a necessity in order to feed them. Our skill in producing washing detergents and such plastic materials as wrappers has produced a vast amount of waste material that cannot be attacked by bacteria, and thus rendered innocuous, or destroyed, with the result that our advances in technology are costing our environment dearly.

Carbon dioxide and toxic gases from industry and motor cars is increasing, and CO_2 in the atmosphere has a "greenhouse" effect. It slows radiation of heat from the earth, particularly at night, while having relatively little effect in blocking entering heat from the sun during the daytime. This tends to raise the temperature on earth, and it has been calculated that, in something of the order of a thousand years, we may expect a sufficient rise in the mean temperature to melt the Antarctic icecap, which would raise the ocean levels by some 300 feet and swamp out all cities within miles of coastlines. Happily, the temperature's rise is not as rapid as it would be were it solely from the "greenhouse" effect of CO_2, because solid pollutants producing our damaging smog, including discharges from jet planes, have the effect of retarding the entrance of heat from the sun, tending to balance out the "greenhouse" effect of carbon dioxide.

In a recent book, Dr. Charles F. Park, of Stanford University, holds that the world population already is well beyond the number that can be supported by our planet. He bases this on the fact that we are consuming our mineral resources at a rate that cannnot be continued for long and he discounts the argument that ample mineral resources can be obtained from the sea because of the excessive cost. Our technological society is consuming metals at an appalling rate, which will rapidly increase if and when the developing countries become industrialized. In 1967, the United States had a per capita consumption of about one ton of iron. He argues that, if global consumption reaches that level by the year 2000, and the population has doubled, the world's annual demand for iron will be 12 times that amount. In similar considerations, assuming the doubling of the population and industrialization of global areas not now developed, he shows that there will be a 16-fold demand for lead and an 11-fold demand for copper over today. These figures are based on the assumption that there will be no increase in per capita consumption by our highly technological society between now and the turn of the century, and such stagnation is highly improbable.

We are, of course, rapidly exhausting our supply of fossil fuel, with the expectation that nuclear reactors will take the place of the oil and coal burners to supply energy. With the shift to nuclear fuels, there will be increasing problems of disposal of nuclear wastes in air, land, and water and this will markedly accelerate dangerous contamination of our biosphere, with threats of increasing malignancy and genetic damage.

It is the industrialized rich countries that do most damage to their environments. Indeed, most serious overpopulation has been defined as occurring in that nation whose people, because of their numbers and activities, most rapidly deplete the ability of land to support human life. The United States is generally regarded to hold first place, in view of our affluence and vast industrial complex, with the pollution it pours forth and its consumption of raw materials. Wayne Davis has compared us to India. He suggests that we speak of our numbers in "Indian equivalents," which he defines as the average number of Indian citizens required to have the same detrimental effect on the land's ability to support human life as would the average American. This value is difficult to determine, but he takes an extremely conservative working figure of 25 and concludes, with this Indian equivalent of 25, that the population of the United States is now equivalent to *at least 4 billion* in its environmental damaging effect compared to India, and the rate of our growth is even more alarming.

There are only two solutions to the social disease of overpopulation. One is to increase death rates and the other is to decrease birth rates. No one that I know advocates the former solution, although an all-out nuclear war would solve the population problem. Fertility control on a rational basis is the only other answer.

The "pill" was developed by the late Gregory Pincus and M. C. Chang at the Worcester Foundation for Experimental Biology and their principal collaborators, John Rock and Celso Garcia. It has proved to be a safe and effective oral contraceptive, despite scare headlines about side effects. Present oral contraceptives are a beginning, and work with other compounds and other methods of fertility control are of great importance in coping with population problems. This is so even while we recognize that ignorance, prejudice, and a variety of social and psychological factors are the primary limiting conditions of population control. In time, governments may find it necessary to act to protect their people against their prolific reproduction. This is indeed an abhorrent idea to most people, but voluntary family planning is not effective in most countries and the present growth rate can be utterly destructive of civilizations if uncontrolled.

How can governments act to control population? Various suggestions have been made.

1. Senator Packwood of Oregon is proposing a bill to limit Federal income tax exemptions for increasing numbers of children. He would have $1,000 exemption for the first child, $750 for the second, and $500 for the third, but none for subsequent children. On the other hand, Paul Ehrlich would eliminate all tax deductions for all children and would add luxury taxes on diapers, baby bottles, and baby food.

2. Various cash awards could be given for couples who refrain from having children for definite periods.

3. Bachelors in many European countries have been taxed rather heavily to encourage marriage and children. This process could be reversed, and single people could be given tax advantages.

4. Late marriages could be insisted upon by governments.

5. Liberalization of abortion laws and encouragement of male sterilization.

6. Temporary sterilization of all single girls by means of an implanted time-capsule contraceptive, and of all women after each delivery, with reversibility allowed only on government approval, based on government decision as to population size.

7. Compulsory sterilization of men with three or more living children.

8. Various studies have shown that education and jobs for women reduce the numbers of children. Therefore, more education and more jobs for women.

9. A Federal bonus could be given for child spacing, and tax and welfare benefits, with penalties for those who do not.

10. It has been pointed out that the development of government social security systems in developing countries can have a marked reducing effect on population growth. In India, for example, the only social security parents now

have is the production of at least one son, who will carry on the farm and provide for their old age, including the necessary religious rituals so that they can join their ancestors. In the past, with death rates very high, it was necessary to have many children to assure that one son would survive to care for the parents. A Government social security system could take the place of the security furnished by surviving sons, as it has done in industrialized countries.

11. Some 15 years ago, the late Indian Ambassador, Mr. Chagla, said that Indian's most pressing need was for a safe, cheap, effective oral contraceptive that could be put in food or drinking water to reduce fertility. Recently, Melvin Ketchel has pointed out the probability that one or more such substances may become available in the near future, and he discussed the ethical and social problems that would be associated with its use. He pointed out that such a substance would, of course, not be used to sterilize a population completely but, depending on concentrations employed, could reduce fertility statistically to predetermined desired levels, including, of course, the level of zero growth.

Considering the furor over fluoridation of water supplies, one can imagine the public uproar that even the discussion of such a procedure would induce. Compulsory government control of fertility in any form is abhorrent to most people, but it may well become necessary for survival itself if voluntary methods fail as they seem to be failing in many countries, including those whose governments accept and promote voluntary family planning. If we do not control our populations, I believe that a nuclear war will do it for us. Man is the only animal that can deliberately control his own evolution. Let's hope he chooses rational procedures rather than a nuclear holocaust.

The American people are deeply enamored of the virtue of growth for its own sake—growth in population, growth in business enterprises, growth in armaments, growth in gross national product. We even measure the virtue of our automobiles by their size—the bigger, the better. It is my hope to see the time when quality of life will be separated from quantity in our value systems and that any politician who boasts that he has ten children would be regarded as too socially irresponsible to be elected to important public office.

If and when this day comes, we shall have control of our population problems as well as most of the problems of pollution.

CONTROL OF POPULATION: NUMBERS

John B. Calhoun

National Institute of Mental Health
Bethesda, Md.

Inclusion of this topic on the agenda of the present assembly reflects a widespread concern about this poorly defined topic. No solution of the human population problem will be reached unless we can gain more appreciation of the item man to be counted or of the meaning of control. Furthermore, neither control nor counting may proceed rationally without placing the human present into an evolutionary perspective embracing both past and future. My intent here will be to consider sequentially (1) evolution, (2) counting, and (3) control. In treating the subject in this manner, it will be apparent that I am forced into a holistic view of the issue. Much of the position taken here is developed in detail elsewhere.[1, 2, 3] I am purposely not trying to summarize the opinions of others relating to population control. Rather, my purpose is to develop a philosophy about the problem that will help us navigate a resolution of the population crisis.

Evolutionary Perspective

Man as a biological species reached his present form about 57,000 years ago. At that time, the world population stood at about four-and-one-half million. This population was formed by nearly 150,000 hunter-gatherer bands averaging 12 adults and 18 children. The average band inhabited a territory nearly 15 miles in diameter. Despite intervening changes in body size, territorial extent of bands, or range of the species, it is highly unlikely that the biomass of man increased more than tenfold during the previous million or so years. During this long period of relative constancy of numbers and biomass, technological innovations of both genetic and cultural origin permitted those groups that were more efficient in extracting resources to supplant less efficient ones.

However, beginning about 40,000 years ago, a drastic transition in population growth occurred. The world population doubled repeatedly. We are now witnessing the tenth doubling, and those now born may see an eleventh doubling—to nine billion—a 2,000-fold increase in only 40 millenia. Each successive doubling required only half the time of the prior doubling. From the first doubling, requiring 20,000 years, the 11th could take only 20 years. Initiation of this unique trend marked the transition from biological-man-only to man-become-human. Unless we appreciate the import of this pattern of population growth, unique in the animal kingdom, solution of the present population crisis will remain illusory.

Up to the beginning of this human era of evolution, man, like so many of his progenitor species, lived in relatively closed social groups, which, on the average, rarely exceeded 12 adults. Through millions of years of evolution, genetically based alterations of physiology and behavior arose that optimized satisfaction from life within such small groups. Conforming to these alterations required maintaining a way of living that would ensure continuance of this small-group existence from generation to generation. Meeting this need for intimate social living became the barrier that prevented biological man from becoming human. Man, like his lesser brethren, could optimize gratification from social living only if contacts with associates continued at the rate that normally occurred within a small group of about a dozen adults.

The capacities for reproduction, with survival into adulthood, regularly caused groups to exceed this optimum number. As the size of any group approached twice the optimum, the rate of increase in frustration from unsatisfactory interpersonal contacts became maximum. At this time, the group would split and one portion would emigrate. Its only chance of survival lay in its capacity to replace a less efficient neighboring group. By this means, technological efficiency increased and total world population was preserved relatively unchanged.

The essential limiting factor to an increase in total population was the genetically built-in need of each individual to experience meaningful contacts with neighbors at a rate consistent with that in a small semiclosed group. Movement within the bounds of the band's territory ensured this rate of contact. Because all available inhabitable space had long since been filled with territories of hunter-gatherer bands, no increase in total population was possible. Increase remained impossible until an essential discovery simulated invasion of uninhabited physical space. It permitted social contact at greater physical density to proceed at the same rate as if the members of each larger group were distributed over a greater physical space.

Elaboration of social roles provided the key to this simulation of physical space. Whenever a group invented or adopted new social roles, its numbers within the former domain could increase accordingly without the necessity for emigration. At the first social breakthrough, some 40,000 years ago, a simple sexual differentiation of roles was sufficient to permit the group to increase from 12 to 24 adults. Within each of these two role-defined subgroups, most members gained most of the satisfactions of social living from within their own subgroup relations. By this tactic, the rate of social contacts was preserved in harmony with that demanded by the genetic constitution of every individual. Each doubling of group size and each doubling of population demanded a related increase in the number of social roles.

Such increase in social roles required comparable increases in the concepts that made social relations more effective. As the population increased in a given physical area, added concepts permitted more efficient mining of resources. Both of these categories of concepts derive from the total information pool available at any time. This total pool of information represents an equivalence to physical space and may thus be termed "conceptual space."

At the time the world population stood at about four-and-one-half million, it was primarily limited by available physical space and by the unmodified natural resources contained therein. Beyond this level of world population, every further increase in the numbers of man required discovery of additional conceptual space. At present, 99.9% of the space we live in is conceptual. (This is not to gainsay the importance of the 0.1% of the space we inhabit that is physical. Our present environmental crisis reveals how mismanagement of this now very small component of our "spatial" requirement threatens our survival.)

Codification of the information forming this conceptual space into usable concepts has proceeded at a rate that makes this total pool of concepts proportional to the square root of the magnitude of conceptual space. We may equate this pool of concepts with human potentiality. Because increases in conceptual space are proportional to the increase in total world population, it follows that with each quadrupling of population, human potentiality doubles; the item that is the average man so doubles.

These trends of increase in population, conceptual space, and human potentiality were possible during the past 40,000 years only to the extent that two related but independent processes exhibited parallel development. The first of these includes such prostheses for information storage, transfer, and transformation as written documents and computers. The second linked larger and larger numbers of individuals into effective sociopolitical networks. Both of these contribute to the rapidity and effectiveness with which we "move" through conceptual space. By sustaining these two processes, man has been creating successively larger and functionally more complex social brains. Brain as a generic concept merely relates to the union of elements that permit the storage, transfer, and transformation of information. Progressive development of social brain largely replaced the evolution of biological brain that terminated as the current, or *sapient,* era of human evolution and population increase began, 40,000 years ago.

This eruptive increase in man's numbers will need to continue only as long as it contributes to the evolution of social brain. Increase in numbers should continue until all adult members of the world population are effectively united into a single sociopolitical network. As we look back over the character of prior unions during the past 40,000 years, a general pattern may be seen. On the average, each doubling of world population has been accompanied by a set of seven of the then extant larger sociopolitical unions coalescing into a new and larger social unity. Thus, each time world population doubled, the geographic extent of the social brain increased sevenfold. Continuation of this general trend has led to an increase in sociopolitical enclaves, or social brains, from a diameter of about 15 miles to the extent represented by larger nations today. This evolutionary course makes a world population of four-and-one-half billion consistent with achievement of a state of union in which seven sociopolitical entities encompass all mankind. Conforming to this evolutionary picture means that by about 1985 there will have arisen seven such unions. In a similar fashion, final world union should be reached during the first quarter of the next century. Attainment of this union will demarcate the level of population beyond which further increase will no longer contribute to evolution of social brain or to an increase in human potentiality.

Upper Optimum Population

If the past trend of population increase continued unabated to this time of world union, the total world population would stand at nine billion, which might be defined as the upper optimum. "Unabated" includes the criterion of a ratio of three children per two adults in the standing population. If this ratio dropped to 2.2 children per two adults, consistent with approaching a stable population, the optimum would drop to about seven-and-one-half billion. However, there are reasons for suspecting that continuing the evolutionary process of enlarging social brain and increasing human potentiality beyond the time of world union will require a continuing slow decline in world population. To initiate such a decline as average life expectancy increases requires a further reduction of the number of children to 1.5 or less per two adults. If this degree of fertility decline is achieved by the time of world union, the upper optimum world population would stand at 6.3 billion or slightly below. Thus, taking an evolutionary perspective leads to the conclusion that an upper optimum world population will contain approximately 3.6 billion adults and enough children to produce either a stable population or a slowly declining one.

Ideomass and the Item Man Being Counted

In either case, the nature of the item being counted has been treated only by implication. We are concerned with the individual as measured by his potentiality for adding to the total information pool (conceptual space) and from its creating new concepts and acquiring and utilizing existing ones. As already pointed out, the potentiality of the individual doubles each time the population quadruples. This means that with each fourfold increase in population, the total capacity of mankind as an assembled species increases eightfold. Multiplying the average potentiality of the individual by the total number of individuals provides a measure, *ideomass*, of the capacity of man as a species. Therefore, when we speak of controlling population, we are concerned with numbers only to the extent that their product, with the average potentiality of the individual, affects ideomass. To place individual potentiality, world population, and ideomass into perspective, we may look at the degree of increase each should have attained by the time of world union early in the next century, in contrast to their magnitudes at the time of transition, 40,000 years ago, from biological-man to man-become-human. Individual potentiality will have increased 45-fold; adult world population 2,000-fold, and ideomass, 90,000-fold. Our present crisis is only partially one of numbers; the main crisis pertains to ideomass in which numbers and individual potentiality are the basic factors.

Over the evolutionary-historical past, ideomass has continued to increase. This increase has been made possible by linking more and more people together into a social network to form more effective brain. But at the time of attaining the upper optimum world population, the linking of people with people, as neurons in a social brain, can no longer contribute to increase in individual potentiality, which can arise only as brain in its generic sense enlarges. This enlargement of brain previously arose from an interaction (product) between expanding networks of people and an expanding repertoire of information storing and processing prostheses. With only this latter factor functioning to increase brain, the rate of increase in human potentiality will be so gradual that it will fail to provide the stimulus for further invention of those informational prostheses required to further expand human potentiality. For all practical purposes, as long as world population is unchanged in numbers, ideomass will remain constant. This means that human potentiality will remain constant. The pattern of life will become ever more rigid and traditional. Complete cultural traditionalism will then terminate the continuing experiment that has been man. His cultural template governing behavior will make him as strictly instinctive as are the simpler creatures controlled by genetic templates that elicit predictable behavior in constant environments.

To remain human, man requires an ever-changing environment and a continual expansion of his own potentiality. If ideomass remains essentially constant, these two requirements may be met by developing a declining population and further expansion of informational prostheses. Each successive halving of world population will probably take twice as long as the prior halving, so that 165,000 years from now, population will stand at about the level reached during the Golden Age of Greece. With each halving of population, individual potentiality can double.

Basic Variables in Population Control

Against this perspective of past and future, we may now evaluate the present period of acute transition between two eras of human evolution. Our concern with the number of men on earth derives primarily from the way that changes in number affect individual potentiality and ideomass. In order to avoid overemphasis on numbers alone, we must examine how fluctuations in other variables have the same consequence as alteration of numbers of individuals.

Social Roles: Any inadequacy in the availability of social roles in relation to an existing physical density has consequences comparable to increasing the size of the population. In the absence of such roles, the frequency of unsatisfactory social contacts will increase. Furthermore, many avenues for fulfilling a potentiality consistent with the population size will not exist. Social disorganization that produces ineffective utilization of existing roles will also have an impact comparable to increasing population.

Linkage: Where individuals, institutions, or various levels of government are too ineffectively linked to facilitate the storage, transfer, and transformation of information necessary for realizing a fulfillment of potentiality, the population will, in effect, be too large.

Prostheses: Inadequate development or utilization of informational prostheses has the same implication as ineffective linkage.

Education and Creativity: The realization of potentiality for development requires acquisition of concepts about role playing, functioning in communication networks, mining resources, and utilizing informational prostheses. To the extent that education fails to provide these, overpopulation exists. Furthermore, becoming more human follows from a continuing expansion of potentiality, which can arise only from the invention of concepts that replace or are added to prior ones. In this sense, the creatively deviant person forms our most precious commodity. In his absence, each person added to the population contributes much more to overcrowding.

Each of the above four variables must be considered in any formulation of population dynamics. They determine the magnitude or capacity of the item, man, being counted. A Kalahari Bushman might be equated with a Norbert Wiener in so far as caloric intake is concerned, but were all members of the world population of two thousand years ago at the level of conceptual development of Bushmen, that world would have been far more overpopulated than it is today, even though at present most roles are not filled with persons of Wiener's potentiality. Simply counting bodies, without considering how each contributes to ideomass, has little relevance.

Fertility: Finally, we come to the variable that, currently, is nearly always the sole consideration in seeking a solution to the impending population crisis. Technological and behavioral means for sufficient fertility reduction to produce either a stable or a declining population exist. Though better means will probably be discovered, the question of means is the least important problem in either the control of the numbers or quality of a population.

Mortality: Termination of physiological functioning—biological death—removes one from inclusion in a population census. If we extend the definition of mortality to include cessation of any kind of functioning, then any individual so characterized becomes subject to removal from the roles of a population census, though he may still survive biologically. Loss of social functioning, in the sense of no longer being able to work toward a fulfillment of potentiality consistent with an existing population level, amounts to death. Fostering such social death is far more inhumane than inflicting physical death. These two kinds of death must be included in our assessment of population dynamics.

An Experimental Study of Social Death

Studies are under way in my laboratory that provide insight into the relationship of population dynamics to social death. The subjects are an inbred strain of house mice. Though they are far simpler creatures than man, they nevertheless give us a picture of man's possible future destitution.

Artificial environments have been designed that consist of modules or cells, each of which represents a socially optimum situation for nine adults (three males and six females). In each cell, accessibility to food, water, and shelter could permit survival of 25 times that number. That is to say, physical resources essentially do not represent limiting variables. Five universes are currently being studied, consisting of one, two, four, eight, and 16 cells, respectively. Each has followed a similar history. Population growth continues until density reaches about 160 mice per cell—eighteen times the optimum. At this maximum density, reproduction essentially ceases.

From the time of initiation of the population in each universe with four males and four females, to the time of maximum density, the number of socially active males per cell rarely exceeds three. "Socially active" for this simple species implies roles in the contention for, acquisition of, or maintenance of a territory. Some young males who enter the social milieu may succeed in replacing a territorial male. Other young males contend for a time before total rejection from the social flux of status striving. All such replaced dominants or rejected contestors became recruits in a socially pathological category of withdrawn males. No longer having access to the sheltered living spaces, these males aggregated in compact pools in exposed public space. The withdrawn males restricted their motor activity to the minimum required to obtain food and water. They exhibit a marked lowering of the threshold of tolerance to disturbance; very minor dislocations of their usual sedentary relations precipitate violent episodes of aggression that culminate in severe mutual wounding without any attempt to escape.

As the population increased in numbers, comparable pathological changes took place among females. Increasingly many females began to exhibit male-type aggressiveness in public space outside the sheltered living spaces. As these shifts in behavior became more pronounced, more and more females lost the capacity to build and maintain nests adequate for sheltering their young. We know very little about the specific ways in which altered maternal behavior or the intrusion of subadults alters the experience of young mice. However, the consequence of this altered experience is drastic and apparent. Both males and females fail to mature socially. Behaviorally, they continue into adult life as juveniles, without the capacity for either mating or aggression; their normal adult roles never emerge. Females move into adulthood with rare, if any, pregnancies. Males return to the nesting boxes where they associate with their nonreproducing sisters, with males like themselves, or with the few remaining reproducing females. We term these males the "Beautiful Ones" because they are physically perfect specimens, free of the wounds characteristic of behaviorally more adequate males, who moved out into the social scene.

At the maximum attained density, about 18 times the optimum, all of the more recently born males are Beautiful Ones; no new contenders emerge. The socially active males senesce and die, as do the large aggregates of withdrawn males living in public space. A few of the older, aggressive, male-like females exhibit territorial defense, but the bulk of surviving adults have lost their potentiality for engaging in the normal reproductive and aggressive roles. After viewing these studies at this stage, Mr. Mayer Spivack of Harvard applied the term "overliving" to these populations. If the current conditions of failure to reproduce persist, the populations will be "overliving" in the sense that surviving individuals no longer contribute to the survival of the population.

Even if some few individuals are able to recoup their normal behavioral roles after many of their associates die, the essential point of the above account remains valid. Where there is no physical escape from a closed environment, and where no capacity for elaboration of roles prevails, social death may mark up to 95%—151 out of 160, approximately—of the population. Had such populations existed in more normal ecological settings, this 95% would have fallen into émigré category, moving to suboptimal habitats, where most would soon have died from predator action or other mortality factors. This raises the question of warfare. Man has often been called the only species that engages in warfare. From a comparative and evolutionary perspective, the term *warfare* must be broadened to include all deaths within a species resulting directly or indirectly from actions by members of their own species. In this light, we may examine where man is now. We recently reached a time when there were 1.2 billion adults, with a life expectancy of approximately 50 years after weaning. During their lifetime, perhaps no more than one million persons per year have died from all kinds of warfare, from simple homicide to nuclear attacks. For adults only, we may thus take 50 million deaths out of a population of 1.2 billion, roughly 4%, as the currently attained warfare mortality. During biological and cultural evolution, warfare mortality has been reduced from 95% to 4%. If world union materializes within the next century, warfare mortality should become even more negligible. As a factor in controlling population, warfare has become less and less important from an evolutionary point of view.

Therefore, we will be faced more and more with the problem of social mortality. Studies on both animals and man suggest that the stochastics of a dynamic social setting will usually produce an 8% to 16% social mortality that correspondingly reduces ideomass. Further decrements in ideomass accompany the social morbidity of realized potentiality falling short of that which should characterize the attained population size of biological bodies. The planet Earth is becoming progressively a closed system within which there is no physical escape. If adequate measures are not taken to provide habitable conceptual space, the prevalence of social death and morbidity will increase. And if we focus excessively on simple control of the numbers of the biological entity that is man, our effort to preserve the human experiment will be endangered.

* * * *

Natality: Here, too, we are circumscribed by a too-narrow concept of natality by restricting it solely to biological birth. Though we may wish to reduce biological natality enough to produce, first a zero population growth, then a population decrease, natality as a broad generic concept may continually increase for the foreseeable future. As a broader generic concept, natality includes both the birth of new ideas, and their spread and acceptance by a larger segment of the population.

Control

This broadened concept of natality encompasses the essence of control, with control implying the design and management of future evolution, of which the number of men, accompanied by an increasing individual potentiality, is a major component. Creation of new ideas is a major essential form of natality to be continually enhanced. Relatively few individuals are involved in this category of reproduction. However, every individual is faced with the need to evaluate emerging ideas, accepting those judged valid, and participating in their dissemination. This process of value change is a kind of midwifery by which ideas are born into an entire population.

"Control of population: numbers" may be realized only through the functioning of an inclusive population policy encompassing a set of interrelated subsidiary policies and their implementation. If we take as our goal a survival of man that enhances the continuation of the human experiment, priorities of policy implementation become apparent:

1. Evolutionary designing.
2. Value change with conflict resolution to foster acceptance and spread of ideas.
3. Education and opportunity for fulfillment of potentiality.
4. Enhancement of the linkage of individuals and institutions into communication networks.
5. Family planning.
6. Promotion of creativity.
7. Development of technological prostheses to extend the capacity for storage, transfer, and transformation of information.

This is my personal assessment of the priorities within a comprehensive population policy.

REFERENCES

1. CALHOUN, J. B. 1964. The social uses of space. *In* Physiological Mammalogy. W. Mayer & R. Van Gelden, Eds. **1:** 1–187. Academic Press. New York, N.Y.
2 CALHOUN, J. B. 1970. Promotion of man. *In* Global Systems Dynamics. E. O. Attinger, Ed.: 36–58. Karger. Basel, Switzerland; Munich, Germany; New York, N.Y.
3. CALHOUN, J. B. 1971. Space and the Strategy of life. *In* Behavior and Environment. A. H. Esser, Ed.: 329–387, Plenum Press. New York, N.Y.

CIVILIZATIONAL CONDITIONS OF
PROLIFERATIVE DISEASES

Julian Aleksandrowicz

III Klinika Wewnetrznych
Akademii Medycznei
Krakow, Poland

This paper considers the problem of the relations existing between the characteristic civilization of some given population, with ethology (as the science of morals) as its essential component, and the incidence of proliferative diseases of various types, including leukemia.

In different social groups the prevailing cultural patterns are, of course, different and so also are their ethical attitudes. There are, moreover, known geographical differences in the occurrence of proliferative diseases.

The evidence assembled to date is sufficient for us to conclude that in various social groups there must be some specific biological, physical, chemical, and psychosociological factors whose combined action influences the rate of occurrence of one or another proliferative disease.[1, 3, 4, 6-10]

A better knowledge of these factors is of tremendous significance for the healthy development of mankind; such understanding would allow us to identify, utilizing scientific principles of investigation, the pathogenic factors in the environment that cause the excessive rate of particular proliferative diseases and to develop methods for controlling these factors. The effectiveness of the prophylactic measures will, however, depend not only on man's intellectual and scientific effort but also on the degree of his emotional involvement in the attempts to eliminate the pathogens. In other words, the effectiveness of prophylactic measures depends on the mode of action and on the methods used to make real the system of values, which has as its highest aim the achievement of the health of mankind in the broadest sense.

I will, in the following discussion, restrict my consideration of the dynamism of proliferative diseases within a set of ecological conditions to one type only, i.e., to leukemia in men and animals. The working hypothesis may be summarized in the assumption that, because the evidence supplied by epidemiological statistics is already sufficient to demonstrate differences in the rate of occurrence of various kinds of leukemia within the same parameters of time and space, it is now possible to pass to the next stage of research and to identify and check leukemogenic factors by scientific methods. The third stage will then be the control and the elimination of these factors from the environment, i.e., prophylaxis.

Historical Variability of Diseases and Their Dependence on
Civilization and Social Structures

Before presenting the current state of knowledge about the forms of leukemia in men and animals in different parts of the world, in different countries, in different districts of a country, and even in limited districts of a town or village, I would like to review briefly the changing pattern of human diseases in different epochs and different parts of the world. We will see that diseases were influenced by the civilization prevailing at some particular period of history and in particular parts of the inhabited world.

When we look back on the past, we see how each new environmental factor, one after the other, influenced the animals living on the earth. With changing climates, the soil changed, causing, in turn, change of the flora. New environmental factors appearing one by one caused the "variability" phenomena, which shaped the bodies of the anthropoid apes. As the forests disappeared from the surface of the earth, the apes were forced to come down to the ground. New species with erect bodies then appeared. Their further development reflected their way of life, determined by the conditions of their habitat, the open plains of South Africa. The local flora and fauna determined the forms by which these animals satisfied their energy needs. In the glacial epoch of the next stage of evolution, the equilibrium between the living creatures and their environment was dangerously disturbed. Our ancestors died in large numbers of starvation and cold. Only individuals capable of protecting themselves from cold and starvation could survive.

Thus, the appearance of *Homo sapiens recens* depended on the influences of climatic factors, such as the amount of sunshine, humidity, winds, etc. For example, in the course of the lengthy process of adaptation, melanin was concentrated in the epidermis of the human skin and protected the body from the deadly effects of ultraviolet radiation. But before a new morphological type completed its development by adapting to environmental conditions, many individuals must have suffered, and many may have died, from diseases caused by insolation and thus, presumably, also from proliferative diseases.

It, therefore, becomes clear that abiotic factors exerted an influence on the modeling of the human body in the process leading to the development of the particular varieties of the human species. These morphological changes and, hence, the psychological structure of man, depended solely on the random action of natural forces lasting for many millions of years.

It is only when mankind had developed the ability to think and to organize consciously the social environment that the morphological and psychological structure began to change more rapidly.

Because of reconstructions made possible by archeological findings, we can now look back over many thousands of years and with a high degree of probability define the environmental conditions in which diseases developed. The outlines of the history of diseases in different periods and different geographical regions may be traced from the information supplied by the study of preserved human skeletons and by observations of certain primitive peoples, whose way and standard of life, even today, often corresponds to the prehistoric epochs.

The most common diseases of the paleolithic civilization were, according to Jean Bernard, those caused by parasites, as is frequently the case among today's primitive nomad tribes, whose way of life is still governed by archaic social structures. In that distant period, diseases and anemias resulting from food deficiency were not widespread, and men suffered mainly from bacterial and protozoan parasitic diseases, ague, schistostomiasis, etc. Paleontological examinations of 40,000-year-old skeletons show that in the distant past man was not afflicted with deficiency diseases. The nomadic and pastoral way of life and cannibalism permitted him to satisfy his energy requirements sufficiently.

In the more recent past of the neolithic civilization, the important change from the nomadic to the settled way of life occurred. Men domesticated some animals and learned to select some plants for cultivation. This settled way of life produced special conditions for the development of a new civilization. It was during this epoch that the first signs appeared of social stratification and of groups specialized in particular handicrafts. This, in turn, produced a situation favoring armed conflicts between the tribes. Ever since, wars have had as their consequence hunger and outright starvation. Characteristic for this epoch are such deficiency diseases as the anemias. This is confirmed by neolithic skeletons, which, in their histological structure, show symptoms of dietary deficiency.

We are, thus, justified in presuming that blood diseases, in which various deficiency symptoms were superimposed on the parasitic invasion, were characteristic for the neolithic period. We may, therefore, suppose that the common diseases then afflicting mankind, especially when crops failed, were the counterparts of what today are known as malnutritional disease, kwashiorkor, tropical macrocytic anemia, etc.

Such nutritional deficiencies may have been associated with the disappearance of cannibalism from the territories of Europe and North Africa. This was, however, a critical period, and mankind, having passed through an all-important revolution, reached a turning point in its moral history.

It seems highly probable that in the distant past men also suffered from proliferative disease of various types; they could have been exposed to such natural oncogenic factors as 3,4 benzpyrene, which sometimes was excessively concentrated in the vicinity of asphalt deposits near volcanoes, or telluric radiation. Nevertheless, these factors must have been incomparably more rare than they are today, when the chemical industry is developing increasing numbers of new products with mutational, teratogenic, carcinogenic, and leukemogenic effects.

Contemporary civilization has been dominated during the last two centuries by the industrial revolutions, which have changed the pattern of prevailing diseases. Progress in hygiene and the growth of the pharmaceutical industry have undoubtedly limited parasitic infections and the advance of factors causing nutritional deficiency diseases. Appearing in their place today are diseases caused by the toxic effects of industrial by-products that often possess carcinogenic properties. Technical civilization, which represents man at a definite stage of his evolution, brings with it both favorable and unfavorable conditions for his further development. It has made possible the suppression of the epidemics that harassed mankind for centuries. It has added many years to human life. By reducing infant mortality, it has produced, along with highly desirable effects, a new demographic explosion and its concomitant alarming economic and social consequences.

Thus, the contemporary epoch is characterized by a steady growth of the number of tumors, exceeded in the developed countries only by premature death from arteriosclerosis, heart disease, and traffic accidents. Tumors and leukemias, which, during the first decades of this century, were not observed in young people and children or in animals and fish, have become much more frequent.

All these observations lead to the conclusion that the extremely rare occurrence of proliferative diseases in the primitive civilizations was due to man's very limited exposure to a very limited number of oncogenic factors. Today, our industrial civilization increasingly exposes man and the higher vertebrates to carcinogenic and leukemogenic factors that accumulate in the environment. They disturb the biocenotic equilibrium, a fact becoming increasingly obvious in the particular links of the water-soil-plant-animal-man chain.

As can be seen from the above discussion, the kind of social structure characteristic of the particular stages of historical evolution, i.e., considered in relation to time, determines the occurrence of specific proliferative diseases. Any change of the social structure obviously leads to change in the environment and, in turn, in the manner that proliferative diseases afflict humanity.

Contemporary Regional Differences in Proliferative Diseases and Their Incidence

R. Doll has published maps showing the worldwide distribution of cancers of the esophagus, stomach, lungs, colon, breast, and liver.[5]

It appears that cancer of the esophagus has three main high-incidence foci: Kazakstan on the northern coast of the Caspian Sea, Bulawayo in Southern Rhodesia, and the region of Transkei in South Africa.

Comparison of esophageal cancer occurrence in various countries shows that in the Guryev district in Kazakstan, its incidence is 200 times higher than in the Netherlands and Nigeria. This phenomenon has not yet been satisfactorily explained. Burrell and co-workers have suggested that the factors responsible for these differences lie in a deficiency of certain soil components and the infection of crops with substances causing the production of nitrosamine.

Cancer of the stomach is most frequent in Japan, the U.S.S.R., South Africa, Chile, and Colombia. Its incidence is particularly low in southwest England and among the white population of the U.S. Such a distribution is attributed by various authors to dietary factors. Associating this form of cancer with substances in decaying foods, they suggest that improved methods of preserving food that are widely applied in the United States determine its low incidence of stomach cancer.

Lung and bronchial cancer is most common in Great Britain, Central Europe, especially Germany, Belgium, Czechoslovakia, Hungary, the Soviet Union, Canada, the United States, Uruguay, and New Zealand. On the other hand, in Spain, France, Italy, Yugoslavia, Romania, and the Balkan States, it is less frequent. Men are more often attacked than women, except in Great Britain, where women patients are more numerous.

A remarkable property of cancer of the large intestine is its apparent association with the Anglo-American way of life and not with race. It is equally common among white and black. Its occurrence is very low among the African population of South Africa and in the Balkans, India, the Philippines, and Japan.

The incidence of breast cancer is highest in the countries of western Europe, South Africa, Australia, New Zealand, and Israel. The further east, the lower the incidence of this disease; it is very low in Uganda and Mozambique, Taiwan, and Indonesia. In the U.S., the rate of the disease among white and black women is the same. Japanese living in the Hawaiian Islands suffer a higher breast cancer incidence than those living in Japan.

The data on the distribution of cancer of the liver refer to man between the ages of 15 and 44. This age group was chosen for observation because in countries with an especially high incidence of hepatic cancer, it appears earlier in that age group. Another reason for choosing this age group was that it ensured the fewest diagnostic errors, because in older persons it is often difficult to distinguish between primary cancer of the liver and cancer of the biliary ducts.

It appears that in this age group hepatic cancer is rare everywhere in the world except in some areas in Africa: the province of Ibadan in Nigeria, Johannesburg, and Mozambique, where it occurs as frequently as lung cancer in Great Britain and is more than 1,000 times higher than primary liver cancer in various parts of Europe. It is believed that aflatoxin, a metabolic product of *Aspergillus flavus,* is responsible for the high incidence of hepatic carcinoma. This mold develops with special abundance on such substrates as groundnuts and foodstuffs kept in a damp and warm environment.

Lymphatic leukemias in men and animals are extremely rare in China and the Far East, a fact pointed out to me during my visits in 1960 to the clinics and hospitals of Peking, Ulan Bator, and Canton. On the other hand, in the Netherlands and the countries bordering on the Baltic Sea, i.e. Denmark; Schleswig, West Germany; and the Polish northern provinces, morbidity from lymphatic leukemia in both men and cattle is particularly high.

Studying morbidity from leukemias in cattle in Poland, Karaczkiewicz found that the rate was very high, sometimes higher than 200 per 100,000 in the northwest provinces, although in the southwest provinces it was as low as about six per 100,000. The geographical difference between these high and low rates of leukemia incidence in cattle corresponds to the partition boundaries between the parts of Poland, which before 1918 were attached to Germany on the one hand and Austria and Russia on the other. According to Karaczkiewicz, this remarkable epidemiological situation was caused by the differences in agricultural techniques in the partitioned parts of the country. It is common knowledge that agriculture was more advanced in Prussia than in the other two partitioning countries; potassium fertilizers were used in much greater quantities, which led to the reduction of the magnesium content of the soil, thus disturbing the biocenotic balance in a direction that favored the proliferative processes in tissues, a problem I shall deal with below.

Cultural and Industrial Causes of Proliferative Diseases

The problem at this point is to find an explanation for the differences described above in the geographic distribution of cancers. Recent epidemiological research has proved that morbidity in the proliferative processes depends to a much higher extent than previously supposed on man's conditions of life. For this reason, nearly 90 per cent of cancers are today believed to be stimulated by human activities, and thus, theoretically speaking, it lies in man's power to prevent them.

Some authors, for instance Higginson, classify cancers according to the nature of the precipitating factors into cultural cancers, industrial cancers, and idiopathic cancers.

The first of these groups are the cancers that may be attributed to traditional customs of a particular population. For example, in India, oral cancer is common, and the cause, in the Bombay region, is the custom of chewing betel leaves. A small piece of the betel leaf, which has narcotic properties, is placed in the mouth between the tongue and the cheek. The leaves are powdered with lime, which causes congestion of the mucous membrane of the mouth, thus improving assimilation of the drug. It is not surprising, therefore, that cancer of the mucous membrane of the mouth is a common disease among these people.

In South Africa, similarly, there is a custom of using snuff made from the leaves of the aloe plant, and this seems to be the cause of cancer of the sinuses.

It is now well known that among cigarette smokers, lung cancer is two to three times more frequent than cancer of the mouth. On the other hand, among peoples where there is a custom of using *naswayam*, a mixture of tobacco, ash, lime, and oil, independently of cigar and cigarette smoking, cancer of the mouth is twice as frequent as lung cancer. In the Kashmir, a common form of cancer called *kangir* develops on the skin of the abdomen; among the peoples of this area, there is, reportedly, a custom of warming the abdomen with a lump of hot clay.

Our second grouping, industrial and occupational cancers, includes asbestos tumors that occur in workers living near asbestos deposits or working in asbestos mines. In South Africa, mesothelioma caused by irritation with chrysolite frequently occurs. Arsenic compounds, to which farmers are often exposed because of the extensive use of pesticides and insecticides in modern agriculture, are also responsible for a substantial proportion of lung cancers.

Case reported that naphthylamine used by the textile industry for dyeing fabrics caused tumors of the bladder in 15 workers employed as distillers. He concluded that everybody was susceptible to exposure to carcinogenic factors when their dose was sufficiently high.

Finally, it is a well-known fact that leukemia and erythroleukemia occur among persons working in a benzene-contaminated environment.

In the last group—idiopathic cancers—Higginson lists different forms of cancer for which an etiological background could not be identified with research methods existing at present.

Ecological Conditioning of Etiology and Pathogenesis of Proliferative Diseases

The evidence presented here to prove both quantitative and qualitative differences between various forms of proliferative disease in various cultural formations seems also to indicate that there must be differences of concentrations of the known and suspected carcinogenic and leukemogenic factors within the particular ecological environments. In an attempt to obtain a clearer picture of these factors, we undertook a group of investigations of small areas, such as districts, towns, and villages, and even individual farmsteads and houses, together with the environment in the homes of cancerous marriages and families.

These studies showed that in specific areas there were centers of cancerous and leukemic diseases and that their distribution was not accidental but was determined by specific factors. In order to show more clearly to what extent environmental conditions, shaped by man himself to a much greater degree than by the forces of nature, are responsible for the spread of proliferative diseases, I shall outline here the present state of knowledge about oncogenesis.

For the development of a proliferative disease, the following hypothetical conditions must be fulfilled: (1) predisposition, (2) presence of the virus infection factor, and (3) presence of cocarcinogens. Predisposition consists of a set of biological phenomena occurring at the molecular level of the genetic structure. It is precisely these phenomena that are responsible for the fact that in the same environmental conditions some persons are attacked by proliferative disease, which suggests a predisposition, whereas others are not.

The differences in predisposition are due to deoxyribonucleic acid (DNA), which carries the structural, and hence the metabolic, code of cells and determines that the daughter cells have the same structure and functions as the parent cells.

The genetic system and the molecular structures of cells are weakened with regard to lysosomes and genes by the absence of some stabilizers, of which so far Mg^{++} ions have been identified, a result of the work of Goldberg and others. A deficiency of Mg^{++} ions facilitates penetration of the genetic structure of cells by viral oncogenic factors; these factors then cause the uncontrolled multiplication of cells, resulting in proliferative growths. Among the carcinogens, the best known are ionizing radiation, some chemical compounds, and mold metabolites and are referred to by the general terms of physical, chemical, and biological cocarcinogens, respectively. Under specific conditions these cocarcinogens activate the nucleic acid molecule, which probably has the same nature as a virus and causes a structural change in the genetic apparatus of the cell. This disturbs the daughter cells and makes them form a clone that differs in structure and function from the parental cells and assumes the form of neoplasmic growth.

This viruslike factor, the particle C, as it has been called, has now been identified by electron microscopy by Dmochowski and others. The particle C seems to be responsible for the vertical and horizontal transfer of the proliferative disease from experimental animals to animals with a specific immunological structure, from man to animals,[11, 12] from one farm animal to another, an occurrence often observed in agriculture, and perhaps from animals to man. In this last case, the direction of the transfer is indicated by some results of prospective experiments to which, of necessity, we have to restrict ourselves.

The aim of our ecological researches—I present here only one of the investigated cases—is to identify the environmental factors that may, in the present state of knowledge, be responsible for oncogenesis due to mutual infection or to exposure to conditions having an oncogenic effect both on men and on animals.

We conducted our investigation,[2] in collaboration with professors of agriculture, University of Krakow, Ewy, Litynski, Komornicki, and Smyk, in the village of Okocim-Pomianowa. In this village there was a "cancerous" house where five persons, three of them relatives, suffered from cancer. Adjacent to the house was a cow shed where for years the cows had had leukemia (leukemia herd). The farm was an experimental one belonging to the School of Agriculture in Krakow. This environment was very suitable for ecological studies on oncogenesis.

We found that in the farm there was (1) a reservoir of oncogenic viruses in cows with leukemia, (2) a reservoir of biological cocarcinogens from molds producing carcinogenic metabolites, i.e., *Aspergillus flavus, Penicillium meleagrinum,* and *Rhizopus nigricans,* in the walls of the "cancerous" house, barns, and soil, and (3) a magnesium deficiency in the Uszwica River, in the soil, and in the wells (TABLE 1).

TABLE 1

Water and Soil Contents of Two Farms

	Water *					Soil	
Farm's Name	Ca⁺⁺	Mg⁺⁺	K⁺	Fe⁺⁺⁺	SiO₂	Mg⁺⁺ in mg/100g	pH
Pomia nowa	53.9	6.9	12.9	0.4	16.0	2.0– 7.0	±5.6
Radlow (control)	67.5	46.5	3.5	11.2	4.0	10.0–28.0	6.0–7.0

* Mean value from May and October, 1969, in mg/1,000 ml.

We have recorded many similar examples where the presence of molds from the group of *Aspergillus flavus* and *Penicillium meleagrinum* and the Mg deficiency in the soil were confirmed in the environment of leukemic and cancer patients.

In the next environment in which the incidence of proliferative diseases was significantly higher, we made the following observations:

In the village of Liszki, about 20 km to the west of the city of Cracow, 14 cases of various neoplastic diseases were reported in the years 1964–1968 (I. Gajda), i.e., an incidence of 180/100,000 annually. The incidence rate of the Cracow province in these years was 136.63. Mapping of the households inhabited by the patients showed that 11 houses were situated within a radius of 250 meters of the house where the first case of neoplastic diseases occurred (K. Janicki). This distribution is statistically significant on the basis of the chi square test with Yates' correction, because theoretically the expected number of cases in the area under consideration is 3.35, and 10.65 outside of the area. The empirical number of cases in the area exceeded the theoretical number threefold. Hence, this tumor "cluster" is statistically significant.

After the termination of this study, in 1970 two new cases of lymphoma occurred in humans inhabiting the foregoing area and no new cases in the remainder of the village. The age of the patients ranged between 23 and 67 years. There was no evidence of family predisposition.

Water from all four wells in the area inhabited by the patients was found to have high contents of K, SiO₂, and very low content of Mg, Cu, Fe. The Mg/Ca ratio was many times lower than in water from a control farm in Radlow situated about 60 km away in a straight line, according to assays at the Chair of Pedology (Prof. Komarnicki). TABLE 2.

The Provincial Agricultural Chemical Station reported high acidity (pH 6.4–6.7) of soil from the area, low values of Mg and Cu. Other microelements were not studied. Fungi imperfecti were cultured from the dwellings of the patients. The five most frequent fungi producing mycotoxins known to possess carcinogenic properties (verified at the Chair of Prof. Smyk) obtained were: *Aspergillus flavus, Penicillium meleagrinum, Cladosporium herbarum, Rhizopus nigricans* and *Alternaria geophila*.

Because the magnesium ion acts as a stabilizer for ribosomes and for the genetic systems (Goldberg, West, and others), and as an activator of RN-ase and the properdine system protecting the cell against invasions of oncogenic viruses and action of mutagenic factors, we hypothesized that feeding the animals a diet deficient in Mg would favor oncogenesis. This hypothesis was confirmed by the work of McCreary, Battifora, and P. Bois, and, in Poland, by Kowalczykowa and Stachura, who caused leukemic proliferation in rats fed Mg-deficient diet.

The experimental results suggest the hypothesis that in countries in which dietary magnesium is high, the incidence of leukemia should be low. This is confirmed by Seelig, who reported that in the Far East, where leukemia in cattle and lymphatic leukemia in men were rare, the daily dose of magnesium was 6–10 mg per kg food. On the other hand, the daily dose of the Euro-American population is less than 5 mg of magnesium and is steadily dropping.

The level of magnesium in soil is determined by the contemporary cultural pattern; modern agricultural techniques are dominated by potassium fertilization; because of the magnesium-potassium antagonism, the magnesium content of soil, and therefore of plants, animals, and men, is steadily reduced.

* * *

In contemporary civilization, the dominant pattern of interhuman relations is characterized by malnutrition of two-thirds of the inhabitants of the world; therefore, the use of chemical compounds for more intensive farming and higher agricultural output is justified, even if it leads to a disturbed biocenotic balance and, consequently, to—among other problems—proliferative disease caused by overuse of chemical fertilizers. The logical consequence of this situation should be, however, such interhuman and international relations that would allow man to preserve the biocenotic equilibrium, thus acting prophylactically to prevent the spread of proliferative diseases.

It remains to be seen whether the attempts to correct the supply of magnesium and probably copper, iron, etc., in soil, water, and plants, will reduce the incidence of proliferative diseases in the districts where the deficiency of these elements and proliferative disease morbidity are high.

In sum, we present the opinion that one of the most important ways to prevent neoplasm is by protection of the balance in biocenosis; it means a conservation of nature. Conservation of nature is conditioned by war prophylaxis. It is obvious that in some situations, war industry pollutes water, soil, and food with carcinogens destroying the biophysical environment considerably more strongly than would industry without military purposes. Hate, fanaticism, and racism are ruining the psychosocial environment connected

TABLE 2

MICROELEMENTS IN WELL WATER IN CLUSTER OF HUMAN CANCER IN LISZKI VERSUS CONTROL IN RADLOW

	Mg++	Ca++	Ca/Mg	SiO$_2$	Fe+++	Cu++	K+Na+
Liszki well 1	42.2	321.6	7.62	26.0	0	0.022	292.7
" " 2	36.2	162.2	4.46	22.0	0	0.009	496.8
" " 3	51.3	260.9	5.08	20.0	0	0.009	388.5
" " 4	30.2	200.1	6.62	32.0	0	0.009	396.9
Radlow well central	46.5	67.5	1.45	4.0	11.2	—	26.5

with the biophysical environment. In this way, the vicious circle devastating human kind is being closed.

We believe that the human being—the highest value—creates new values and is dependent on them. So, the fate of humanity in our own hands.

REFERENCES

1. ALEKSANDROWICZ, J. 1969. Estudios sobre la influencia del Mg++ en la pato-genesis, terapeutica y prophylaxis del sistema linforreticular. Folia Clin. Int. 19(4): 3–4.
2. ALEKSANDROWICZ, J. & B. SMYK. 1971. Mycotoxins and their role in oncogenesis, with special reference to blood diseases. Polish Med. Bull. (In press.)
3. BATTIFORA, H. A., P. A. MCCREARY, B. M. HANEMAN, G. H. LAING & G. M. HASS. 1968. Chronic magnesium deficiency in the rat. Arch. Path. 122: 610.
4. BOIS, P., E. B. SANDBORN & P. E. MEISSIER. 1969. A study of thymic lymphosar-coma developing in magnesium deficient rats. Cancer Res. 29: 763.
5. DOLL, R. 1967. Prevention of Cancer, Pointers from Epidemiology. Nuff. Provinc. Hosp. Trust.
6. DELBET, P. 1944. Politique préventive du cancer. La Vie Claire. Montreuil, France.
7. SEELIG, M. S. 1964. The requirement of magnesium by the normal adult. Amer. J. Clin. Nutr. 14: 342.
8. STRIEBEL, A. 1966. Über das Calcium-Magnesium Verhaltnis in Urin von Gesunden und Krebskranken. Oncologia 20(3): 209.
9. VOISIN, A. 1964. Les Nouvelles Lois Scientifiques d'application les Engrais. Quebec, Canada.
10. TROMP, S. W. & J. H. DIEHL. 1955. A statistical study of the possible relation-ship between cancer of the stomach and soil. Brit. J. Cancer 9: 3.
11. ADAMS, R. A. 1968. Cancer Res. 28: 1121.
12. LIKNAITZKY, D. G. 1969. Cancer 23(1): 94.

RECENT ADVANCES IN CHEMICAL SCIENCES

Ernst D. Bergmann

Department of Organic Chemistry
Hebrew University, Jerusalem, Israel

To be a prophet is a thankless task, even for an Israeli—although Israel has been the land of the prophets. Especially in science and technology, it is probably more reasonable to extrapolate into the future existing knowledge, trends of development that characterize the present. One may then be faced with the possibility that a "breakthrough" will upset the forecast and the planning; however, the history of science appears to teach us that breakthroughs are rare indeed and that, on the whole, there is an inherent logic in the development of science and technology.

Methodology. The period in which we live is characterized by a weakening of the delineation between the various scientific disciplines. As far as chemistry is concerned, we are witnessing on one hand an increasing influx of mathematics and physics into the science of chemistry in the classical sense and, on the other, an expansion of chemistry into the life sciences. The meaning of "molecular biology" is, in fact, only the transfer of certain areas of biology into chemistry.

The mathematical methods of quantum chemistry have been so refined and the use of high-speed computers so perfected that it is almost a routine task to calculate the physical and chemical properties of known inorganic and, especially, organic molecules and to predict the properties of molecules that have not yet been prepared by the chemist. Of particular interest is the application of these methods to molecules of biological importance; it has very recently become possible to calculate the various *conformations* of biologically active compounds and to determine the relative energies of such conformations.

Physical methods of analysis have made it possible to detect quantities of chemical compounds several orders of magnitude smaller than hitherto. I am referring to such methods as nuclear magnetic resonance and mass spectroscopy, on one hand, and to the chromatographic methods, on the other. It has recently been shown, for example, that we can make a quantitative analysis of the purine and pyrimidine bases of the nucleic acids of a single brain cell by chromatography on a quartz fiber.

Computers make it possible not only to supervise industrial processes "on stream" and to operate a factory almost without human labor but even to design, in the most rational way, complex chemical syntheses. Corey has recently given a most impressive example of this possibility by applying it to the synthesis of patchouli alcohol.

As for the expansion of chemistry into biology: in the past we explored mainly the bulk constituents of cells, e.g. proteins, sugars, and lipids, but we are now engaged in the elucidation of the structure of those compounds, which are so enormously active that the cell needs them only in minute quantities. Here, of course, the modern methods of analysis that we have referred to are of extreme importance, especially because many of them are of nondestructive nature. The study of enzymes, vitamins, and hormones has thus become a chemical rather than a biological activity. Let us mention here the elucidation of the complete structure of the enzymes lysozyme and ribonuclease or the isolation and identification of the insect pheromones, which belong to the most active compounds known: the pheromone of the female silk moth (hexadeca-*trans*-10-*cis*-12-dien-1-ol) is active in quantities of 10^{-10} micrograms, so we may well suspect that only one single molecule may be necessary for the biological effect.

Within chemistry itself, the classic distinction between organic and inorganic chemistry has become blurred. This is because of the rapid development of the study of organometallic substances, which not only are of considerable theoretical interest but appear to have many practical applications. In particular, homogeneous catalysis of many reactions has thus become possible, such as hydroformylation, decarbonylation, hydrogenation, and dehydrogenation, or isomerisation of unsaturated hydrocarbons. Perhaps most spectacular is the ability of certain of these organometallic complexes to bind molecular nitrogen (in the same way as the isoelectronic carbon monoxide) and thus to permit nitrogen fixation in a manner that might well prove to be a model of the method employed by nature. Another example of such "hybrid" chemicals is ferrocene, which is the parent substance of a whole series of interesting "pseudoaromatic" substances.

Raw Materials. Before the end of the century, we will apparently be faced with the fact that the raw materials on which our present technical civilization is based have gradually been exhausted. This is true for the ores we are exploiting, and even more for the fossil fuels we are employing both for our machines and as the starting point for the petrochemical industry. We will thus be forced to utilize poorer ores than those presently mined, and we will have to develop methods for the enrichment of the important constituents of such poor ores without greatly increasing the price of the final products. On the other hand, we will have to develop to a greater extent synthetic materials that can replace the fundamental materials we use, even for construction; this will be an additional burden on the petrochemical industry, which today is the main source of synthetic products.

There is, however, one area that may yield large quantities of ores, viz. the ocean, and it is likely that chemical and industrial oceanography will attract increasing attention in the future.

Whatever the outlook for new sources of fossil fuels may be, there is no doubt in my mind that we will gradually abandon such fuels as the source of energy, so that more and more of our present resources can be made available for the petrochemical industry. Even so, the amount of crude oil and natural gas is finite, and the question will arise: is there a raw material that is available at least in the same quantities as natural oil and gas or, if possible, in really unlimited quantities? It would seem that there is only one such possibility that can be taken into consideration, and that should be explored even if it appears unusual at first sight. One can transform, by fermentation, carbohydrates (sugar, starch, cellulose) into chemical compounds containing two, three, or four carbon atoms and thus equivalent to the fundamental substances, ethylene, propylene, butylene, isobutylene, and butadiene, which the petrochemical industry produces by cracking and then transforms into its final products.

The scheme in FIGURE 1 gives an idea of these possibilities.

The sugars from which this scheme starts are indeed available in unlimited

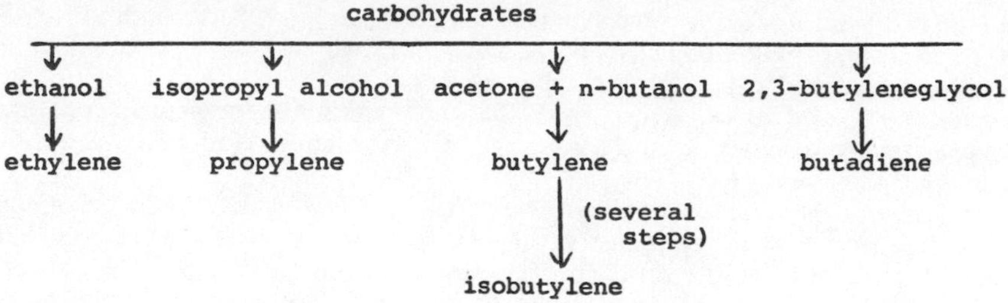

carbohydrates

ethanol	isopropyl alcohol	acetone + n-butanol	2,3-butyleneglycol
ethylene	propylene	butylene	butadiene

(several
steps)

isobutylene

FIGURE 1. The fermentation of carbohydrates.

quantities because they are reproduced by nature year after year. We would have to learn to look to agriculture and the fermentation industry as an alternative to the cracking industry. The implications of such a change of view are obvious: they may be of particular importance to the planning of the developing countries in the tropical and subtropical areas of the world and therefore capable of producing large quantities of carbohydrates in various forms.

Naturally, this approach to agriculture not only from the point of view of food production but also from that of industry is by no means new. The utilization of agricultural products, including agricultural wastes, for industrial purposes has been and still is one of the foundations of the chemical industry. However, it might be useful to consider that, from a more philosophical point of view, the proposed scheme (for which the name chemurgy was coined during the Second World War) is highly satisfactory. Our machines burn fuel to carbon dioxide, which is also the end product of the oxidation of food in our body. Carbon dioxide is assimilated by plants and transformed by them into sugar. In prehistoric periods, geological processes converted plants into fossil fuel and thus created a potential cycle that no longer exists. The fermentation of carbohydrates produced by plants creates a bypass to this blocked cycle and thus re-establishes the economy of nature. This is indicated schematically in FIGURE 2.

In general it can be foreseen that biotechnology will become a major constituent of the chemical industry. As examples, I would like to mention the production of glutamic acid by fermentation of sugar in the presence of ammonia, the technical synthesis of lysine by a mutant of *Corynebacterium glutamicum*, or the use of various microorganisms to modify the structure of steroids to give them higher biological potency than they have in nature or to induce biological or therapeutic effects previously unknown. Perhaps one can expect that similar methods will become available by the use of tissue cultures for the modification of chemical substances of biological or industrial importance.

I would like now to refer to a number of areas in which the chemist will be faced with new and, to some extent, urgent problems that will have to be solved in the near future.

Food Production. If we assume that at the end of the century the population of the globe will be of the order of 6 billion, larger areas will have to be made available for food production. This will be possible only if we extend agriculture to the vast desert lands that are not utilized today. The problems of plant physiology (and physiology in general) under desert conditions will thus become a major objective of research, and we will have to extend our knowledge of and experience in applied genetics to adapt useful crops to the conditions of arid zones. Let us recall that the breaking of the genetic code has opened many avenues to such unknown fields as adaptation to "unnatural" environments.

Whatever the outcome of this research, two predictions can be made safely, namely, that the fertilizer industry will see an unprecedented growth and that we will have to solve the problem of seawater desalting. Water is such an essential commodity that economic considerations will have to take second place in the planning for this important area. The combination of nuclear reactors and reverse osmosis on a very large scale will be the most practical solution of the problem, if we can find an answer to the chemical question of how to produce membranes sufficiently active and mechanically stable for this purpose.

This is not to say that other methods to increase water supplies will not be investigated, among them the recovery of used water, more economical methods of crop irrigation, the avoidance of excessive evapotranspiration, and, of course, artificial rainmaking by cloud seeding.

The most important problem in food production is that of protein supply. Today, one can define developed countries as those in which protein is the nutritional mainstay and underdeveloped ones as those in which starch plays the central role in nutrition. Furthermore, there exist in many areas of the world religious reasons for not utilizing animal protein. In any event, new methods of protein production must be discovered. Only recently, processes for the production of protein from crude oil by certain bacteria or the production of protein by algae have been developed, along with that of protein-rich meal from fish that are not fit for human consumption. In any event, present methods of production of meat, namely the feeding of plant protein (mainly industrial-waste protein) or the products just mentioned, to animals and its recovery as meat or other animal products permits a utilization of the protein

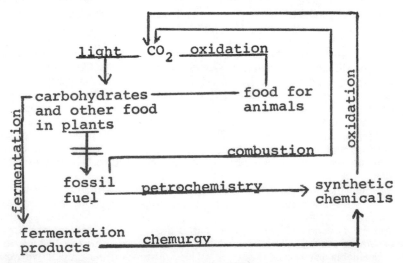

FIGURE 2. The fermentation of plant carbohydrates.

nitrogen in the order of magnitude of only 10 percent. On the other hand, there is, perhaps for genetic reasons, a fundamental difference in our ability to assimilate and utilize vegetable and animal protein: the former generally has not the amino acid composition that we find in animal products, and its enzymatic hydrolysis in the body is less easy and less complete. One can, therefore, predict that we will have to have recourse to hydrolysates of vegetable proteins (which obviate the difficulties of enzymatic digestion) and to the enrichment of such hydrolysates with synthetic amino acids.

A second, not less important, problem is that of food preservation, which is most urgent in the countries with high average temperatures. We will witness rapid development of the methods of food irradiation that have shown such outstanding results in the preservation of potatoes, onions, some fruits, fish products, and the like. No doubt a certain resistance to such irradiated food will have to be overcome by a more reasonable appreciation of the advantages of this procedure.

Insect Control. In the field of insect control we are at present witnessing an even less rational approach. There is no doubt that the use of chemical insecticides has disadvantages and creates dangers, but against this we have to weigh the advantages that these chemicals have given the human race in its fight against insects; let me mention only the control of malaria by DDT. We have to reckon with the fact that insects gradually develop resistance to the insecticides we apply so that we must use larger quantities of the known compounds or develop new ones that are more toxic and not only to insects. New methods will have to be developed, and it is rather surprising to realize how little we know about the chemistry and biochemistry of those insects with which we have lived in an uneasy symbiosis for so many thousands of years. Several new methods, I believe, will emerge. The first will be the rational utilization of the sex attractants (pheromones) of insects, and it is gratifying to see that for a number of insect species these highly active compounds have been isolated, identified, and even synthesized.

A second important possibility is the sterilization of insects by ionizing radiation or chemosterilants. This method has not been utilized as fully as it deserves; it shares with the application of pheromones the feature that it applies only to an individual species and therefore will upset the equilibrium in nature only to a minimal extent. Third, if we can find a characteristic step in the biological processes in insects that has no counterpart in higher animals, we may be able to develop a "magic bullet," in the sense of Paul Ehrlich, that will interfere with this specific process. Although on the whole the biochemistry of the insects is not very different from that of higher animals, there are indications that our search will not be in vain. One observation seems to me of interest in this respect, namely, that insects require cholesterol for their development but are unable to synthesize it themselves.

Pollution of the Environment. Here again we will have to arrive at a less emotional approach to the problem before we can look for a solution. The question is obvious: What are we willing to pay for the comfort of modern society? It seems to me that the use of atomic reactors will greatly reduce the amount of air pollution and that we will soon see cars run on electricity instead of liquid fuel. However, the chemical industry will also have to modify its processes, even if its products become more expensive. This applies not only to industrial wastes but also to what I would call societal wastes—namely, packaging materials. In spite of recent suggestions, there does not yet exist a proper solution for the question of what to do with the vast amounts of packaging materials employed today by modern society, apart from plastic materials, especially glass and metal. Perhaps we will have to renounce completely the use of inorganic packaging materials.

Medicinal Chemistry. Undoubtedly the field of drug research will continue to attract the attention of the organic chemist. Already a rather empirical approach is being replaced by a more rational one, for the design of active molecules on the basis of a reasonable mechanism for the process that the drug is intended to inhibit. This "chemotherapeutic" approach, an outcome of Ehrlich's original model of drug action, has found its justification in the development of the sulfa drugs. Another example is the antimetabolic action of certain organic fluorine compounds, such as fluoroacetic acid or 5-fluorouracil, which are antimetabolites to the corresponding natural hydrogen compounds. In this case, it appears that the antagonistic action is based on the great similarity in size of the fluorine and the hydrogen atoms, between which the living cell is not always able to differentiate.

The general approach to medicinal chemistry is thus to find the relationship between chemical structure and activity, chemical structure including the spatial structure of the substances. We are looking for the structure of the active site of enzymes or of the chemoreceptors on which the drugs act, and we are trying to elucidate these structures by a systematic investigation either of the inhibitors of the drug action or, more directly, of the difference in the fine structure of active compounds and of related inactive analogues. One of the most elegant examples is that of acetylcholinesterase, its poisoning by organic phosphorus compounds, and its recovery by antidotes.

In this, a hydroxyl and a negatively charged group form part of the active site of the enzyme, and indeed, phosphorylated serine could be isolated from a hydrolysate of the inactivated enzyme. Another characteristic example is provided by the morphine series. One has developed a picture of the morphine molecule, which appears to be able to envelop a certain part of the chemoreceptor; this conclusion has been reached by a study of morphinelike active and inactive compounds. In the case of the drug methadone, e.g.

$$CH_2-C(C_6H_5)_2-CO-C_2H_5$$
$$H-\underset{\underset{CH_3}{|}}{\overset{}{C}}-N\begin{array}{c} \diagup CH_3 \\ \diagdown CH_3 \end{array}$$

we have found that if one gives the activity of its racemic form an arbitrary value of 100, the dextrorotatory form has an activity of 10 and the levorotatory one (which has the same configuration as D-alanine) a value of 180.

As a result of such elaborate studies and by the use of antibiotics, we have been able almost to eliminate bacterial diseases; we remain faced with various metabolic disturbances and with viral diseases. They will be the target of research in the future. It is gratifying that some antibiotics have recently been discovered that are antiviral agents, but, more important, some deeper insight has been gained into the biochemistry of viral action, of which we know much less than that of bacteria: it has, for example, been shown that the cancer-producing viruses require arginine for their reproduction.

Two other medicinal areas that will be of increasing importance are those of psychopharmacological drugs and contraceptives. Both areas are in a state of active development, both are undergoing some regression because of side effects that surely should have been expected. However, there is no doubt that the development of such compounds will continue, the former not only because of the necessity to find cures for mental diseases but also to neutralize the deviations from the optimum use of our mental and psychological facilities that the present structure of society entails. If we knew more about the mode of action of the psychopharmacological drugs, it would be easier to educate the public and to prevent the present abuses. As to population control, this is the second aspect of the rapid growth of the population of the globe, the first being, as pointed out already, an increase in food production. In both cases, it is interesting to note that we are returning to the habits of primitive people. Hallucinogenic drugs have been used since time immemorial in religious rites, and we are told that morphine had been known and used in the Stone Age. Fertility control was not carried out in primitive society by contraceptives but rather by abortifacient plant products, and it seems not unreasonable to predict that we will give greater preference to this aspect of population control, perhaps because we instinctively believe more in natural products than in synthetic ones.

Bulk Synthetics. In this field, too, the main progress made and expected to be made is based on a better knowledge of the relationship between the atomic and molecular structure of the compounds and their physical and chemical properties. We are approaching the time when we can speak of "tailor-made" molecules. This is true for metals and metal alloys used in magnetic devices and electronic components, and also for organic synthetic materials, although our knowledge in this particular field is semiempirical. We are concerned with disordered or partly ordered aggregates of molecules, an extreme case being glass, a field in which impressive progress has been made in recent years. In the field of plastics and rubber, the knowledge based on our experience is great, but it will have to be deepened and enlarged. We know about plastics that are resistant to high temperatures, about polymers that have a low viscosity index (temperature dependence of viscosity), about plastic materials of high mechanical strength. However, as already pointed out, we will have to proceed in the direction of structural materials of synthetic origin to replace the metals now in use.

The art of organic chemistry can produce molecules of quite unexpected properties, such as organic semiconductors, photoconducting substances, or compounds that can serve as the "memory" of computers. The great variability of organic structures makes the production of materials with desirable, specific, and well-defined properties a matter of time rather than of principle.

It would not be wise to conclude without noting the field of military chemistry, not only that of propellants, which for obvious reasons will continue to be developed, but also that of chemical warfare agents. It can be hoped that we will reach a point at which the danger from the use of such chemicals will cease to exist, but in the meantime it is obvious that we have to expect further development, especially of incapacitating agents that will probably belong to the group of psychopharmacologically active drugs. Another area that will have to be developed further is that of protective agents against ioniz-ing radiation and against gas warfare.

The foregoing is a wide panorama that, of course, has much to do with practical problems with which humanity is faced or will be faced in the future. Nevertheless, it should be obvious that in order to reach these goals, a great deal of basic research is necessary. In fact, it seems to me that the dichotomy of which so much is being made today is an imaginary one. Basic understanding is necessary if we want to design applied-research projects and carry them into effect. Scientists and planners alike will have to show great wisdom in apportioning manpower and funds to these two avenues of scientific and technological development; very often the best approach to the solution of problems of great urgency is the fundamental, the long-term approach.

PRODUCTION AND DISPOSAL OF WASTES
SOCIAL CONSEQUENCES—POLICY IMPLICATIONS

Jacob Feld

Consulting Engineer, New York
Past President, The New York Academy of Sciences

Introduction

Every living thing, vegetable or animal, produces waste products, the emission of which is necessary for the continuation of life. Production of goods similarly is accompanied by the discarding of waste products because no process makes complete use of its raw materials. Developed industrial nations produce a surplus of goods that are exported to less efficient or less developed nations and accumulate the wastes resulting from such production. The surplus goods are exchanged for more raw materials, including much waste to be disposed of by the more productive nation. The natural result is for the more efficient producer to bury himself in his own waste.

These wastes alter the soil, the water, and the air that are necessary for human existence. With small population concentrations, a group of men could abandon the area of activity when the waste pile got too large and shift to another area for gathering food and carrying on their minor production of goods. With more concentrated populations and intense industrial production, capital investment in cities and factories deterred the abandonment of occupied areas, and pollution of air, water, and soil became a recognized problem.

As early as 1306, a royal proclamation was issued by Edward I against the use of sea coal, because it filled the air with pestilential odor. A century later, further regulations to control smoke and odor production came from Richard II and Henry V. In 1661, John Evelyn published his "Fumifugium—Inconvenience to the Aer and Smoke of London Dissipated." Evelyn was instrumental in the building of St. James Park and other open areas around London, planting trees and flowering shrubs to purify the air and counter the concentration of such industries as the slaughterhouses, tallow-rendering works, and brickyards to the south side of the Thames River. As these plants increased in size, the chimneys were carried higher to dissipate the smoke above London Town. Early experience and recent research show the failure of dispersion and dilution from taller chimneys.

Affluence and Waste

Man upsets nature's balance of production and deterioration and successive use of refuse as food. The unbridled expansion of production without use or storage of the waste products is suicide.

Affluence comes from overproduction of the necessities of life and is therefore directly responsible for waste and pollution. The size of the household garbage can is a measure of the affluence of a people and of the severity of pollution in the area occupied by that people.

111

Life generated by, and existing within, a set of natural conditions can tolerate change in the natural conditions by time-dependent acclimatization, such as living in an air-conditioned environment. But every such change requires the production of goods and of power, with an offshoot accumulation of wastes to pollute the air, the water, and the soil, and even the natural light and heat from the sun

Except for the very few "nature children" or bush people in the center of Australia, no one wants a return to the good old days before electricity or even before steam. And to earn the benefits of the power added to the energy of man and to that of his domesticated beasts, of extensions in the provision of food, shelter, clothing, mobility, interchange in word and thought, we must pay the price of disposal and control of waste production. If the waste problem is not solved, we do not earn the benefits desired and obtained from any development in production.

Every living thing continuously produces wastes and contributes them to the air, the water, and the soil. Are they necessary replacements of food for other forms of life, or are they surplus waste and therefore pollutants? Each level of life produces wastes for the beneficial use of other, not necessarily lower, levels. How can the balanced ratio of production and use, in effect the elimination of affluence, be determined and controlled?

Source of Raw Materials

All raw materials for production of food and goods come from the earth's crust, from the waters that hold desirable compounds in solution or suspension, from the chemicals in the air, and from the waste products of vegetable and animal life. Waste products are disposed of in the soil, the water, and the air of the environment. All material gained from the removal of resources must be balanced by restoring the used-up volumes. As a slow but continuing geological agent, man depletes available resources and must find substitute products with the remaining sources of ever-scarcer materials and completely new designs. Many of these scars in the crust—abandoned worked-out mineral, fuel, salt mines, cavernous deep vacancies of oil and gas production, gullied surface coal mines, and even depleted natural water courses—become limitations to the farming and urban uses of the land. Continuous increase in population pressure, at least in direct proportion, intensifies such an imbalance in the natural state of the earth's crust. Replenishment of natural resources by recycling of waste and obsolescent products must become a normal way of life, or progress will be throttled.

Chemical fertilizers have largely replaced animal products for restoring the food-producing qualities of farm lands. Vine-covered arbors were seen along farm roads in China before 1940 as an inducement for the traveler to leave his animal wastes for use as fertilizer; they are not now considered a proper source of that raw material. To the native farmer in India, the cow is not a problem but a necessary link in the chain of the living cycle. Children as young as three, with flat baskets on their heads, follow the animals to collect the droppings. Flat cakes of dung are collected, dried, and stacked in front of their huts as the only source of fuel and fertilizer. The size of the dung pile is a measure of prosperity and also of family size. Eliminate the cows and you eliminate the life of the millions of peasants. When India can provide substitute fuel and fertilizer, producing new kinds of waste pollution, the economic necessity for large families will be reduced, and the control of population may make more sense to these people.

In many parts of the world, organizations of well-intentioned men have for many years preached the gospel of careful use of our natural resources, with some success in providing for future generations. Some, like the Sierra Club, are chiefly interested in keeping the earth habitable for those who live in it. They warn man to stop devastating forests and swamps, stop lining and diverting streams, stop the use of pesticides, eliminate air pollution and supersonic noise. Return life to early natural conditions—but how far back in the progress of mankind's technological development would the present generation be satisfied to go?

Wasteful Use of Resources

Older civilizations drifted into more efficient and less laborious production of food, clothing, and shelter, and without planning for the future were buried by their wastes. Can we control our destiny, or will the skyscraper be a mysterious symbol of our civilization in some future archeological explorations around present centers of population congestion?

In 1966, the U.S.A., with 6 percent of the world's population, consumed 34 percent of the produced energy, 29 percent of the steel, and 17 percent of the timber. And the lumber industry, in fear of competition from aluminum, steel, plastic, and other substitutes for their housing market, continually pressured the Government to allow larger timber removal.

In 1966, we also dropped 48 billion empty cans, many of them rust-proof, and 26 billion bottles. The production of 800 million pounds of trash each and every day is enough to cover the entire land area soon. Of one billion pounds of paper produced, one-third is reclaimed for new paper or other goods; except for the small amount consigned to library and commercial records, the rest is waste. Again in 1966, 9 million vehicles were discarded.

As to food use, the mark of our affluence is the large garbage can. This was expressed in 1948 at a reception in Rotterdam by the Deputy Burgomeister, who had been in the U.S.A. before the war, and who said: "Do you still have those large garbage cans?" Europe learned by necessity, during and for some years after the war, that food can be more efficiently used without detriment to production or health. However, by 1960 the lesson was forgotten and the old habits of wasteful food use have returned. The "haves" are only a small island of plenty in the entire world population of relative scarcity, poverty, and hunger. As national groups become developed, they tend to copy the wasteful use patterns of the minority.

In his December, 1862, message to Congress, President Abraham Lincoln included the biblical quotation: "One generation passeth away and another generation cometh, but the earth endureth forever." The basic laws of mass and energy restate that truth but do not stop the alteration of the natural resources, without replenishment, so that a generation can rob all future generations of their source of many necessary materials and of power. In 1952, President Truman's committee to "study the broader and longer-range aspects of the nation's materials problem as distinct from the immediate defense needs" issued the Paley Report in five volumes of valuable information.

Most important was the warning that in the 1940s the U.S.A. passed the point of no return, consuming more natural resources than could be produced from available sources. And of 72 strategic and critical materials listed by the U.S. Munitions Board, 40 had to be entirely imported and some of the others were partly imported. The wasteful use and lack of recycling of natural resources will be the deterrent to future progress.

Pollution Production

The available air, water, and soil are limited not only in total quantity but even more in usable volume. If their use is limited, natural cycling of change will not denude the available resources. If used without thought of the future, any material can be exhausted, even if present sources are plentiful. The disposal of wastes into natural sources not only diminishes their usable volume but also affects the availability of other resources. Annually, some 160 million tons of solid particles are emitted into the air from the U.S. and some 800 million tons around the world, loading the atmosphere with dust. The effect is not only degraded air but change in the normal trade winds, change in cloud formation, change in the daily and annual temperature cycle, and change in the rainfall pattern. Rainfall upwind and downwind of the Gary steel mills in Indiana varies by almost 50 percent.

Total precipitation per day is given by Gale Young, of the Oak Ridge National Laboratory, as $340,000 \times 10^9$ gallons, 21% on land, and by G. P. Kalinin and V. D. Bykov in the UNESCO report as 520,000 cubic kilometers per year—about 10% higher than Young's figure—with 21% on land. Of the rainfall on land, the only present usable source of water, two-thirds evaporates, leaving some $24,000 \times 10^9$ gallons/day (7% of the total rainfall) available for use. Much of this falls in unpopulated areas. Some water is from glacier melt. Even for a population 4×10^9, there is an available total of 6,000 gallons/day, or a hundred times present average requirements. All of the waters eventually go into the seas, and, unfortunately, they are not well distributed over the earth. The Amazon River supplies 20% of the total runoff, and the next 20 largest rivers—the Hudson River is the 21st—total only 19% of the runoff. The runoff of the Amazon into the sea exceeds the total U.S. rainfall.

The runoff, without help from man-made wastes, carries soil particles into the sea; evaporation does not return them to the earth, and there is a continuous loss of available land for food production.

Man's use of the land adds wastes and pollutants directly to the sea and to the waters that travel to the seas. In the U.S., water heavily polluted by household wastes totals 20×10^9 gallons/day, and industrial-waste water is twice that volume. Materials are disposed of in solution and in suspension and are not easily altered or absorbed by marine organisms. Farming adds herbicides and pesticides, 10 million boats contribute sewage and oil wastes, storm water from paved and covered land brings silt and salts used for snow removal, and a substantial amount of abnormal heat removal further denudes the dissolved oxygen of the receiving waters. By 1980, a sixth of the freshwater runoff in the U.S. will be needed for cooling in power-plant operation, with a possible doubling of that volume by the year 2000. The effect of such large thermal change is unknown but must be determined. For example, the

Milwaukee River, in an area of winter temperatures consistently below 0° F, does not freeze over.

The large volume of the oceans makes them appear to be a logical and safe disposal pit. But it does not work that way. Disposal by estuary and outfall into the oceans merely concentrates the fine residues along the shores. The trend of movement is shoreward rather than seaward. Ocean movements tend to concentrate the waste materials towards the shores rather than into the ocean deeps beyond the continental shelf.

In the period of industrial development, the cities occupied infertile hills surrounded by farmland; in this century, population concentration has caused the expansion of urban areas to cover and wipe out large farm developments. For example, most of the potato farm area in Long Island has been covered with suburban developments. The construction industry also encroaches on food production for supplies of its raw materials. Sand and gravel beds are being depleted at the rate of 6 percent per year (11 percent in the Mid-Atlantic section) in the U.S., leaving large denuded areas for rapid erosion. Some have been developed as artificial lakes for recreation and park use. Pits for ornamental granite in the islands of the Penobscot Bay in Maine were dug to such depths that stress release cracked the natural rock beds and abandoned scars were left, surrounded by huge piles of shattered and seamy rock fragments. These abandoned man-made geological changes in the earth's crust must be utilized as storage for sorted waste to become future mines for usable and necessary resources.

Accumulation of mine wastes in unstable piles has resulted in slides with serious consequences; one such was the loss of 200 lives when a conical pile of coal waste slipped into Aberfan, Wales, in 1966, engulfing an occupied schoolhouse and 18 homes.

Were it not for the rapidly increasing yield of farmlands, due to technological developments and improved seed, the available farmland would have become a critical control of human life on earth. There is not too much total land available, and there is none to be turned into wasteland. Of the 22.2×10^9 acres, 55 percent is arid. Main crops and grain cover 2.3×10^9 acres, with minor crops on about 50 percent more area. Pastureland used 6.4×10^9 acres (29 percent of total land), some of which could be upgraded for more intensive food production. The total irrigated farmland is 0.3×10^9 acres, and only serious famine and higher food prices would significantly increase the area. Of the arid regions, fully a third are located more than 500 miles from any source of water that can be processed for irrigation.

Energy production and energy release into the environment are becoming more significant pollutants with the rapid increase in power requirements. David A. Berkowitz, of the Mitre Corporation, estimates that the energy now released artificially by man equals 1/2,500 of the radiation balance on the entire earth's surface and the release is increasing at 4% per year. The environmental compatibility of power generation is associated with the heat balance of the earth as well as with pollution of air, water, and soil. The estimated power consumption in the U.S.A. for 1970 is 1.5×10^{12} kilowatts, with a 50% compounded increase for each decade. Solid, liquid, and gas fuels need expensive sulfur control for their combustion gases. An oil-desulfurization plant, with the capacity to produce 100,000 barrels of low sulfur fuel oil per day, costs $150 million.

Atomic energy plants have gas, liquid, and solid wastes. Gas wastes can be precipitated and added to the solid waste. Solids are shipped to fuel processors, treated, and stored. The AEC estimates that by the year 2000 there will be 800,000 cubic feet of solid waste stored in 700 acres of abandoned salt mines. And some 200,000 gallons of tritium-bearing water is disposed into deep wells each year. The Hanford plutonium plant has 20 tanks, each containing 500,000 to a million gallons of stored liquid solution of waste products. Waste storage becomes a major item of the production program and may be the incentive for further development of solar energy to meet future demands.

Solution of the Problem—Use the Wastes

The solution of the problem is not elimination of technological production but better production controls, with major deletion of wastes by using them. Some 20 years ago, Swift's new hog-slaughter and packing house in Chicago advertised the complete use of the animal; everything but the grunt was packaged into salable items. But air and water were still polluted by the process. Air pollution from the power plant can be reduced by chemical and physical filtration of exhaust gases or, if the problem of atomic waste disposal can be solved, by the use of atomic energy. Water pollution can be reduced by liquid recirculation and reconstitution. Costs will be increased but protection of the environment for future generations is a small addition to the cost of raising and training each successive generation.

The technology for water pollution control is available; for air pollution control, it is still under study; for solid pollution control, it is just being analyzed. The solution of these problems is the most urgent task facing the present generation.

The return of large fractions of solid wastes to use in the economy must become common practice in every industry and must be a national objective for immediate implementation. There are now some 10 to 20 million discarded motor vehicles stacked in junkyards. In 1966, some 6 million cars were returned to the steel mills, but another 6 million were added to the junk pile. The process must be speeded up to clear away the 50-year accumulation of wastes. A system should be developed to collect, sort, transport, salvage, process, and recycle all worn-out materials. The system must be more efficient than that of the flea markets of many old countries, such as the one in Old Delhi, where continuous trading in trash, junk, and waste is in progress.

The Bureau of Mines has collected much valuable data on waste disposal and reuse. Included in their reports are case histories of reclaimed metals from magnesium foundry waste, aluminum dross, electrolytic recovery of zinc from sal skimmings, extraction from S-816 alloy scrap, cobalt-nickel from superalloy scrap, vanadium scrap, lead and copper from blast furnace matte, silver from waste photographic solutions, copper from auto scrap and from molten iron, aluminum and fluorine from leached potlining residues, titanium from residues, minerals from mica tailings, and gold from electrolytic solders. Automobile and refuse disposal has many recovery possibilities.

The Reynolds Aluminum Company, to reduce solid waste disposal and stretch the supply of natural resources, is active in the reclamation program. At present, 30 percent of its aluminum products use reclaimed or secondary material. In the last three years, Bethlehem Steel Company has spent $67 million, or 5.6 percent of its capital investment, in environmental control improvements in connection with their production.

Economic utilization of industrial wastes, scrap metal, mining wastes, and municipal refuse must be planned to produce usable products and recycled resources. Production costs may require subsidy, and such funds should come from a charge against the products or process that produced the wastes. In this way, solid wastes become a source of raw materials and can be converted into useful building materials, insulating fibers, mineral wool, mulches for agricultural application, and fertilizers. Ferrous and nonferrous metals must be recycled. The technology for such development is available. The concept of charges against the producer of the wastes is not novel; many jurisdictions make a charge for liquid-waste disposal as a percentage addition to water-supply charges. Some examples of waste utilization indicate economic solutions without subsidy. Above the Conowingo Dam, on the Susquehanna River, the accumulation of coal dust from the washing operations upstream is pumped as a slurry and injected into a furnace for steam operation and electricity production. The only cost of the fuel is the pumping cost; production is very economical. Arid soil at Phoenix developments becomes lush gardens when household garbage is buried therein. Sawdust accumulations at lumber mills make good bases for roadways that cross deep marshes. Glass fragments have recently proved good mineral fillers for asphalt-mix roadway surfaces.

Water salvage in industry is one method for reducing pollution of the streams. With an average precipitation of 170 gallons/square yard per year and a national average household use of 60 gallons/day per person (100 gallons in the Southwest), better water harvesting from rain catchment areas and reduction of seepage and evaporation losses should protect the necessary supply for many generations. Such plants as Kaiser Fontana Steel save, recover, and reclaim water for use; it uses 1,400 gallons per ton of steel produced, as against a 65,000 gallon average for the entire steel industry.

The estimated 1980 water requirements are: Industrial, 243×10^9 gallons/day, with 224 reusable; seepage losses, 760 with 25 possible saving; irrigation, 136 with 20 saving. Additional reductions in present uses in control of household supplies, and correction in conduit losses can further help to eliminate shortages.

Control of dust and fumes from industrial plants, buildings, and vehicles can be accomplished by expenditures that will increase the cost of production, climate control, and transportation. We dare not refuse to spend the extra price for protecting the atmosphere. Some industries have actively incorporated necessary protections. Asphalt-mix plants normally belch dust and fuel-oil smoke from their chimneys. Adding sonic water atomizers in the chimney removes 98 percent of the dust particles at 3-to-15-micron size and 80 percent of the particles smaller than 3 microns. Scrubbing water is atomized by a blast of steam or of compressed air, and a second blast of air produces high-sound energy to induce vibration in the droplets, providing beneficial microscopic turbulence and aiding in the surface collection of solid particles by the water droplets. The added cost to the finished product is minimal.

Conclusion

Production of wastes is a necessary condition of any life or of manufacture. Control of the wastes to reduce deleterious effects on the air, water, and soil that sustain life must be a recognized duty of each producer. Reuse of salvageable resources in refuse and cast-off goods must be planned and executed as a future source of raw materials. These are not public duties and should not be considered the province of government, but a planned step in the production of any item using natural resources.

PSYCHOLOGICAL ADAPTION IN A WORLD
OF PROGRAMMED MACHINES

Robert Cancro

Department of Psychiatry,
University of Connecticut, Hartford, Conn.

> *What is a man profited, if he shall gain*
> *the whole world, and lose his own soul?*
>
> —Matthew 16:26

Perhaps the most important mental operations that man performs are those that mediate his psychological adaptation to the environment. Adequate individual psychological adaptation is an essential prerequisite for species maintenance. It requires the establishment of a viable dynamic equilibrium between the demands of an individual's internal and external worlds. As with most higher biological species, psychological adaptation in man relies partially on social institutions that are transmitted through the medium of culture. An individual, therefore, must adapt not only to the physical and interpersonal environments in which he finds himself but to the existing social institutions as well. Any radical alteration of the environment—including its social organization—necessitates a change in the dynamic equilibrium between the total environment and the individual.

These theoretical formulations suggest the value of studying the predictable effects of programmed machines on both social institutions and the physical environment, in order to anticipate some of the probable consequences on man's efforts to adapt psychologically. This paper describes a conceptual approach to the modeling of the future effects of full-scale automation on work—a single social institution, important in psychological adaptation, that will be radically altered. Although the study of a single social institution will provide very limited insight into a complex problem, the approach can serve as a paradigm for the study of others. Much has been written and much thought has been given to the probable effects of full-scale automation on employment, especially the negative consequences for the employee. The primary concern has been safeguarding the economic welfare of the workers—both those displaced and those not displaced. One approach to the potential economic problems of automation has been a guaranteed annual income independent of the individual's personal productivity. It may be feasible and even necessary to pay a man a decent salary not to perform a job. Another economic approach to the problem has been to generate an adequate number of jobs for the work force. This point of view was presented quite succinctly by A. H. Raskin[7] in a most sensitive and perceptive article. He stated that "Today we have two very distinct problems in trying to make sure that everyone will have a useful place in our work life. One is to protect the man or woman who already has a job and who worries about being nudged out of it by a robot. The other is how to make sure that there will be enough new jobs to accommodate the vast number of youngsters who want, need and must have a creative outlet for their energies and their aspirations and the freshness

119

of thought they can infuse into our jaded society." Apart from our degree of concurrence with Raskin's view of the jaded nature of our society or its potential salvation by nonjaded youth, there can be no denying that this is a more sophisticated approach than a guaranteed annual income. This approach explicitly recognizes that in our present society there are functions served by and values derived from employment that go beyond the simple economic returns.

There is a third approach to this problem that recognizes not only the existence but the overriding importance of several human factors. There are, for example, psychological benefits derived from real work that are at least as necessary to the maintenance of our social system as are the economic benefits.[1, 3, 5] In order to take a systems approach to this problem, we must place automation in its appropriate social context so that we can examine and understand its potential effects—positive and negative—on work. A systems approach is necessary if we hope to understand such highly complex and interrelated issues as the environment and society. If we ignore the multiplicity of interrelationships and deal with problems in solitary isolation, we run the danger of naïvely-planned, rather than spontaneous, disaster.

A traditional approach to problems at this level of complexity has been evolutionary—to muddle through and hope for the best. Through a time-consuming process of trial and error, followed hopefully by appropriate adjustment and correction, a satisfactory steady state is achieved in the system. This approach, with its underlying faith in the beneficence of an almighty Lord, may be adequate for a system only slightly perturbed or slowly evolving. The effects of automation, unfortunately, promise to be both profound and rapid. The probability is remote that we will have the luxury of enough time to muddle through.

Proponents of an evolutionary approach invariably make inspirational statements about man's striking adaptability. No such statement would be complete at the time this is being written without a reference to the flight of Apollo 11. Although the reference is emotionally moving, we should not ignore the simple fact that the astronauts brought much of their earth environment with them in rather elaborate space suits. Man's adaptability is indeed remarkable, but it has both limits and deficiencies. The environment in which man evolved over eons of time, and which produced its genetic selection pressures, differs so sharply from the environment of a technologically advanced society as to be almost unrelated. We are not saying that all deviations from the conditions of a prehistoric jungle are bad but that rapid and massive changes from traditional environmental sources of sensory, perceptual, cognitive, affective, and physiologic nourishment are dangerous.

In a very real sense, man's adaptability is unfortunate. An organism that either survives or fails to survive in a given environment presents a much simpler case than does man. For example, if air pollution wiped out a major part of the earth's population in a short time, man would be forced—if there were enough men left—to take effective measures of control. It is because most people are able to adapt to foul air with minimal immediate and/or obvious damage that the pressures that are brought to bear on the sources of pollution are not particularly great. Paralleling this adaptive "gift" of man, we find, however, a sharp increase in a variety of chronic degenerative diseases. Coupled with the endemic diseases of an advanced society—ulcers and hypertension—these are parts of the price man pays for his ability to adapt partially to environmental changes. The unfortunate medical reality that ulcers—gastric and duodenal—are found increasingly in children suggests that the role of a negative environment in this disorder is being extended to additional segments of the population.

This trend is recognized by a number of outstanding scientists. Professor René Dubos,[2] at the 1969 World Health Assembly in Boston, commented on a number of these particular issues. He stated that "the evolution of biological mechanisms is far too slow to keep up with the accelerated pace of the technological and social upheaval of the modern world. . . . Today, everything changes so quickly that the processes of adaptation do not have time to come into play . . . the more a population is exposed to modern technology, the more it appears to be subject to certain forms of chronic and degenerative diseases . . . [Man] probably cannot adapt to the toxic effects of chemical pollution and of certain synthetic products, to the physiological and mental difficulties caused by lack of physical effort, to the mechanization of life, to the presence of a wide variety of artificial stimulants." In an interview following the meeting, he added: "Our *wealthiest* [author's emphasis] suburbs are so impoverished in the stimuli which they give to their children—the stimuli are so narrow, so objectionable—that these children will never develop their full potential."

As we have emphasized elsewhere,[1] man's ability to manipulate and thereby alter the environment to which he must adapt has developed much more rapidly than any changes in his psychological adaptive mechanisms. His technological skills have outstripped his adaptive abilities. This does not mean to say that man cannot survive at all in this artificial environment but, rather, that he survives poorly. There is no evidence that increasing per-capita income, life expectancy, and material comforts have produced a concomitant increase in happiness, well-being, or the distinctively human virtues. Although it can be argued that the goal of technological development is not to produce human joy, it does seem quite remarkable that none of these victories has been accompanied by a greater sense of generalized human satisfaction and contentment. If anything, the opposite seems to be the case. Man appears to be better equipped to struggle against natural forces than he is to triumph over them. It is as if in this arena, too, man wages war more effectively than peace.

An opposite approach to such highly complex problems has been revolution; that is, to follow a charismatic leader who believes he has the answers to questions that have not yet been articulated. Answers in the absence of questions are not often productive but can serve to reduce anxiety. The charismatic leader traditionally is able to guide his followers only into deeper difficulties, although he is characterized by never losing his certainty, his "cool," or his charisma.

In spite of the inherent limitations of man's mental apparatus, closely reasoned, logically determined change offers the preferred and the only real alternative to both evolution and revolution. Many interactions will not be anticipated and others will be estimated incorrectly. The models we construct will of necessity be inadequate and unrealistic, but they are still preferable to anything else available. There are a number of qualitative and quantitative changes in the nature of work that will be produced by automation. This paper describes a schematic model in which the most immediate effects of automation on employment are postulated. From each of these, the subsidiary consequences—both secondary and interactive—are derived. By continuing this process, a model of probable consequences and interactions is constructed that highlights potential problem areas. Theoretically, values can be assigned that express the strength of certain of these relationships, thereby producing a variety of numerical results. The present state of knowledge concerning the measurement of the social variables in question—and their interrelations, is so meager that although such an assignment of values is feasible, it is premature. However, even a simple qualitative examination of the schema can be productive.

The effects of automation on employment are related to a variety of other factors, including education, population size, and feedback effects; ideally, they should be included in the model. Initially, it may be wiser to de-emphasize the consideration of feedback features in the model until further data are gathered. This technique of pyramidal schematization need not be restricted to work but can be used to model conceptually the effects of programed machines on a variety of social institutions that impinge on man's psychological adaptation, thereby ultimately creating a large, complex, interacting model that may adequately simulate the real system. The predicted consequences are not inevitable but only probable in an economy that has not been planned to anticipate them. The goal of modeling this system is to increase society's ability to produce rational options with known costs and benefits. The critical feature of this approach is the emphasis on the quality of mental functioning; in that sense it represents an attempt to introduce traditional humanistic concerns into our scientific deliberations.

It should be emphasized that we do not consider the industrialization and increasing mechanization of the first half of the 20th century in the United States to be automation. These industrial changes are historical precursors of automation, but they differ from it qualitatively and not just quantitatively. Nevertheless, they shed light on some of the probable effects of automation. Our concern is with computer-based automation, which has only begun to make its impact felt. Total production processes, through the packaging and invoicing of the final product, can be automated with the present state of the art. We must look carefully at selected industries, e.g., the petrochemical, in which large-scale automation has begun, for prodromal signs of the anticipated disorders.

The first level of predictable consequences of automation on the quantitative features of employment will include changes in productivity, work hours, and number of jobs. We can predict, with a high degree of confidence, an increase in both individual and national productivity in a fully automated society. There is clear evidence that productivity has been increasing in most technologically advanced Western societies for a number of years, and automation should enhance rather than reverse this particular trend. With this increase in individual productivity we can safely predict that the purchasing power of the individual worker will also be increased. The weight of the evidence is quite clear that the American worker will continue to receive a major share of any increase in his productivity or in the gross national product. The bargaining and political strength of American labor unions and the general trend in our society toward a more equitable distribution of wealth should insure this outcome. This increase in purchasing power will open a number of new options for the use of available leisure time.

Probable consequences of an increase in national productivity are an increase in industrial sources of environmental pollution and a further drain on limited natural resources. Pollution is a good measure of the gross national product of a society and, therefore, of its technological capability. It is not difficult to anticipate that, as personal income rises, individual sources of pollution and resource consumption will also rise. An American living in our present value system who receives greater wealth will spend it on additional items that consume power and other natural resources, e.g., summer homes, automobiles, snowmobiles, boats, etc. Obviously there are a variety of possible social, political, and scientific counterforces to this consequence, such as pollution control and the development of new sources of energy. Many pollution problems are undoubtedly soluble, but they will not simply solve themselves. The effect, if any, of this increase in productivity and purchasing power on population size is uncertain. Financial success has not been consistently associated with family size in the past; more recently, the relationship between socio-economic status and number of children has become more complex and includes additional variables. There is no a priori reason for assuming that the "average" person will have a larger family if his real income goes up. On the other hand, an increase in purchasing power in conjunction with some of the factors that will be described later may tend to influence family size.

To return to the first-level consequences of automation, the second quantitative feature to be considered is the length of the work week. The basic work week, as opposed to the number of hours actually worked, has been going down slowly since the turn of the century. It is not certain that automation will, in itself, produce a shorter work week for all, but the probability is high that there will be displaced workers who are unemployed, underemployed, or paid to stay home. When automation's effects are coupled with the pre-existing trend in our society, a shorter work week becomes highly probable. The objection can be raised that individual workers are at present working more hours than they did ten or twenty years ago through overtime and moonlighting. This argument rather begs the question that the basic work week will continue to diminish. If a man chooses to increase his working hours beyond the basic week, through a variety of means, we must ask if his motivation is solely economic. This is a separate question that will be explored later.

The predictable diminution in work hours will have certain secondary consequences, such as an increase in the amount of leisure time. Some workers will choose to use this "leisure" for additional work; others will not. The need to develop satisfying and growth-promoting uses of leisure time is very likely to become increasingly important. The reduced number of hours spent on the job results in a corresponding diminution in time spent in relatively structured activities; less structured activities will probably take their place, e.g., recreational. Man's psychological adjustment is affected in highly individual and varying ways by the ratio between structured and relatively unstructured activities in his day.[6] The trend toward less work, in combination with prepackaged leisure-time activities—e.g., television—can combine to produce an increasing homeogeneity of perceptual inputs, both in variety and intensity. That this alteration of perceptual-pattern input may play an important role in mental adjustment is suggested by sensory-deprivation studies.

Any sharp reduction in the number of hours worked weekly or in the total length of the work life must be accompanied by a diminution in the quantity of work-related sources of satisfaction and work-related opportunities for the acceptable discharge of drives that are otherwise socially unacceptable. Clearly, there would be a complementary increase in the amount of time for leisure-related satisfactions and leisure-related drive discharges. Unfortunately, Western man has more experience with the use of work as an adaptive device than he does with the use of leisure time. In our technologically advanced, Judeo-Christian society there are very few people who know how to use leisure time productively. Enforced early retirement is usually a euphemism for imminent death. On the other hand, the increased purchasing power available will make it economically possible for people to utilize this leisure time in a greater variety of ways, some of which might support adaptation.

The final first-level consequence of automation on the quantitative nature of work is the number of jobs. Full-scale automation will result in a reduction in the number of jobs per unit of population and in an even greater reduction in the number of particular types of jobs. The unique combination of psychological and physical frustrations and satisfactions that a particular job offers cannot be matched so easily by another. An enormous asset of the broadly based technological society in America has been the great variety of jobs available to meet the differing dynamic combinations of economic, physical, and psychological needs of workers. At a time when the population is increasing both in size and in heterogeneity, we can ill-afford any loss in job diversity. Yet the most unskilled and the most physically taxing jobs will be automated out of existence, as will be a variety of clerical, managerial, and so-called white-collar jobs.[1] This not only will lead to further structural unemployment but will markedly reduce the range of job satisfactions. The selective diminution in the range of skills necessary for the running of a highly automated society means that there will be a number of individuals who are unemployable.

If we concede that there will be a reduction in the total number of available jobs per unit population, a highly probable consequence will be the further deferment of young people's entrance into the work force. A "traditional" method of limiting the potential work force has been discrimination. As we move toward a more equal society there will be great pressure from ethnic and racial-minority groups for their fair share of the job market. An even larger group who have been systematically discriminated against are women. There are a variety of social and political pressures developing for equal job opportunities for women. The success of these efforts will only intensify the problem because any reduction in employment discrimination will increase job demand without increasing supply.

The draft has been an effective means of temporarily postponing the entrance of many young men into the work force. For some, this has been a relatively permanent postponement, because certain individuals have chosen to remain in military service. There are powerful influences in our society to reduce the size of standing armies and to terminate the draft. If these efforts are successful they will contribute to the imbalance between available jobs and available labor.

The raising of the general educational level—education lasts longer and is available to a larger proportion of the population—has been an important means of delaying entrance into the job force. Obviously, if this is to be an educational experience and not professional baby-sitting, there will be a need for drastic revamping of curricula. It is unrealistic to expect that the form of higher education that evolved to satisfy the needs of a small, élite subset of the population can be extended, basically unaltered, to large numbers. If education is not to be a transparent device for frustrating young people's legitimate desire to enter society as full and productive members, it may have to be radically altered, made interruptible, and perhaps even shortened for many.

Automation will create new jobs, but the weight of the evidence is that it will produce fewer than it removes. There can be no question that the jobs it creates will be of a totally different character from those it terminates. Economically, the problem of the displaced worker is primarily one of maintaining his income at, or above, its previous level. Experience indicates that only one out of three such workers is successful in doing this.[8] With the increased productivity that a fully automated society promises, it may be possible to maintain income without creating an undue economic burden on the society. On the other hand, we do not know how large a nonproductive population any country can tolerate without endangering its economic viability. The economic realities of industrialization necessitated an increasing urbanization of society. The diminished need, in an automated society, for the geographic concentration of population near the means of production, along with the reduction in work force and working hours, will make it economically possible to achieve a more rational distribution of people over the available land. The psychological aspect of the displaced-worker problem is more complex than the economic. Some workers can be retrained, but this is not always feasible or even theoretically possible.[8] It is incorrect to assume that anyone can be trained to perform any task. The physical and mental limitations of the displaced worker must be considered, especially because a large percentage will be from the lower half of the I.Q. distribution. It is unrealistic to assume that they can be retrained, with our current knowledge of educational methods, to perform the needed and prestigious jobs of the automated society. The danger of producing a new social class of individuals who are able to understand and thereby control automation is quite real. This raises the potential problem of creating a new—technological—aristocracy.

Certain experiences with job retraining have proved most discouraging, though some have been more successful. When the job for which a worker is being trained does not differ substantially from the one he previously performed, the task of retraining, including psychological adjustment, is minimized. Retraining a man who has done hard manual labor to sit in a chair watching a gauge is not easy. It is not just a matter of his intelligence and ability to comprehend the new technology but also of his psychological needs, value system, sources of external gratification, role identity, etc.

These considerations suggest the importance of examining some of the effects of automation on the quality of work. Automation will continue the trend established by industrialization to reduce the physical effort expended by the worker on the job. There is a relationship, the exact nature and strength of which is unclear, between muscular effort and health—both physical and mental. To maintain good psychological adaptation, man needs to use his muscular system.[6] Menninger[5, 6] argues that work is a uniquely valuable social institution for the constructive use of aggressive drive-energy. The element of struggle implicit in dealing with problems on the job may permit at least partial discharge of aggressive drives. In the job situation, it is often acceptable to become quite angry at an inanimate source of frustration. To curse a tool or to kick a machine that fails is not seen as a sign of illness but is accepted and even applauded. The probability that much of the socially acceptable discharge of anger that takes place in these work situations owes its origin to nonwork-related problems must be considered. If work serves as a socially acceptable outlet for aggressive energies, then the loss of this outlet would result in an increase of aggressive behavior. Evidence in support of this position includes the relationship between unemployment and violence in American cities and the fact that strikes are frequently the occasion for violence. Clearly, a more rigorous examination of this question is required.

In addition to the hypothesized aggressive drive-discharge function of work, there are a variety of more highly visible psychological benefits. Hard manual labor has been a traditional source of masculine pride and self-confidence for many men. There is clinical and census-data evidence to suggest that men who are unemployed or underemployed often compensate for this by refusing to practice birth control and fathering large numbers of children who are visible evidence of their manliness. The problem of a selective increase in population pressure as a result of a loss of the traditional sources of support for masculine identity for certain socioeconomic groups must be considered. The value system and ethic that sees man as earning his right to exist and to consume through the sweat of his brow must change, or else man must continue to derive substantial narcissistic support from the experience of working. Our value system has changed enough so that most men can be proud of working with their minds, although in some groups this is still seen as less masculine. These changes are not enough. The values inherent in such roles as breadwinner and/or head of the household may not be compatible with the realities of a highly automated society.

Some of the psychological satisfactions that man gains from work derive from task mastery and the resulting sense of competence and completion.[4] The solution of new problems is obviously a source of much satisfaction to many workers. Performing a task in its entirety—i.e., carrying it through to completion—is also an important source of satisfaction. An unfortunate consequence of the assembly line, with its repetitive performance of the same subtask, has been the removal of much of the challenge and the sense of completion in work. There can be but little feeling of task mastery in the absence of both a challenge and a task. A man who is challenged in his work and is successful feels more satisfaction than a man who is not challenged at all. Needless to add, a man who is challenged by his job and consistently fails is in an even more unfortunate situation.

The various ways in which work currently serves to support a man's psychological adjustment must be better understood if we are to avoid the consequences of naïve action in this area.[1] Automation threatens to disrupt the dynamic equilibrium that presently exists between people and their work. This paper has described in considerable detail some of the highly probable effects of automation on both the qualitative and quantitative features of work in an economy that has not been planned to anticipate its disruptive potential. The consequences are presented as a schematic model because this arrangement not only highlights some interesting relationships but theoretically permits quantitative manipulation as well. The computers that will create the problems offer the technological basis for their possible solution. Electronic computation is a powerful tool whose potential for the solution of problems in these areas is real but undemonstrated. It is certain, however, that the mind of no given individual can retain, let alone manipulate, the number of variables required for the modeling of complex social systems. Computers, with their virtually unlimited memories and remarkable capacities for high-speed calculation, can be programmed to do the cost-and-benefit analyses that are necessary for the production of socially viable options. Finally, man not only must create the models that computers will use but must imbue them with his value system. It is not enough to provide for man's continuance as a biological species; we must insure the preservation of those distinctively human qualities that we treasure.

References

1. CANCRO, R. 1969. Automation: the second emancipation proclamation. American Journal of Psychotherapy 23: 657–666.
2. DUBOS, R. 1969. Reported in *Medical Tribune* August 4.
3. ERIKSON, E. H. 1964. Insight and Responsibility. Norton. New York, N.Y.
4. ERIKSON, E. H. 1963. Childhood and Society. Norton. New York, N.Y.
5. MENNINGER, K. 1942. Love Against Hate. Harcourt. New York, N.Y.
6. MENNINGER, K., M. MAYMAN & P. PRUYSER. 1963. The Vital Balance. Viking. New York, N.Y.
7. RASKIN, A. H. 1965. Automation and the War Against Poverty. The City College Papers, No. 3. New York, N.Y.
8. U. S. Department of Labor, Bureau of Labor Statistics. 1964. Experience of Workers After Layoff. Bulletin No. 1408. Washington, D.C

RECENT ADVANCES IN ENERGY SYSTEMS

Francis Perrin

High Commissioner for Atomic Energy (France)

Collège de France
Paris, France

Within the frame of the Conference on "Environment and Society in Transition," this report will consider the recent advances in those energy systems that convert energy from primary sources into thermal, mechanical, or electrical energy available for utilization. Advances likely to become practical, or even those which appear to have some possibility of becoming practical within the next ten or twenty years, should also be considered. The main object of this presentation is to determine how much these advances can help meet the growing requirements of mankind, in advanced or in developing countries, and how they may improve or worsen the various pollutions that result from the conversion of the primary energy

The Importance of Energy for Mankind

Men, like animals, have always had food as their fundamental source of energy to keep them alive and internally warm and to enable them to move or do mechanical work. The mastery of fire was so important for early men that it identifies the rise of the human condition above that of animals. Fire meant warming in winter, lighting at night, and the possibility of improving food by cooking. The primary source of energy was then the chemical energy that could be released as heat by burning vegetable or animal fuels, such as wood or fat; this energy, as that of all foods, was the energy of solar light stored by living organisms. No other form of energy was mastered by men for thousands of centuries.

It was only during the neolithic period, less than 10,000 years ago, that men discovered the possibility of supplementing their own strength by that of large domesticated animals, such as cattle or horses, to plow fields, transport heavy loads, or move themselves at an increased speed. At that time began the use as fuel of dry dung—still the principal source of energy of large sections of mankind. This archaism is worth remembering at a time when nuclear energy is becoming prominent in other parts of the world.

At the end of the Roman period, the development of watermills for grinding grain was a more important factor than Christianity in the general abolition of slavery. But during the Middle Ages, only the windmills added something to these early sources of energy, though the use of black powder as a propellant in firearms in the early 14th century should be mentioned; its use for blasting in quarries appeared only in the 17th century.

Burning coal instead of wood to produce heat, which began very early, was done on a significant scale only after the beginning of the 18th century; it was then also that coal was first used to produce iron in the blast furnace. The great importance of coal as a fuel started with the development, during the 19th century, of the steam engine, which partly converts heat to mechanical work.

The coal-burning steam engine was the main factor of that large increase in industrial production during the 19th century called the industrial revolution. From the middle of the last century, the discovery of the fundamental laws of electromagnetism lead to the possibility of converting, on a large scale, mechanical energy into electrical energy, cheaply transported and distributed to scattered small or large consumers. Electrical energy is the most versatile form of energy, because it is easily converted to light, heat, motive power, or chemical energy. First produced and used essentially for lighting, then for traction in city transportation (streets cars and metropolitan or underground railroads), its production as an intermediate between any kind of primary energy and utilization has become more and more general, leading to the disappearance of small stationary steam engines. It is still developing very fast.

It was also during the second half of the 19th century that petroleum production began, with lighting as its main utilization. Since the beginning of the 20th century, the development of automobiles has been associated with the use of gasoline from petroleum. Today petroleum is mostly used as a source of gasoline or of diesel oil for truck, car, or ship propulsion, and of fuel oil, competing with coal for power production or in homes for central heating. It should be noted that the fast-growing production of natural gas, which can be transported over long distances by pipe lines or by tankers (as liquid methane) and distributed for local heat generation in factories or homes (for warming or cooking) is becoming a more and more important factor in the total primary energy derived from fossil fuels.

Forecast of Future Energy Consumption

In the past, the raising of the level of living of any population has been closely connected with the increase of its consumption of energy. It is certain that improving the standard of living in underdeveloped countries will be possible only with a large increase in the amount of energy available in these countries, both for industry and for individual consumption. Even to maintain their poor present condition, their consumption of energy should at least follow the rapid growth of their populations. In the case of the most advanced countries, such as the United States, one might think that there would be no great pressure to increase the utilization of energy; but it is so much easier to make possible such an increase in these countries that no saturation is yet apparent, and power consumption in the advanced countries will probably continue to grow at the present rate, at least in the near future.

To speak of the growth of the energy requirements of mankind has no real meaning. A fundamental need for much more energy already exists in the large population living in underdeveloped nations. The problem is to estimate the probable development of the effective consumption of energy in different areas of the world and the future availability of primary energy supplies necessary for this development.

Analysis of the factors governing the increase in energy consumption is very difficult, especially in underdeveloped countries. These factors, like availability and "cost" of capital for investment, education, and training of specialists, general industrial development, etc., are very intricate, and one must remember that the amount of energy consumed per capita in any population is both a measure of its wealth and by itself an important factor in the possible growth of that wealth. Attempts based on such analysis have not been very successful in the past, and global predictions by extrapolation of the observed growth of energy consumption in countries having reached different stages of industrial development seem more reliable.

In fact, during the past 50 years, the consumption of energy, in the different forms delivered to consumers (fuels, electricity, gasoline, has grown exponentially (constant rate of yearly increase) in the whole world as well as in more or less industrially advanced countries. This trend has been strikingly well followed by the consumption of electricity, which has been, throughout this period of half a century, multiplied by two every ten years (five periods of doubling, meaning a factor of more than 30). The consumption of gasoline for automobiles is also approximately doubling every ten years, while the direct consumption of fuels, as such, has been growing more slowly, doubling in about 25 years. Thus, the total consumption of energy is increasing faster now than 50 years ago, its period of doubling having decreased from about 30 years to about 20 years.

It is thus very tempting to estimate future consumption of energy by extrapolating these exponential laws, but it would be very hazardous to do this for more than one or two doubling periods. Obviously, the present rate of increase cannot continue for a very long time: electricity production doubling every 10 years for seven centuries would require, according to Einstein's law, the annihilation of the whole earth, its entire mass being converted into energy!

Forecasts deduced from exponential extrapolation during the next 20 years are probably not far wrong, and decisions may be based on them. Here we will just note that, in comparison with present figures, the electrical energy produced in the whole world will be multiplied by four at about the year 1990, while the total consumption of energy at the same time will be multiplied by three.

For the more distant future, the uncertainty increases rapidly, but we may consider a fairly good prediction that, by the end of this century, the annual world production of electricity may be six to eight times what it is in 1970, serving a world population that may have doubled. It should be remarked that a total energy consumption proportional to the world population should not be expected; on the contrary, it is likely that a smaller rate of increase of the world population would correspond to a higher rate of increase of the total energy consumption, because the standard of living in underdeveloped areas would be higher if the rate of population growth in these areas was reduced. Thirty years from now, the standard of living of a large fraction of mankind may be much lower than that of western Europe a century earlier.

What the situation will be one generation later, around the year 2030, is extremely uncertain, though it is the near future for mankind, which most of the children now living—our grandchildren, for men of my age—will see. If there is no social catastrophe resulting from an unchecked population explosion, or from atomic war, a large increase in the demand for energy is still to be expected during the first decades of the 21st century.

Recent or Expected Advances in the Utilization of Energy

This heading covers not only the increase in the global efficiency between any primary source of energy and its final utilization but also new significant use of energy contributing to the general welfare of mankind.

A first, very important example of a considerable increase in efficiency is shown in the substitution, for warming houses, of central heating for open-hearth coal burning. In this case a multiplication at least by three of the efficiency has not led, in general, to a diminution of fuel consumption but to an increase in comfort. Similarly the consequence of a very large increase in the yield of electric-light bulbs is more light rather than less electricity consumed.

On a larger scale, the development of district central heating has, on the contrary, an important influence on the consumption of fuel used as a source of heat, first because it makes possible the use of low-grade or even apparently worthless fuel, such as garbage, which must be burned for disposal, and second, because the size of the heat-producing unit is sufficient to make the use of back-pressure turbines profitable, associating the production of hot steam, adequate for heating purposes, with power production. The use of back-pressure turbines to provide heat in large chemical plants may also result in an important saving of fuel.

In the future, added fuel saving will result from further improvement in the efficiency of heat conversion into electric power. Some progress in the steam cycle is still possible besides the diminution in the cost of conversion due to larger and larger production units of more than one megawatt per generator. Still greater efficiency may be attained when technological progress makes it possible and advantageous to start the conversion cycle at a much higher temperature, either by using gas turbines (with open or closed circuits) or by introducing a lithium-vapor cycle preceding an ordinary steam cycle.

The so-called direct conversion of heat at high temperature into electrical energy, requiring no moving gear of any kind, may also be used in the future to improve the total efficiency of heat conversion. For large power stations, the only direct conversion principle that seems promising is based on magneto-dynamics (MHD). It might become economical to use a MIID cycle to precede a conventional steam cycle if the price of fuel increases markedly.

Technical progress, which is thus making cheaper and facilitating the conversion of primary energy and its distribution, is also the principal factor increasing the demand for more energy because of the new possibilities created for raising the standard of living. Among these possibilities, the development of small, efficient cooling units that can be installed in any house connected to a power grid is particularly significant for the underdeveloped countries, most of which are located in tropical regions. In warm climates, this development has not only led to an improvement in the quality of foods, with refrigerators keeping them fresh and wholesome until they are eaten; it has also made possible air-conditioning, which should be, and will eventually be, considered as important for the inhabitants of tropical regions as heating in winter is for those of temperate regions. A large increase in the demand for electricity will thus certainly result from the extension of air-conditioning, when the wealth of the population in question is high enough.

The Conventional Sources of Primary Energy

Until the recent past, the only important sources of primary energy were:

—*Waterfalls or Waterchutes,* essentially used for electric power generation,

—*Fossil Fuels* (coal, petroleum, natural gas), with general utilization for heating purposes or electric power generation, and special uses for which no substitution by electric power is practical, the most important of these being the use of coal, as coke, in blast furnaces and the use of gasoline as a power source in automobiles.

To these principal conventional sources of primary energy may be added a few others, although it does not seem likely that, in the near future, they will have more than local importance·

—*Energy from Ocean Tides:* A large pilot power station, using the exceptionally strong tides of the Mont Saint Michel Bay in France, has been in operation for the past few years, with an annual production of some 400 million kilowatt-hours. In the same region, with a very long dam, up to 50 times more energy could be produced, but it is not likely that this project will be carried out before the end of the century. In a few other places, as the Bay of Fundy, between Maine and New Brunswick, projects of similar magnitude are possible, but the total world production of electric power of tidal origin will remain relatively very small.

—*Energy from the Wind:* Small-scale experiments have shown that modern windmills would be nearly competitive for generating electricity in regions where the wind is regular and strong; but no significant development is to be expected in any predictable future. Although a large number of small aeolian pumps are used to provide water for irrigation in many places, winds will contribute little to the world supply of energy.

—*Geothermic Energy:* The rather high temperature of rocks at depths within reach of modern drilling makes the idea of using heat from inside the earth for power generation attractive. Until now, this has been practical only in a very few places in the world, in Italy, or New Zealand, where some kind of volcanic activity brings molten rocks not far from the surface. The only good experience is that of Lardarello in Tuscany, where a power station of several hundred megawatts has been operating for many years. This source of primary energy may become important, but certainly not before the end of the century.

—*Solar Energy:* Of course solar energy is, and always has been, of fundamental importance for mankind; all the energy in our food comes from it, and burning wood for heating or cooking is still essential for populations living outside cities. The possible development of this use of solar light, through photosynthesis in plants, will not be considered here; neither will the solar heat that evaporates water in salt-marshes (thousands of kilowatts per acre), nor that which warms greenhouses. The question here is: to what extent will solar energy be a substitute for fuels?

From this point of view, it may be noted that sunlight is already beginning to be used for heating water for private homes in North Africa and for cooking food in India and that the use of solar heat for desalting water, which is still only experimental, may become important in desert areas, but without producing a large amount of energy; a large production of electric power from solar energy is not in sight.

From this brief survey, it may be concluded that, within a century or so, the conventional sources of primary energy that may be of significant importance are hydroelectric power and fossil fuels. Considering the possible sites for new hydroelectric power stations throughout the world, some of which would be very large, it appears that their contribution to the satisfaction of the fast-increasing demand for energy will continuously decline, though they may be of considerable local importance.

Thus, if the conventional sources of primary energy alone were available, the future energy requirements of mankind would have to be satisfied essentially through an increase of the consumption of fossil fuels. It seems that the increased production of fossil fuels, mainly petroleum and natural gas, necessary to face this situation, could be met during the next few decades by the probable discovery of new oil fields. But it would lead, within a century or so, to the exhaustion of all the natural reserves of fossil fuels, and long before this complete exhaustion, a scarcity would appear, entailing an increased price for fuels and consequently for energy delivered to consumers. This would jeopardize the increase of energy consumption in the underdeveloped countries; the hope of raising the standard of living of the very large population of these countries to a level comparable to that current in advanced countries would be very dim.

Nuclear Energy

The discovery, some 30 years ago, of the nuclear fission by neutrons of uranium atoms and of the possibility of nuclear chain reactions has opened a new era for mankind by making available to man a practically inexhaustible new source of primary energy.

Fission is an explosion of a heavy atomic nucleus, usually caused by the absorption of a neutron, releasing a very large amount of energy and emitting or re-emitting, on an average, more than one neutron; this last feature makes possible the development of divergent nuclear chain reactions. A nuclear reactor is a system in which such divergent or self-sustained nuclear chain reactions are possible, with a corresponding release of energy.

Fission is not possible for all uranium atoms. Only the atoms of the rare isotope uranium-235 (only 0.7 percent in natural uranium) are "fissile"; the absorption of neutrons by atoms of the dominant isotope uranium-238 is followed by their transformation into atoms of a new element, plutonium, which does not exist in nature.

Nevertheless, divergent fission chain reactions may be obtained with natural or slightly enriched uranium, if neutrons are slowed down by a "moderator" between their emission by fission and their absorption by uranium, with fission of uranium-235 atoms or transformation of uranium-238 atoms into plutonium atoms, which happen to be fissile like those of uranium-235.

Because of these facts, the first models of nuclear reactors developed for industrial power production, whether operating with natural uranium or with enriched uranium, cannot utilize more than 1 percent of the initial natural uranium for generating energy. But one must remember that about

3 million times more heat is generated by the fission of any mass of uranium than by the combustion of the same mass of coal. In fact, nuclear power stations in which heat is generated by reactors with this low efficiency of uranium utilization are now competitive in most places with fuel-burning power stations, at least for units of more than 500 megawatts. In the advanced countries, where the use of such powerful units is possible, nuclear power stations already constitute an important fraction of all power stations under construction.

The rapid development of nuclear power stations would lead, within perhaps 25 years, to the exhaustion of all the rich deposits of uranium ore that may exist in the world. If low efficiency were to continue the price of uranium might become prohibitive. But it is already known that, by using the plutonium produced in the reactors of the present first generation as nuclear fuel in fast neutron reactors, operating without a moderator, more plutonium will be produced by conversion of uranium-238, than would be consumed by fission to generate energy. Such reactors, known as *breeder reactors*, will eventually make possible the generation of energy by fission of all the atoms contained in any quantity of natural uranium, either directly or after their conversion into plutonium atoms. The amount of energy generated with a given quantity of uranium will thus be much larger than at present, so that the price of uranium would increase more slowly with much less influence on the cost of the electric power produced. Eventually the rather expensive uranium that may be extracted from ordinary granite, which is a very poor but very abundant uranium ore, might be used without much increasing the price of the electric power produced. The corresponding reserves of uranium are so enormous that an annual energy consumption a hundred times larger than the present one could be satisfied for more than 10 million years.

The Problem of Pollution

The combustion of fossil fuels, a necessary process in the generation of energy, may entail very serious pollution of the atmosphere, depending on the quality of the fuel and the completeness of the combustion.

In the case of power production by fuel-burning central stations or of heating for industrial activities, the worst possible pollution derives from the sulfur contained as impurity in the fuel. The combustion of this sulfur produces toxic sulfur dioxide. Even the use of very high smokestacks to increase the dilution of combustion gases, will not be sufficient in the future, when the amount of fuel burned in populated areas will be much greater, unless the fuel consumed has a very small sulfur content such as in some coals or fuel oils. But it is practically impossible to desulfurize coal and difficult, that is, expensive, to desulfurize heavy fuel oil. For these, the rather costly purification of the combustion gases before they are dispersed in the atmosphere should be considered. On the other hand, desulfurization of natural gas is easy and even profitable, because of the value of the extracted sulfur.

The limitations on the sulfur content of fuels used for power production, which will have to be imposed by law for health protection, will undoubtedly increase the average cost of production of electric power produced by fuel-burning power stations. This will favor the development of nuclear power stations that do not pollute the atmosphere. This factor has already led to the choice of nuclear power stations instead of fuel-burning stations in Switzerland and, recently, in New Jersey.

The other very serious cause of atmospheric pollution is the incomplete combustion of gasoline by automobile motors, especially when they are throttled down. Such incomplete combustion leaves a certain proportion of carbon monoxide, which is extremely toxic, in the exhaust gas. This means that there is already very serious pollution of the air in the heavy and slow traffic of large cities. A considerable diminution of the percentage of carbon monoxide in exhaust gases is technically possible without being very expensive. Here again, strict regulations will have to limit this percentage, as is already the case in California.

In automobiles, dangerous pollution of a different kind results from the extensive use of tetraethyl lead as an antiknock compound in high-octane gasoline. Every year, a hundred thousand tons or so of lead oxide, in the form of fine dust, are ejected into the atmosphere by automobile exhausts. It is in large cities that the consequences of this pollution may be most serious, with slow lead poisoning of the inhabitants a possibility. High-octane gasoline could be produced at slightly higher cost without the addition of tetraethyl lead; regulations could suppress this pollution by forbidding the use of lead compounds in gasoline. In the United States, according to a recent decision, such a regulation will be effective in 1975.

Radioactive Pollution

In normal operation, a nuclear power station produces practically no radioactive pollution of air or water. In case of a minor accident, any contaminated water could be held in special tanks before being released, after decay or purification according to the strict regulations universally in force, into the cooling river or sea. The air contaminated by fission products in such a contingency could be contained within an airtight reactor building and exhausted into the atmosphere through ordinary charcoal filters that retain the fine radioactive dust and radioiodine. Only radioactive inert gases, with no possible fallout, would contaminate the air exhausted through a high stack.

A major accident, with destruction of the reactor building—which is very unlikely—might mean radioactive contamination by fallout of the neighborhood of the power station. This would not imply an immediate danger for the inhabitants but might lead to evacuation for a more or less long time. Because of the difficulty, hazards, and cost of evacuation of a large population, nuclear power stations should not be situated in a densely populated area.

When large, fast neutron-breeder power stations, of a thousand megawatts or so, are in operation, a major accident would be more hazardous because of the large inventory of plutonium (a few tons) necessary in each of them. Because of the high radiotoxicity of plutonium, it would be wise to situate these powerful breeders in scarcely populated areas, with little or no population within a distance of some kilometers. Economical transmission of large electrical power, above 1,000 megawatts, at very high tensions, 750 or even 1,000 kilovolts, for long distances up to 1,000 kilometers has been already considered, and this should not prevent extensive development in the 1980s of nuclear power stations using fast neutron breeders.

The reprocessing of nuclear fuels after they have been used, necessary to recover the residue of the initial fissile uranium and the plutonium that has been formed, leaves as waste a large amount of highly active fission products. As concentrated solutions, these are usually dumped into underground double-shell stainless-steel tanks. This is temporarily safe enough, but in the future, when nuclear energy is used extensively, such dumping will no longer be acceptable. Because the tanks are vulnerable, they might be destroyed in time of war or by sabotage, leading to very great radioactive pollution of underground water in a large area. To avoid these possibilities, quite safe methods of dumping have been devised; for instance, the fission products may be incorporated in a special glass, invulnerable to water, that can be very simply buried underground. Such methods, which are beginning to be used, are a little more expensive than the present one, and their general use would mean a small increase in the cost of electricity produced by nuclear power stations, which would not seriously impair their expanded use.

Thermal Pollution

The so-called thermal pollution of rivers by steam power stations, whether fuel-burning or nuclear, limits the amount of heat from condensers that may be dissipated in the river used for cooling them and consequently limits the total electric power that may be produced along a given river. This last limitation is significantly lower for most of the present types of nuclear power stations than for fuel-burning power stations, mainly because of their lower thermal efficiency (lower steam temperature). But for other types of nuclear reactors and for the future breeders, this difference does not exist.

Far from the sea and if no large rivers are available, cooling towers, in which heat is dissipated by water evaporation, may be used; it is a little more expensive but does not change the temperature of the river from which the small amount of water to be evaporated is drawn.

Controlled Fusion

The release of nuclear energy may result not only from the fission of heavy atomic nuclei but also from the *fusion* of light nuclei. Such fusion, essentially the thermonuclear fusion of four hydrogen atoms to form one helium atom, is the origin of the energy radiated by the sun and, in general, by the stars.

In the H- or thermonuclear bomb, energy is produced by the fusion of heavy hydrogen atoms initiated by the explosion of a fission bomb that raises the temperature of the thermonuclear explosive to a hundred million degrees. It is extremely difficult to "control" such a fusion process to obtain the more or less steady flow of heat necessary for industrial power production rather than a violent explosion. At present, the only possible technique appears to be the heating of heavy hydrogen at low pressure by an electric discharge, creating a very hot plasma (fully ionized gas) that might be contained away from any solid wall by a strong magnetic field. To avoid instabilities in such containment, the magnetic field must have a very special configuration. It has not yet been possible to contain a hydrogen plasma (or any plasma) for sufficient time to obtain self-sustained thermonuclear fusion, even in the most favorable case of a mixture of deuterium and tritium. Nevertheless, it seems that, within less than ten years, demonstration of the feasibility of controlled fusion may result from the operation of large experimental devices now under construction or from still bigger ones of the following "generation." When the feasibility of controlled fusion is well established, it may take only 20 years more to develop a fusion reactor. It is likely that the heat generated by the first large fusion reactors will be several times more expensive than the cost of the heat generated by a fission reactor, in spite of the fact that the cost of the fusion nuclear "fuel" (deuterium from heavy water and lithium, from which neutrons regenerate tritium) will be negligible. It may take more decades to develop a competitive fusion reactor and to be able to use it especially if its minimum economical operating power is very large.

The main advantage of a fusion reactor in comparison with a fission reactor is that it would not produce large amounts of long-lived radioactive waste. Controlled fusion may thus be of great importance for mankind, but we should not expect to be able to use it on a large scale before the middle of the next century.

General Conclusion

Because the production of electricity by nuclear power stations is already economically advantageous, at least in certain regions of the world, large development of the use of nuclear energy may be expected in the near future. The extensive use of nuclear energy will help to avoid a rapid increase of atmospheric pollution.

In the somewhat more distant future, the possibility of producing very large amounts of cheap electrical energy anywhere in the world may create new conditions of development, such as the agro-industrial complexes conceived by the Oak Ridge team.

In the long run, without nuclear energy there will undoubtedly be a scarcity of fossil fuels, which would increase the price of energy, limiting the amelioration of the average standard of living in underdeveloped countries, and perhaps even start a certain decline in the material condition of mankind.

The fundamental importance of nuclear energy is that it may prevent such a decline, not only in the next century but for millions of years to come.

ANTHROPOLOGY, SOCIOLOGY AND PSYCHOLOGY

Chairman, Margaret Mead

Cochairmen, Gardner Murphy, Hans Speier

MEAD: I got a new insight from reading Gardner Murphy's prepared paper, and *that* possibly is going to distinguish the kind of contributions that the different sciences or different disciplines make to their whole field: there is a difference between the people whose scientific materials are precious and those whose scientific materials are extraordinarily cheap and virtually boundless; a difference between those who, at least until very recently thought their materials were boundless—such as oxygen—and who deal with a great number of reversible processes in many cases, and those disciplines whose materials are precious, rare, or irreplaceable, or who are observing irreversible and possibly unrepeatable processes.

One of the things that is happening in our whole attitude toward this planet at present is that we are beginning to think of it as small, isolated, and if not unique in actuality, quite unique in particularity, because chances of our working with other species in outer space within our lifetime seem rather small at the moment. And we are beginning to think of how many irreplaceable materials there are in the many irreplaceable processes.

If we once lose people with special sorts of experience and education, we are not likely to find them again, certainly not in our lifetime and possibly not at all. So that one of the differences between treating human beings "as guinea pigs" (one of the accusations frequently made against the social scientists), or not, is not so much that we treat them as animals, because I don't think anybody does treat them as animals, but that we treat them as if, in a sense, we have an inexhaustible supply.

Anthropology is one of the human sciences that deals *par excellence* with rare and precious materials, with vanishing people, rare fossils, bits of prehistoric man—types that survive with only a very few fossils in their midst.

And this gives us, perhaps, a different attitude, one that can be more easily transferred to our own present ecological crisis, in which we feel that our situation is unique, the state of the planet is unique, and the steps we have to take are unique, historical steps related to a particular moment.

This panel differs a little from the others at this Conference, because, although we have discussed fragmentation several times, much of the fragmentation we have been discussing has been within a science, but nevertheless one whose relationships to other sciences at least were historically regular.

We represent three branches and, if we include education, four. I do not like to think of education as separate from the other human behavioral sciences. It should never have been separated, and has been only by perverse historical circumstances perpetuated by perverse human beings. If we had not been so separated, if we had been working with the whole body of materials that are appropriate for each of these three disciplines, we might have a great deal

138

more to say than we have to say today. We are the outstanding charge of irregular and inappropriate fragmentation.

We now have batches of people who are utterly illiterate in other branches of their subject: Sociologists will publish an article without one reference to an anthropologist—and anthropologists do the same, and psychologists.

Given the fact of this fragmentation, we have practitioners with different styles of behavior and different bodies of material on which they draw. In my formal paper I attempted to emphasize the point that anthropologists have historically dealt with wholes, starting with the whole of mankind, his history, his language, his past and his present, including the development of his body as well as of his mind and culture.

Because we have had to narrow our sights to do particular work, we have always dealt with small, manageable wholes. We have preferred very small, preliterate societies because it was possible to control the material. Language is a preliterate material; the linguist was the first person who ever put language in writing, and he has very good control over his material in these circumstances.

We can be compared somewhat to those paleobiologists and ethologists who deal with species on very small islands, where they command the entire flora and fauna, and who work within bounded systems that can be completely known to the extent that we know what we do not know. This is, I think, one of the ways in which we perhaps differ from the sciences that attempt to take some aspect of society, like economics, and study it without a systematic inclusion of the unknown.

Because we have worked with small, bounded wholes that can be encompassed within the study of living human beings, we are able to treat wholes in terms of what is known and what is not known. This is rather easy to extrapolate to the planet, or the solar system, or whatever size universe the scientific community is focused on. One can move from a small, known society on an island to the planet, to the solar system without a change of posture, and rather easily.

The other thing we realize very accurately is that we have lately moved from the use of human beings as complex instruments for the study of human behavior. We have recognized that the only instrument we had that was worth anything until very recently was the whole human being, and by having the same human being study several cultures, if possible, and integrate the material in his head—and I think we can still agree that the human brain is the best computer that we know anything about—we have been able to arrive at the ordering of materials that we have to date.

At the same time in history, we were no longer able to find students who could make this kind of integration because they had grown up in a world that is fragmented and so multiple that they did not have the kind of experience of one culture that makes it possible to learn another.

We now have methods of recording and analysis that can supplement these orders of human intuition. Heinz von Foerster has been scolding me about this point, because he says the young people do want to integrate these things. This is true. I am not saying they do not. I think we have a whole group of students that would like to look at everything at once.

What I *was* saying is that the older type of anthropologist is disappearing and we now have, instead, young people who can use the new methods of reporting and analysis and computerizing that will replace, for our more complex world, what the old forms of anthropologists were able to do in their heads. We have whole groups of people who are working very hard to develop units appropriate for the analysis of complex wholes that can be handled mathematically and with computers.

There is a new science of kinesics, which deals with culturally patterned human behavior, with the unit called the kine.

There are the minor subsciences of anthropology called proximics, which deal with the interrelations of human beings in space with the proxime as the unit.

There is the use of videotape and the development of the action, by which we can analyze very complex videotape records of behavior in the relationship of human beings and the other objects in space. These are all units that relate to situationally placed, historically determined human behavior, and that can be dealt with with the sort of materials that we work with. These always are unrepeatable events; no event in the study of culture that is simple enough to repeat is valuable enough to study, on the whole, but if we can record it satisfactorily, then it can be studied over and over again by new theories, new forms of analysis, et cetera.

So I think the ways in which anthropology is going to be able to contribute to our common enterprise is its habit of dealing with wholes, and it is a very significant contribution when compared with working with those social scientists who have necessarily had to limit themselves to one kind of data or another within complex societies.

But this habit of dealing with wholes seen as the behavior of living peoples, and the tremendous interest today in the development of minute or micromethods of analysis that are appropriate for the computer will mean that as our material escapes the grasp of single human minds, the computer can substitute for a large number of these activities.

I think it even possible that adequate methods of retrieval will mean that individual human minds can deal with materials from many different disciplines as well as from many different observations; not only won't they have to go through all the steps, but in many cases they won't have to go through the same masses of data that used to be necessary, or sit in the same room for 12 years before they can talk to each other.

In both cases, simulation models of wholes into which we can feed material that has been adequately analyzed will replace the old forms of interdisciplinary work, which primarily consisted of catching a group of well-disposed social scientists and keeping them in one room long enough so that they could learn to talk to each other, which is a slightly crude method.

We will still need some of that talk, which is what we are doing here, catching a group of well-disposed people from many countries. We are not here using any materials that have been reduced to a form that could be used by everybody here, and I think this is something that we can hope for.

MURPHY: I would like to underscore a few of the things that Dr. Mead has said.

If one asks what can be expected from psychology as a social science and wonders why some of the results that we would expect are not forthcoming, we must remember that psychology came into existence less than a century ago.

In the effort to create a new biological science using experimental methods, essentially those of experimental physiology, I would say the environment is one of the first things a psychologist would have to consider. If, however, you visit a modern laboratory in which a room has been established for the study of the environment, you will find beautiful studies of air currents, temperatures, moisture, and light and shadow, and possibly something as complex as the perception of movement. But anything like the culture within which the professors and their subjects are operating would be automatically ruled out as a disturbing variable.

The rules of the game are established in such a way as to make broader environmental studies increasingly difficult because of that urge toward specialization that Dr. Mead has pinpointed.

I think we may say that our prime problem today is to focus on those studies of psychology that enable the experimental psychologist to do what he can do and at the same time find himself in a companionship of kindred sciences that take a larger course, notably cultural anthropology, which is disciplined to conceive of larger processes within which the specific psychology of the individual has membership character.

With this point of view you automatically include as part of your problem that which has previously been excluded as irrelevant. As Margaret Mead suggested, the Yale sophomore coming for a psychological participation experiment is supposed to park his culture outside. Not only the sophomore, however, but very often those who designed the experiments carry out the same operation. Within this setting, what can we do? We have to catch them young. When we do catch them young, we find this holistic, this Gestalt tissue, a broadly human preoccupation that has not yet been stifled.

I encounter them at all levels, hungry for a kind of bread that is not allowed, literally driving their insights to a period after their doctorate, as if they could possibly preserve this first broad flush of excitement about humanity until after they have been subjected to the nose-grinding process of undergraduate specialization and four or five more years of graduate study. During these years, the status, all kinds of social reinforcement, economic ones as well as those of prestige, are determined largely by this skill in avoiding the general, not getting too philosophical.

I think the issue is not hopeless if one realizes that the answer is to encourage whatever we can in the way of humanness when it first appears in the young. I am speaking of all those I see first exposed to psychology, sometimes even freshmen, certainly very commonly sophomores, and very much so juniors and seniors. They are preoccupied with something that will both be theoretically exciting and likewise have some relation to the ghastly situation in which the human race finds itself.

Is it not a little preposterous, for example, that one finds, as a superspecialty later in psychology, the question of whether it is too late already, given our weaponry, to expect to make some contribution that will be relevant? For example, the elementary problems of vertical communication that students begin to raise relatively early: Why is it that all peace research that has been done, with all the able people devoting themselves to the psychology of international relations and the threat of war, has not developed a channel that would enable us to make possible vertical communication with those in positions of political responsibility?

Those young people begin to discover that is not a part of scientific psychology. In fact, ways are usually found, relatively early, to deflect their enormously intense, passionate craving to participate in a psychology that will make a difference, in which a conference like this, or any serious conference of professionals, will work toward communication that will be evaluated in research terms, so that one learns what is effective and what is relatively ineffective, so that the whole professional wisdom of the social sciences be- comes a matter of creating a magnificent system of broad human understand- ing and takes on a form, a character, a structure that will make possible further continuation of humanity on this planet.

These are the kinds of questions, both theoretical and practical, that young people want to have partially solved, and in the further solution of which they hope to have a part.

It will take, however, an enormously hard, long pull. Psychology as it stands today is not ready to do much. It is ready to contribute, along with the biological sciences, toward a fuller understanding of the human central nervous system by chemical processes, et cetera, which, of course, have their continued pertinence to social issues.

But to see the ecological issues broadly will take a long period of reedu- cation of the psychological profession, particularly of the eager students who, if encouraged, will grow into the men and women who are concerned with that type of applied research, research on which the human future will depend.

THE EXPANSION OF SCIENTIFIC KNOWLEDGE

Derek J. de Solla Price

Yale University
New Haven, Conn.

Science is explosively filling the world at an exponential rate faster than that of any population explosion. It has been doing this for the last few centuries, so that at all times past we could look around and see that almost everything known in science had been discovered only recently. During the centuries, more and more countries erupted into the scientific age. Now, something is suddenly new, and we have witnessed for the last decade a steady and mounting deceleration of the largest and most developed countries.

The "overdeveloped" countries, the U.S., the U.S.S.R., and perhaps the U.K. and a few other nations, are suffering from a sort of ecological disaster of information pressure. There is an overpopulation of scientists and scientific papers manifested in a flood of publications, of abstract and index journals, and of bibliographies of indexes *ad infinitum*. We witness also the cybernetic damping effects caused by lack of national funds, lack of motivation toward the hard and basic sciences, and, to crown it all, an onset of machine murder, neo-Luddite rumblings, and a craving of the young to escape by way of mysticism, drugs, and contemplation.

Already the general effect of this onset of a ceiling to efforts has gone far, much farther than the more recent recession in U.S. Federal research and development funding would suggest. My guess is that both the U.S.A. and the U.S.S.R. have now only about half the scientific stature they would have had in the absence of saturation.

The result of this overdevelopment and scientific crowding is that the world is not filling up evenly with science; the little countries are now growing more rapidly. Though the big powers still grow absolutely, they are declining relatively, and the decline becomes faster each year. It is rather like the vote in the United Nations, where the crowd of small nations may now have an absolute majority that a single great power cannot overcome.

Not yet visible, but inevitably approaching, is the fact that before we are even used to the blessings of a world where scholarly and diplomatic discourse can get along with very few big languages, those biggest languages must now decline. English, which had constituted much more than half the world's philosophical and technical language, is now slipping back by about one percent per annum as the little languages rise. We must expect Japanese to mount in importance as that nation's technology approaches and, perhaps, will soon rival that of the U.S.S.R. and the U.S. not long thereafter.

We must recognize that science is a world task, growing on a world scale and less and less amenable to organization by the superpowers. Perhaps the first moral is that the smaller nations should make even more efforts to integrate their university science with their primary and secondary education by moving away from English and Russian and toward getting decent home-grown books in their national languages. We had taken an opposite line, against nationalism in science, but it may now be good tactics to reverse course.

One need not worry too much about the difficulty of communication. Scholars do not read much in any case, and virtually nothing in foreign languages; to be more precise, each language looks at foreign literatures through dark glasses that let through only about one-tenth of what is there; the smaller the language, surprisingly enough, the darker the glasses. The probable reason for this is that it is vital for scientists to feel themselves part of a highly interacting scientific community that can be evaluated and possesses the autonomy and competence to facilitate its work by providing reasonable conditions and checks.

The most oppressive influence of the science that continues to grow on the world scale is not the unevenness of that growth from one country to the next but the stupendously continuing rate of accretion. We have almost exhausted our resources by straining so that all might be published and all indexed, and now it begins to appear that this was not quite the right line to take. By having a hierarchy of merit among journals and by having editors and referees sit in conscientious judgment on each paper, we thought we had improved quality. In fact, the gatekeepers only delay and shift publication; they do not prevent the flood. By supporting journals that could not stand a cruel marketplace we maintained valuable forums—but for what?

It seems we might have done better to take immediately the line that is being forced on us. Any journal that can die in the marketplace should be encouraged to fold up forthwith; any paper that can be aborted or strangled should not be set in type. Perhaps we should take another look at the vigorous and productive Information Exchange Groups that were suppressed by the scientific establishment because they spelled danger to the canonical procedure.

Let it be realized at last that *publication is not a duty consequent on time and money spent; it is not a right consequent on a mere desire to speak: it is a privilege consequent on there being some person who really wants to listen.* We write to communicate with real people, not with a faceless archive. Most of the so-called information crisis is caused by technologists wanting to read something different from what the scientist writes for his own purposes of scholarship.

Part of the reason we might be pushed to this realization is because we have now developed a considerable disillusionment with the potential of the computer in the particular capacity in indexing and setting in retrievable order the science that is embedded in the gigantic literature of the archive and the research front. Order exists, but that which is in papers is not quite an ethereal phlogistonlike fluid of "information" that may be stored and retrieved. It seems that a large graininess is built into indexing methods of all conceivable varieties, and in the long run the computer does not help fundamentally.

We have only two radically new and efficient tools. One is the citation index, which revolutionizes the research front and enables us to move forward in time instead of backward when we search bibliographically. The other is the taxonomically indexed data bank, such as can now be managed in chemical compounds and nuclear constants. It is about time that papers containing such data be fed straight to a central data bank rather than via a journal.

Though we can do fabulous things, what we cannot do is provide a direct substitute for the ingenious packing down of knowledge that goes by the name of scholarship. It is a most clever process: How else could we leapfrog a graduate student right up to the research front when that front is, by cooperative effort, running much faster than any one established individual scientist can run?

Perhaps it is not commonly recognized that it is in this matter of the existence of tight-structured research fronts that science sets itself apart from all other sorts of scholarship. The difference between science and humanistic scholarship is not one so much of methodology or concern. It is in the sociology of interaction of its workers. If scientific communication is wounded, science may die.

Lastly, a word about the use of this knowledge, i.e., the reason we need it. Scientific learning grows, in the best model we have, like a giant jigsaw puzzle in the process of cooperative solving by a team of players.* Each player strives to find and add pieces; sometimes one area is hot, sometimes another. One cannot pick a piece and decide that this must be fitted next; it may or may not have its natural "turn" at that juncture. In the knowledge industry, we all play against Nature, and she acts as if there is, in truth, only this one world lying there waiting for its pattern to be discovered. Technology is not like knowledge; it is things and processes that society wants an knows how to make. Each nation and culture can choose and buy the technology it wishes to have and reject what it wishes—within reason. Science one must take or leave. It happens, however, that to work with modern highly scientific technology one needs a population educated to some degree—often the highest—in matters scientific.

To put it bluntly, *science is the only means we have for investing in future technologies.* By attempting to train for the technologies perceived today we can get only products that are and must remain a decade stale. Moreover, if we demand "relevance" to needs and wishes, to techniques and knowledges that are visible and aware today, we are also in jeopardy of an investment in only the present. Science and technology move too fast for that, and even the most developed nations do not have enough control of the world to declare a moratorium. We must all plan for this future by a certain return to faith in the virtuous necessity of basic scientific knowledge and full education to all the research fronts, no matter what their apparent relevance—or lack of it.

* Oddly enough, there seems to be an absence of research on the psychology of jigsaw-puzzle solving and no time/motion studies of it.

CULTURAL RELATIVISM AND PREMARITAL SEX NORMS: RESEARCH BASES FOR MORAL DECISION

Harold T. Christensen

Purdue University
Lafayette, Ind.

This paper focuses on premarital sex. It attempts to build an empirically based theory explaining social change and describing cause-and-effect relationships as background for rational decision making in a scientific age. Although at this stage of development the analysis is exploratory and its data are somewhat limited, it can illustrate how the sociologist works in applying himself to problems of this sort and also point to the kinds of research needed for clarifying and perhaps eventually resolving the issues.

By *cultural relativism* we mean that things are relative through time and space according to their cultural settings. It has been commonplace among social scientists to observe variations in belief and behavior. To this we add the relativity of consequences. By *norms* we refer to the mores, or established moral standards, of a society. These are the prescriptions and proscriptions that the group considers of great importance to its welfare and so surrounds these norms with positive and negative sanctions—rewards and penalties. When norms become internalized within the personalities of group members they are called *values*. Values are the individual counterparts of group norms.

By *moral decision* we refer to choices based on criteria that are deemed important; that is, on values and norms. Although morality is sometimes used in the narrower sense of a "good" or "right" assumed to be intrinsic in the nature of things, it is applied here to connote any system of propriety, whether its standards are thought to emanate from divine or human sources. Thus, in the context of this analysis, moral decision making can be thought of in relativistic terms.

Science, which by definition is value-free, cannot in and of itself make a moral decision. But, by laying bare the varying consequences attached to alternative lines of action, it can provide the needed elements or *research bases* for rational choices aimed at maximizing the net positive effects. Our present task is to apply these concepts to the phenomenon of premarital sex.

NEW FREEDOMS AND CONFUSIONS

Social scientists know of no society, past or present, where sexual behavior has been completely unregulated. The quality of imposed restrictions has varied greatly, and does now, but always there have been restrictions; never has there been absolute sexual promiscuity as a standard practice. Another way of saying this is that all known societies have some form of marriage designed to exert some degree of control over adult sexual relationships.

There has developed in the Western world a tradition known as the Judeo-Christian sex code, which has for many centuries reinforced marriage as a regulator of sex. According to this code, sex is for marriage only, and even there it exists chiefly for purposes of procreation. Strict chastity has been promoted as the norm, backed up by the belief that this is an external principle commanded by deity.

But with the development f science and technology, this tradition has been exposed to severe challenge and is now losing ground. In his search for facts, man has become less willing to rely on faith alone. Furthermore, with greater understanding of the processes of and advancements in contraceptive and medical technology, many of the earlier risks have been greatly reduced— risks of detection, infection, and conception. On top of this, there have been court decisions that have legalized all but the hardest of hard-core pornography, and recent developments in the mass media mean that almost everyone is exposed to an almost constant barrage of erotic stimulation. Today the lid is off: There are topless, and even bottomless, entertainers; there is complete nudity in the legitimate theater; there are underground newspapers and magazine, book, and motion picture portrayals that leave almost nothing to the imagination; there are "filthy-speech movements," "free-sex movements," "group-marriage movements," and the like.

The question of whether premarital coitus has increased significantly in America during recent years is in some dispute, because of a lack of relevant data. That attitudes have liberalized everyone is agreed, but practice may be another matter. The earlier Kinsey data suggested a large jump in nonvirginity during the 1920s and then not much change during the 1930s and 1940s. Influenced by this finding, Ira L. Reiss and others have concluded that there has probably been no significant recent increase in premarital coitus. But the Kinsey data are too old to throw light on the 1950s and the 1960s, and no other data suitable for a trend analysis have been reported for these decades. Nevertheless, some of my own data are relevant here.[a] As part of a larger study, students in a midwestern university were sampled, first in 1958 and again in 1968 (Christina F. Gregg gathered and analyzed the 1968 data); they were asked to indicate their personal approval of and experience with premarital coitus:

		Males	Females
Percent approving:	1958	46.7	17.4
	1968	55.4	37.7
Percent experiencing:	1958	50.7	20.7
	1968	50.2	34.3

The results give rather clear indication of three things: (1) approval went up for both sexes but the increase was proportionately greater for females, which means a trend toward intersex convergence; (2) coital experience remained at about 50 per cent for males but moved from nearly 21 to more than 34 per cent for females, which again spells out an intersex convergence; and (3) attitudes liberalized in greater magnitude than did behavior, so that, for both sexes, more experienced premarital coitus than approved of it in 1958, while the reverse was true just one decade later.

[a] For a more complete reporting, see Harold T. Christensen and Christina F. Gregg, "Changing Sex Norms in America and Scandinavia," *Journal of Marriage and the Family*, 32 (November, 1970).

Whether this new freedom over sex is a long-term trend or but a temporary pendulum phenomenon is not clear from available evidence. Only time will tell But there can be no questioning that it is a radical departure from the past— at least the immediate past in this country—and that it is resulting in con-siderable confusion, tension, and conflict. So rapid and far-reaching are the changes that the movement is often dubbed "The Sexual Revolution" and its outcome "The New Morality."

A revolution, very likely; but a new morality, not yet! It takes more than freedom to constitute a morality system. The Judeo-Christian sex code has been yielding ground, but, in many circles at least, there have not developed new standards to take its place. Although beliefs and practices are in radical transition, discussions concerning what is and what ought to be remain largely at the level of polemics. Science has had a great deal to do with creating the present environment of freedom and confusion within the sexual arena. It also has a responsibility—and offers the best hope, perhaps—for lessening the confusion and easing the decision-making process.

Deriving Values from Research

It is not within the reach of science to decide on questions of right or wrong in any ultimate sense. By definition, science must confine its work to measurable data and let its generalizations spring from these data alone, avoiding value judgments. But this is not to say that the scientist cannot *contribute* to the development of a morality system. By clarifying a problem in terms of its elements, the processes at work, and the varying effects found to be related to alternative lines of action, the scientist certainly can add something. He can, that is, providing the primary underlying assumptions give credence to the scientific method—which is generally the case today. Choosing to rely on science is itself a value judgment, but once this decision has been made, the way is open for objective testings of cause-and-effect rela-tionships, and the empirically derived knowledge that results can then be used as the raw material for one type of morality system. It is in this special sense that I refer to the possibility of values' being derived from research.

Alternative Value Positions

The positions that people take concerning questions of sexual morality are many and varied, differing primarily because of the nature of their basic life-organizing assumptions. If they assume the existence of transcendental authority, their position is likely to be absolutistic; if they think in pheno-menological terms, relativistic; if the pleasure principle is given priority, hedonistic.

Here are brief characteristics of these three basic value types: (1) The *absolutistic position* is that the ultimate sources of truth and right transcend the reaches of man, that there are unchanging guidelines emanating from deity and intrinsic in the nature of things. This is a "morality of commandment," and its followers typically do not feel moved to test or prove anything, only to believe and obey. In its purest form it has been ascetic, encouraging self-denial and expecting sexual abstinence outside of marriage for both sexes and self-control even within marriage, acknowledging only the need for reproduction. This has been the traditional Judeo-Christian code mentioned earlier. (2) The *relativistic position* is that human behavior must be seen within the context of cause-and-effect sequences, that acts may be wise or foolish in terms of their positive or negative results but not necessarily "right" or "wrong" in any ultimate sense. This is a "morality of consequences." Its followers are interested in rational decision making, relying chiefly on understanding the relative effects of acts according to time, place, and circumstances. It has been called "situation ethics," but its full development will require more than supporting arguments submitted up to this point. It will require the careful and somewhat detailed testings of science. (3) The *hedonistic position* is that personal enjoyment is its own justification, that what is pleasurable is good, regardless of religious pronouncements or social norms and (in its most extreme form) regardless of consequences other than the pleasure achieved. This is a "morality of indulgence." It sometimes has been called "fun morality." When hedonists pay attention to religious or social norms in addition to their primary emphasis on the pleasure principle, as they sometimes do, their position then loses some of its uniqueness and blends with one or both of the other two.

The absolutistic, relativistic, and hedonistic positions on sexual morality, as I have outlined them, are what Max Weber would speak of as "ideal types"; that is, they are logical constructs seldom found in pure form but useful in conceptualization and in the interpretation of data. In real-life situations, any given individual is apt to stress one type above the other two, but not necessarily to their complete exclusion. All sorts of combinations are possible.

THE RELATIVITY OF CONSEQUENCES

We shift now to a focus on the relativistic position, for here is where science can make its greatest contribution.

There have been a dozen or so serious research attempts in this country to determine if premarital coitus in any way affects the outcome of the marriage. In general, the approach has been to use presence or absence of premarital coitus as the independent variable and some measure of marital success as the dependent variable. Overall, the finding has been that marriage turns out to be more successful when premarital chastity has been maintained. But some of the reports have been inconclusive on this point, and in others the purported relationship was found to be of relatively low magnitude.

We can speculate as to why the studies so far undertaken are, in the first place, somewhat consistent in the direction of their findings and, in the second place, somewhat inconclusive because of the low magnitude of the reported relationship: abstinence from coitus before marriage is found to be associated with marital success probably because chastity has been considered the norm in this country and to remain chaste is in line with expectation. The statistical relationship between chastity and marital success, where found, has turned out to be of relatively low magnitude because of certain confounding variables that the studies have failed to take into account. It has not been demonstrated, for example, that premarital coitus is *in itself* the negative influence, as against an explanation that would argue that basic instability of the personality can be a single cause of *both* premarital sex experience and postmarital friction. Furthermore, since individual differences are concealed within group statistics it seems reasonable to assume that the magnitude of reported relationships may have been reduced by the fact of opposite effects in different marriages that cancel each other out.

In my own research, about to be summarized, I hypothesized that the values people hold and the norms of the cultures within which they live act as intervening variables to *affect the effects* of premarital coitus—with the greatest negative effects showing up where there is what I have labeled "value-behavior discrepancy."

This hypothesis was tested by comparing samples from sexually permissive Denmark, moderately restrictive midwestern United States, and highly restrictive Mormon country in the Rocky Mountain area of western United States. The study was made in 1958 and repeated in 1968. The latter set of data is only partially analyzed at this time. Nevertheless, results available from the more recent research fully support the following summary points for 1958 (insofar as they parallel each other):

1. Attitude measures supported prior expectation, in that the Danish respondents showed the greatest liberality of viewpoint on sexual matters and Mormon-country respondents the greatest conservativeness, with midwestern respondents in between.

2. Behavior measures, including rates for premarital coitus and premarital pregnancy, fell into the same cross-cultural pattern; that is, highest rates were for Denmark and lowest rates for Mormon country. The consistency of the two cross-cultural patterns demonstrates that behavior tends to line up with attitudes, though not always.

3. The discrepancy between how one behaves and what he believes was found to be proportionately lowest in Denmark and highest in Mormon country. In other words, norm violation varied directly with norm restrictiveness; the most restrictive culture (Mormon country), though having the fewest violators of chastity considered in the abstract, had proportionately the most who violated their own standards because their standards were stricter.

4. And it was this value-behavior discrepancy that was found to be associated with negative consequences. Proportionately, Denmark showed the lowest and Mormon country the highest: (a) guilt and other negative effects following premarital coitus; (b) forced or hurried weddings following premarital sexual activity; and (c) divorce-rate preponderance of premarital over postmarital conceivers.

To illustrate, TABLE 1 gives some comparative data on selected items from the 1958 survey.

Thus, it becomes clear that it is not the sexual act alone that determines the consequences but, more importantly, how this act lines up or fails to line up with the standards held. Norms and values are intervening variables, not only influencing the behavior but determining in significant ways the effects that this behavior will have. And it is value-behavior discrepancy, in all likelihood more than either values or behavior taken alone, that causes the difficulty.

In the research just reported, comparisons were across societies and concern was centered on cultural norms. But personal values are the individual counterparts of cultural norms, and we can assume that value variation among individuals within a society will show the same relationship to behavior and behavior consequences as does norm variation across cultures. Additional research is needed to fully test this out.

Some Next Steps in Policy Development

Policy, of course, entails value judgment. It is not the same thing as science, which strives for objectivity or value neutrality. Yet, properly utilized, each of these endeavors can be made to serve the other. Policy, when based on tested knowledge, is more likely to be successful in achieving the chosen goals. Science, when in line with broadly accepted values, is more likely to receive support and to have its findings utilized by society. Looking ahead to the

TABLE 1

ATTITUDES TOWARD PREMARITAL COITUS AND INCIDENCE

		Denmark	Mid-western U.S.A.	Mormon Country
Attitude (What one values)				
Percent claiming no preference for marrying a virgin	Males	61.0	17.6	5.3
	Females	74.1	22.5	10.8
Behavior (How one acts)				
Percent having had premarital coitus	Males	63.7	50.7	39.4
	Females	59.8	20.7	9.5
Percent Premarital Pregnancy (births in 1st 6 months of marriage)		24.2	9.4	3.4
Value-Behavior Discrepancy				
Ratio of experience to approval in premarital coitus	Males	.68	1.09	1.69
	Females	.74	1.19	3.28
Some Measurable Consequences				
Percent feeling guilt or remorse over premarital coitus	Males	4.3	12.1	29.7
	Femlaes	2.0	31.0	28.6
Ratio of divorce from couples having birth within 1st 6 months of marriage compared with 2nd-year birth cases		1.09	2.17	3.25

possibility of further policy development in the area of premarital sex (and perhaps other areas also), I see the following major steps as necessary:

1. *Delineation of society's core values around which there is reasonable consensus.* Contemporary American culture is characterized by value pluralism, and perhaps to some extent this always will be so. It would be unreasonable to expect mankind ever to be in complete agreement with itself. Yet, all is not chaos, and with clear thinking coupled with research there could be a reduction in the confusion now existing. In the first place, we need to pin down the values on which there is general agreement.

As a start, let me suggest that most Americans today place high value on these five things: (1) the preservation of human life, (2) the maintenance of mental and physical health, (3) the search after truth and progress, however the latter be defined, (4) the right of the individual to freedom and self-determination, and (5) the serving of a common good through love and cooperation. The list may not be complete, and of course it needs elaboration and needs to be checked and rechecked frequently by survey research to keep it accurately descriptive of the *existing core values* of our society.

2. *Investigation of the consequences of alternative environmental and behavioral conditions.* After a dependable set of core values has been determined, this can be used as the reference point for research into cause-and-effect relationships. The core values would be considered as dependent variables, and the attempt would be to test out how various other factors either contribute toward or detract from the achievement of these agreed-on standards.

In my research into premarital sex, for example, it was first assumed that mental health and marital stability are accepted social norms and that such things as guilt or remorse over an act, feeling pressured into marriage or to hurry the wedding, and breaking up a relationship that was entered with an idea of permanence are "negative effects" —negative, that is, not in the context of ultimate or absolute judgment, but only in the relative sense of going against generally accepted norms. (They would seem especially to detract from items two, four, and five of our tentative list of core values. Of course, it already has been agreed that the list needs further testing, elaboration, and continual updating. There also is need to determine objectively whether and to what extent these specifics actually relate to the broader value scheme; divorce, for example, is becoming increasingly accepted.) It then was demonstrated in a small way that values and norms regarding premarital sex are intervening variables, operating to modify the consequences of the behavior. According to the criteria adopted, value-behavior discrepancy was shown to have negative effects.

Although this research represents but a start toward objectively getting at the dimensions of an extremely complicated problem, even this one tentative finding can be of some use in personal decision making and to the helping professions.

3. *Control over the physical and social environments to maximize their positive effects.* Once there has been built up an adequate body of dependable data describing relationships between environmental and behavioral factors and between both of these and an accepted set of core values, the next logical policy move would be to try to control the environment in line with desired goals. This is not the time or place to discuss such things as the depletion of natural resources, the pollution of our physical environment, or the pressure of population expansion—for they are given more detailed attention elsewhere in this volume by experts in these fields. Nevertheless, the problems are relevant to what we here are saying: man's potentials for achieving his goals are to a considerable extent dependent on the environment within which he operates. An important part of this, of course, is the social environment: norms and programs imposed by the society; cultural embellishments that both attract and detract; the quantity and quality of man's interactional patterns. Here again, I can only mention the phenomenon and leave its elaboration for other times and places.

Improving the social environment from the standpoint of better aligning premarital sexual behavior with social goals would include attention to such matters as pornography and sex education. Both of these subjects are highly controversial at the present time, and research evidence remains meager and inconclusive. Pornography is regarded by some as a pollutant in the social environment; but, although this school of thought sees it as debasing sex and as a stimulant to undesirable behavior, there is another school that regards it as a catharsis or safety valve actually reducing antisocial behavior. As matters stand, there is no reliable body of data to settle the argument. Similarly, sex education is seen by some as an erotic stimulus akin to pornography and by others as a needed background for intelligent self-control. But there has been little evaluation research to determine which claim is correct or, more realistically, which *kind* of sex education is negative and which kind positive in its results. It should be obvious, therefore, that even before the environment can be improved in terms of social goals there is need for more and better sociological research.

4. *Encouragement for personal decision making responsibly carried out*
Unless one takes either an extreme, absolutistic position (which surrenders individual choice in deference to deity and thereby projects the responsibility upon supernatural powers, however perceived) or an extreme hedonistic position (which goes all out for individual choice but dodges responsibility by stressing personal pleasure over obligations to others), he must somehow or other face up to personal decision making responsibly carried out. A responsible decision is one that is made in the light of evidence, with a willingness to accept the consequences, and with due consideration for others.

Responsible decision making by the individual is in line with democratic ideals and entirely compatible with the scientific frame of reference. In a society such as ours, indoctrination can be self-defeating. It may stunt the child's growth and leave him brittle and unadjustable. Learning is more than memorizing; it is analyzing and actualizing—in terms of carry-through for improved living. Rather than imposing preconceived values on the child, leaders need to teach the valuing process. This means listening to him, giving him a clarifying instead of a preaching or commanding type of response, helping him to weigh alternatives in light of their consequences, and assisting him to identify and develop his own set of values in the light of this ongoing learning experience.

CONCLUSION

Research underlies policy, or at least should do so in a scientific age. It provides the relevant raw materials needed both to formulate and to implement the policy structure of a society such as ours. Science, as has been noted, cannot go beyond the limits of empirical data, and so cannot make value judgments. But the social scientist can *study values as data* and, taking these and other relevant data into account, can demonstrate the varying effects of alternative lines of action. As scientist, he will avoid questions concerning ultimate worth; but the insights he can bring concerning the consequences of acts make him a discoverer of *relative worth*—and hence a collaborator in the formulation of policy.

Elsewhere I have described the major point of this paper as a *Theory of Normative Sexual Morality*. It is normative in the sense that its takeoff point is values and norms. It represents a morality system in the sense that it provides a basis for personal choice. But it must be remembered that the values that can be derived from social science research are *values* only in a relative sense. We are talking about a morality of consequences, not a morality of absolutes handed down from above.

Finally, if our central proposition is correct—that value-behavior discrepancy is productive of negative effects—it then behooves policy makers to find ways to reduce that discrepancy. The decision whether to try to liberalize values to bring them in line with behavior, whether to attempt stricter control over behavior to make it conform with the values, or whether to work for certain quantitative adjustments in each is for the policy maker, not the scientist. Once such a basic policy question is settled, however, the scientist can again apply himself to testing alternative ways for moving toward the chosen course of action.

Research on family size indicates that more important to the marital satisfaction of the individual couple, more than just number of children, is how that couple is able to control the number of children according to its desires—in other words, how the couple's values and behavior line up or fail to line up. And when the couple have more children than they want, or fewer children than they want, they are more likely to be unhappy than if they were able to control the number of children in line with their desires.

It is parallel to what I have been dealing with in the area of premarital sex, because precisely the same thing was found there. It is when you have a value-behavior discrepancy, when the premarital sex, in this case, or the family size, in the other case, is in violation of the standards held, that you seem to have the negative consequenes.

I would argue from this that social planners need to take into account the values of the individuals, the norms of the cultures in which they live as real data, and if the planning involves any particular goals, such as reducing the birth rate in population control, a goal to which I fully subscribe, ways must be found to modify the values that individuals hold relevant to these things.

Or, if you get the control without altering the values, you are going to get basic dissatisfaction in the people you are dealing with.

* * *

MEAD: I would like to point out that we need to tie together our recognition of the population explosion and all its ecological consequences with the fact that this is the first time in human history when societies have been able to afford to release women from a primary obligation to homemaking. In order to bear enough children who would survive, most women have had to give their whole lives to homemaking and reproduction.

This bears on what Professor Murphy has said in his paper, but not in his speech, about our attitude toward creativity, because we will now have socially available women, as well as men, who can give their full time, or almost their full time, to contributions to society, and this will, in turn, have tremendous effects: for the first time in history, society will have to give women public roles that are worth having.

When private roles and private recognition, and support and dignity, are withdrawn from women as primarily homemakers and the mothers of the next generation, we will then have to work out some way by which they can make comparable intellectual and artistic, philosophical and religious contributions to public life that we have committed to only a few men while all the other men were busy helping the women be homemakers.

* * *

MEAD: We have so far presented you with rather odd pieces of a picture puzzle, partly from two disciplines and partly on two subjects. So far we have been approaching the general problems of the environment rather circumspectly, with a bit of sex behavior, the information explosion, problems of population control, and Professor Murphy's hope that we will get psychology geared into the study of the environment.

We are now going to have a number of discussants, and first will be Dr. Ralph Tyler, Director Emeritus, Center for Advanced Study in the Behavioral Sciences, Stanford University.

TYLER: Some of you know that the Center was established by act of the Ford Foundation and leases land from Stanford University, but it is not actually part of Stanford University.

Also, you may have learned to your sorrow that about one-fourth of the Center was burned down the other night, when students and others, trying to burn the ROTC building at Stanford University, were repulsed by sheriff's deputies, and the only unguarded place around was the Center, so fire bombs were thrown there.

Coming back to the subject of the paper that was assigned to me, Education Balancing, Conditioned Response, and Responsible Claims—and although the formal paper outlines a proposal for dealing with both conditioning and with more flexible means of learning as a way of making a more effective transition in these days of rapid needs for transition in our society—I think we ought to be reminded that, by and large, education, up to this point, for the mass society, has not meant the schools.

The schools throughout the world have been primarily sorting agencies concerned with the problem of bringing that number of people who are able to compete for positions of the social-political-professional-business elite to the number of positions open in terms of the shape of that economy. So that when I was born, 61 percent of the United States labor force was unskilled, and only 5 percent of the U.S. labor force was engaged in managerial and professional occupations.

It was understandable that the school system used methods that are known—by those who study learning—to be inappropriate for learning, namely, having everybody moving at the same rate and using a grading system based on the lower part's always being given low grades; they then were discouraged and dropped out, so that by the time, in my day, children got to the sixth or seventh grade in the state of Nebraska, they could drop out of school; the number that dropped out was about equal to the number of unskilled that were needed.

With the increasing changes taking place in the composition of the labor force because of science and technology, today only five percent in the United States, and a somewhat larger but comparable figure in other developed nations, are unskilled. The number in the United States who are engaged in meeting nonmaterial needs, health services, education services, recreation services, social services, accounting, administration, and engineering and science, now total 60 percent of the labor force.

So we have a demand for a much larger number of educated persons, and the educational system—that is, the formal system—does not serve the needs. Even in the United States, somewhere between 15 percent and 25 percent of our children do not reach a literacy level that enables them to engage in more than unskilled labor. The problem we face is developing a school system, a formal educational system, that makes use of what we know about learning rather than continuing a tradition that was, unconsciously perhaps, based on the business of regulating the number of people to the opportunities available to them in the social structure.

But, although at the primary level teachers, by and large, are more adept at influencing learning and at the undergraduate and graduate level are very little adept at influencing learning, by and large, persons who are called teachers are not very effective stimulators and guiders of learning. Therefore, in modern society the chief means by which mass groups attain their education or their learning are through their peer groups, their family, through the mass media and the like, rather than through the formal arrangement of the school. We find, for example, to take one illustration from Dr. Emily Mudd's presentation, that children develop not through any formal effort of the school but through the family, the peer group, through the fairy tales and other things, but by three or four years of age they have already developed the notion, on the average, that they want three or four children and that having a good many children is a fine thing.

Thus, the kind of goal we require is either massive systematic intervention or a very slow adaptation to the new expectations that come from other sources. But I am primarily concerned with trying to suggest a means of balancing conditioning, which is one form of learning, with a much more flexible and responsible process of learning. And it is the hope that, as we become more concerned with the effectiveness of our formal institutions, there will be more effort to use what is known about learning in the educational institutions, although they now are largely sorting institutions.

MEAD: Our next speaker is Dr. William Simon, Director, Sociology and Anthropology Department, Institute for Juvenile Research.

SIMON: I want to wander around and raise a number of points that have been discussed here. Two themes that to some extent have been repeated in a number of papers were the old familiar notion that we really need more interdisciplinary collaboration and that we become responsible to the environment in the broadest possible sense.

The first part is like the annual brotherhood luncheon; we ought to love each other more. In fact, we really do. I grew up both enjoying and suffering Margaret Mead and the work of Gardner Murphy and his students. I cannot conceive of a sociologist who has not. As one goes around to conferences, one finds in the inner recesses of the disciplines, people doing their own little nitpicking thing, and that is good, because it keeps them out of our way. That is why I really want the journals to continue, and I look at the people in our science, and thank heaven they are doing that, or they might get jobs that will do us some harm. I think that the crucial differences may not be differences of disciplinary loyalty, but I think there may be a substantive distinction that cuts across disciplinary hang-ups. I have a kind of conservative-radical commitment. I think one of the great gifts of anthropology is that it really traumatized the 18th and 19th centuries by convincing man that his time and place did not express any principles or truths but were simply a moment in the sociocultural process that might really be going nowhere.

All our disciplines have helped us to that understanding, and we have always managed to walk away from it: namely, just as society changes in a very fundamental way, man changes, and in very fundamental ways, and he may be able to change very rapidly. And, confronted with the trauma of that realization, we invent things like ecology. And now, very clearly, the socio-cultural moments are changing to the eternal truths, and the softer psychologies somehow have invented love for us, forgetting that love is a very recent inconvenience in the history of man. The human-potential movement, half religion, half science, somehow goes back to William Blake to conceive for us an image of man full of all the basic instincts that somehow get crudded over by layers of civilization. We now have to get back to him.

I think if we are going to think about the future—and I think we have to think about it realistically—we must realize that the kinds of decisions we make will basically change the human capacity, the human sensibility. We have to accept the responsibility that we are designing men in very fundamental ways, accepting freedom as well as the responsibility it affords us.

I think people are going to go into the study of the social structure, the environment in the larger sense, and I think our colleagues have an instinct for the dollar; as more people go into environmental research, there will be an exodus—we have all seen them closing Washington agencies—to find out where the program really lies. I think the worst part of it is that we are going to drown in ecological and environmental research. In this situation, I think we ought to moderate our plea for more research lest the hordes descend on us and begin to pick the people who think they can do it without turning into a major industry.

I have already suggested one of the things we have to get way from is part of the impasse in this environment, and that is the great culture efficiency—which, if we have any kind of world, is going to kill us. We have to begin to learn how to sustain and enjoy inefficiency, because this is the 20th-century version of the Protestant ethic, which had little to do with whether people leaped into beds with one another but had to do with something that is implied in the emerging generation gap, and that is merely the capacity for pleasure, the capacity to turn one's activity into existential correlates and not consume one's life in the pursuit of abstractions. I think, as a terribly critical kind of point, the meaning of the Protestant ethic really was that the moral, prudent man demonstrated his prudence and morality by acquiring the capacity for pleasure and then doubled the debt by being doubly prudent and moral: by not spending it, he simply handed it on to his children, "I gave it to you; don't you spend it either."

I think we now have a whole generation, at least in middle-class America, possessed of continuous prosperity all of their lives, for whom things like cars are not great symbols. We have a group of kids who want to cash the check, and they are prepared to do what I believe will traumatize most of us, and that is to commit themselves to lives of inefficiency and lives that are, in many ways, frighteningly asocial. This worries me a little bit. The direction is healthy; there is a beautiful line that quotes one of the French students during the May riots, and he said that any revolution that demands you to sacrifice yourself is Daddy's revolution, and increasingly the kids won't have it; they are willing to put the fuse in on this existential level. And the worst thing we can do is maintain our need for "integrated identity," our need for stable relationships, our need for privacy and space; they represent some of the expressions of some kind of eternal truths. A couple of lines I used earlier deal with precisely the need for identity; let me end this with a few lines from Leonard Cohen in one of his songs. He has the following verse:

> "Come with me, my little one, and
> we will find a farm/and grow us
> grass and apples and keep all the
> animals warm.
>
> "But if by chance I wake by night
> and I ask you who I am/then take me
> to the slaughter house, and I will
> wait there with the lamb."

MEAD: Professor Klineberg, of the International Center for Intergroup Relations in Paris.

KLINEBERG: I am still convinced, in spite of the rather pessimistic comments here, that the social sciences do have a contribution to make in our attempt to understand and control the environment.

I feel, for example, that at least in three directions there is a very intimate relationship between the concerns of this meeting and the concerns of the social sciences. I think there are certainly data that are useful, particularly the knowledge of other cultures. I think that there are certain techniques that can be used of sampling and public opinion studies that have shown their value. And I think, perhaps in the first instance, of a point of view that needs to be added to the reflections of the demographer—that of the economist. The economist is also a social scientist, but in this case I will put him on the other side of the fence for a moment. We need also the biologist, and all the others who are attempting to understand our problems.

Nor do I quite feel as negative about the possibilities of psychological, sociological, and anthropological data's being used even in the highest places in the land. I would remind Professor Murphy that on May 17, 1954, when the Supreme Court decided that enforced segregation of Negro children in the United States was against the principles of the U.S. Constitution, part of the justification for that decision—and you will find everything I say documented in the footnotes to the Supreme Court decision written by former Chief Justice Warren—part of the arguments used by the Supreme Court were arguments that came from a social science statement signed by anthropologists, sociologists, psychologists, all over the country, indicating two major reasons why enforced segregation was unjustified: one, that there was no evidence for the lack of capacity of Negro children to learn in the same way and to the same degree as white children; and, perhaps even more important, two, the fact that enforced segregation and any form of discrimination has a deleterious effect on the personality development of Negro children.

This is specifically mentioned, together with a very remarkable compliment to psychology in this Supreme Court decision, namely, that the reason for taking a different position today from that taken in 1896 (when it was stated that separate but equal accommodation fulfilled the obligations of the Constitution) was based on the developments in the field of psychology since that time. Now, this is one case where—and remember, I am not suggesting that this was the only reason for the Supreme Court decision—there was specific use of social science data, psychological, sociological, anthropological data, and it was not only read, but taken seriously by the United States Supreme Court.

I leave it to future historians to decide whether this particular act functioned as well as it should have, whether we have made enough progress in that field since that time, but the point is that here was the possibility of a movement all the way up from the social sciences to the Supreme Court, with the consequences that I have indicated, consequences that led many people, as you know, particularly a number of lawyers, to criticize the decision as being a sociological rather than a legal one. I think it was perhaps a little of both. I use the term "sociological" here the way the French do, to cover all the social sciences. But it is that term that was frequently used in criticism of this decision. So the first question I would like to raise with Professor Murphy and others is whether this experience gives us any possibility of expansion or extension to other areas, whether there may be other possibilities of moving a little further along in the influencing processes, or a least in the supplying of information to people in a position to make the responsible decisions. When we move a little more specifically to some of the questions raised, I would say first that the problem of relevant research as contrasted with basic research seems to me to be a false dichotomy, and one that I hope Dr. Price will have an opportunity to defend, especially if I misunderstood his position.

In France, as many of you know, at the time of what the French called the event of May, 1968, among the desires expressed by the students was, first, the desire for more interdisciplinary approaches to the various problems and, second, for relevance. And many of us were disturbed and unhappy because the desire for relevance seemed to exclude basic research. As a consequence, many people jumped to the defense of basic research, as if that were all that existed and as if relevant research were not a respectable activity.

I would like to propose that we keep both in mind—and here we may have to come down to the personality differences between social scientists; some are concerned with the most rigorous kinds of data collection or theorizing, without any notion of what this might do to the society in which they live; others are concerned with the social relevance of their work; many have been working with the techniques of basic science—Margaret Mead, Gardner Murphy—and many others have been concerned with applying the knowledge they have obtained to a better understanding of the problem of society.

So I would suggest that, instead of choosing between these, we regard both as legitimate exercises of the human mind; just as Roentgen was unconcerned about any practical application of the x-ray, so there may be people in the social sciences and elsewhere who are similarly unconcerned; perhaps today their data may be used. There are others who are impatient and want to see the research directed toward the specific problems of our day—now—today.

I see no reason why we should make the choice, speaking about science in general or social science in general. There is room in the mangers of science for both. I spoke of a point of view in connection with the contribution of the social sciences. We talked earlier about the population explosion, and I was struck by the almost exclusive emphasis on techniques of contraception, on the one hand, and, on the other, with the extrapolation of figures based on past experience to the notion of the doubling of the population within a given period of time. I heard no reference to a fundamental aspect of this problem, the attitudes of people—and the attitudes of people do change. The culture changes, and people change with it, or help to change it, and until we know something about where the movement of attitudes is going, I don't think we can say anything about the numbers of people that we will have in the future.

There is no easy way to introduce any statistics with regard to attitudes, but when I spoke of a point of view, what I meant was that these problems need to have added to their consideration that old, slightly outworn phrase, *the minds of men,* the attitudes and values of men and women about this whole area. Until we can find some way of introducing this concept of changing values, changing attitudes, changing positions with regard to contraception and population, we will not be able to make any predictions at all.

Dr. Mudd raised another point, quoting Margaret Mead, and I borrow the quotation from Dr. Mead as a very important one, in reference to the generation gap in the sense that our children are growing up in a world that we did not know, and they are telling this to us. It is difficult to estimate the rate of change, but as I look back now on a fairly long life of observation, at least of technological change, technological revolution, I would like to raise this question: Is this phenomenon that Dr. Mudd and Margaret Mead have described a temporary one or a lasting one?

I wonder whether, now that we have made these tremendous developments based on technological and scientific discoveries, the gap between our children and theirs will be anywhere near as great as the gap between us and our children. It seems to me that the changes that have occurred in our generation have moved us so far along certain lines that the relative changes in the next generation will not be so great.

Have I any right to say this? I don't know. But if it is true, then the problem of the generation gap or conflict or differences in the position of the young and less young in our present society may be a temporary phenomenon and not one we need to assume will be a continuing process in the future. Margaret Mead talked about the emphasis on wholes as the characteristics of the anthropologist, and there again I agree wholeheartedly. But the notions of wholes reminded me of a problem in which my colleagues and I have done some research recently. Our research touched on the attitudes of African students in six countries toward their national as contrasted with their tribal identity. Our query was, to give a specific example for those of you who know Africa, does a Yoruba of Nigeria feel closer to an Ibo of Nigeria—I might say our work was done before the African crisis—than to a Yoruba in Dahomey? In other words, does the tribal identification predominate in the minds of the students, or does the national identification predominate?

We found tremendous variations from nation to nation and from tribe to tribe, and we found variations between individuals within each tribe. But the general movement did seem to be from tribalism to nationalism, this occurring at a time when many of us are concerned with the implications of nationalism in our own, older society, and a time when many of us are concerned with the excesses to which nationalism leads.

We try to cover many of our criticisms by speaking of chauvinism as contrasted with healthy nationalism, but as far as I know, no one has clearly drawn a line of demarcation between what might be called a healthy nationalism and chauvinism. Now we see in the other parts of the world that the building of nations is becoming a tremendously important aspect of their activity, a tremendously important concern to those responsible for the future of those countries. Here again we have a kind of anthropological warning: The things that worry us in one society may be the opposite of the things that worry people in other societies.

This problem, whether our concerns and our techniques and our values can be extrapolated elsewhere, that is, the need to understand the values, the beliefs, the desires of other peoples, seems to be of the most fundamental importance. Not very long ago, in Paris, we were talking about productivity and pollution. This is of great concern to us, obviously, and it is of great concern to this particular conference. But the developing nations have said: Pollution or no pollution, we want productivity. They want industrial and agricultural productivity, and they are literally at this particular moment—this may change later; I am speaking a little too dogmatically—much more concerned with increasing productivity than with reducing pollution. In other words, just the opposite position from that we are taking in our society. And whatever plans we make for the future must take into account these different value systems and different concerns. To quote one of my favorite lines from George Bernard Shaw: "The Golden Rule reads the wrong way; it should state, 'Do not do unto others what you would like others to do unto you; their tastes might not be the same.' "

MURPHY: I am very grateful that Otto gives us the constructive memories of 1954. What has happened, however, in regard to the many threats to the continuation of the human race reminds me of Alice in Wonderland, in which no matter how fast you run, you have to run faster than that to get anywhere. That is the issue we face with regard to the long series of Pugwash Conferences, which have reached a high level of clarity and competence but have not had any effect on policymakers vis-à-vis the actual suicidal threats that are mounting.

Will talks on arms limitation be influenced by science, particularly by social science, and particularly by the kinds of social science we have been talking about today? I think the inspiration of 1954 needs to be recaptured and intensified; the threat is greater today; there is a massive obligation of the social sciences to think about that kind of environment that makes possible the continuity, the only one that does make possible the continuity, of the human race.

DE SOLLA PRICE: A great deal has been said about the desirability of cross-disciplinary interaction. As a physicist cum historian making like a social scientist, I am very conscious of a particular difficulty, and that is that when social scientists talk about science, they mean their own science, and very rarely do they have any idea that there are fundamental differences between the behavior and texture of knowledge for a physicist, a chemist, an astronomer, or an engineer and for an anthropologist, a sociologist, or a psychologist. While there is a large difference affecting the very life of society, as in the relation between the basic research and the applied, there is the question: How many of your trained people does society want to hire for whatever they can do, apart from teaching, which gives you a feedback?

Sometimes we train anthropologists so that we can teach students to become anthropologists to train anthropologists. Society has a hole in the head if it pays for that alone. It is a very nice hole, and we would like to have it around, but what society is paying for is education. When, however, society pays for the same process for the physicist, it is paying not quite so much for general education but for the production of physicists who are—or until last year or the year before—hired in fairly large numbers to add to the profits.

What we now have as a crisis in the social sciences is that we are training rather large numbers of these people because they are useful in universities, because they are rather soft subjects. You can get 5,000 unemployed, trained sociologists in Paris alone. It is nice coffee-house conversation. In a way, that is the difficulty, to make coffee-house conversations out of physics—would that there were more of it. I am not trying to distinguish between brands of scholarship or slur my colleagues. I count myself a social scientist. But the communications differ, the places in society differ. They are as different as the Yoruba and Ibo tribesmen. There is a fundamental difference in application, and I am a little worried by those who would suggest that the social sciences should perhaps have an aim and go after relevance. This can be just a great boondoggle, and we do it at danger to the university and to civilization if we take money to do things they want us to do, rather than the things we know are possible. We still must, willy-nilly, do the research that has to be done in the fundamental, basic knowledge in sociology and anthropology and in the other social sciences, and we cannot subvert it to all the claims or relevance and do the jobs to cure the ills of society any more than the physicist or biologist can give us things that we want. All they can do is to work at what is there, and this is very difficult.

MUDD: I think Dr. Klineberg's question was: Will the generation gap be as great for our children's children as it is for us? Earlier in this conference, one of our speakers said that there probably were living species on other planets, and that we can already communicate as far as the sun. If this communication should be developed in the lives of our children's children, it is conceivable that the generation gap might be very much greater.

Another of our scientists said that perhaps the major discoveries, the revolution in modern science, had been made in the last 45 years. If this should be the situation, there is something quite different for our children's children.

Dr. Christensen spoke of moral values and gave something of his research on premarital relations and their effect. I would like to leave you with a conversation with two brilliant young university students who are training to be teachers and are in the A level of teacher training and have been living together for the last two years, quite constructively, not married, but learning how to buy food, and to cook, and so on. They have told their parents and their grandparents just what their living situation is, and they feel not a bit ashamed of it. I asked a young man, the brother of the girl, "What do you think about the way your sister is living?" and he said, "It's not for me to judge; but if she did not love Bill sufficiently to marry him, she would not be living with him."

MEAD: In a sense, Emily Mudd said that if we have anything as great as getting into communication with other planets, this will probably create a generation gap. I was thinking that we might set up a colony on the moon, and the earth would destroy itself. If we were hit by a meteor and knocked off our axis, it might create a comparable gap. But otherwise this gap is unprecedented; it cannot occur more than once, without external intervention, that the earth becomes one for the first time and that we are all in communication with each other.

The other thing I think may happen is that we may learn how to communicate with our children. We may learn to communicate across lesser lines so there will be less of an extraordinary break, especially in value systems, than there has been in the past.

UNIDENTIFIED DISCUSSANT: I was going to put a question to Dr. Mudd, but I would like to connect it with what Professor Klineberg said; I think I am in much greater disagreement with him than with Dr. Mudd, because I think that maybe we will, in the future, be more bold and not tolerate dictators on either side of the pyramids; that is, we want to have human rights for black children in Nigeria as we have for white children in Europe, and we will not tolerate this kind of nationalism, not even in Africa. Perhaps we will want a greater freedom for women, not only in the northern countries, but also in the Moslem countries, and we will have birth control not only among the poor in North America but also in South America, regardless of what the Catholic Church may proclaim.

So my question is really, putting this more ambitious goal at the forefront, Dr. Mudd, what is your idea, for instance, in terms of liberation of women, of the most useful means of controlling the birth situation and the population explosion; how should it be implemented in countries where the Government and other authorities are working strongly against it? We cannot be content with changing the situation in Europe and North America alone; we have to consider the whole globe, even when the governments are working actively against it.

MUDD: I return—and I would like to hear many other comments—to the point that family planning is essentially a personal problem. I think it is going to be a personal problem no matter what position the government or the great religious bodies take. I do not believe that anyone can force people to have children or not to have children. I think it comes down to a personal decision, and this is why I believe that we have to use, as Dr. Tyler said, educational methods in ways that I don't think we have even thought of in terms of communication in a mass situation. If you have a government that is opposed to this, I certainly do not know the answer to that except to go back to the fact that you can bring pressures on people and give bonuses not to have more babies. All the maneuvering of men and governments in this connection has to go back to the feeling and attitudes and values that the women of child-bearing age—and their husbands—have. This is where you have to do your communication.

MADDUX: I want to state very briefly a parable, and then I want to make a plea.

I am at the World Bank now. Prior to that I was at the Pentagon, and philosophy is my discipline, so that does not make me a scientist of any kind. When I was at the Pentagon—the Pentagon is the largest building in the world; there are 30,000 people in that building—and I was looking through the phone book one day, which is a very good way to get an education about modern America. I found out there was something called the Advanced Weapons Research Unit. There are the sort of people there you would expect. But to my tremendous satisfaction I noticed there were two behavioral scientists listed there. And I was working in the office of the Secretary of Defense, so I called these people and asked them to come up. This was about 1967. I told them there was some interesting work being done on the nature of aggression—Konrad Lorenz and people like that are writing books, and what are we doing here in the Pentagon about understanding what aggression is?

They looked at me rather blankly and said, "We are not doing anything." I asked them why, and they said, "You know, it's very difficult to get those kinds of projects through the military." I said, "Look, we are spending $60 billion a year allegedly to take care of aggression. Could we spend maybe $6,000 to find out what it is?" They said, "Look, we don't disagree with you, but it is not technically possible to do this. We don't have the staff." I told them, "All right, we don't have enough people here in the house to do it, but let's go out to the universities and get it done." They said, "No, we can't do that either, because nobody in the universities wants to do it, because everybody is so down on the war they don't want to take any of this money. This has become blood money."

How could anybody be against taking money to tell us what aggression is so we can stop it? Maybe we are committing aggression. They said, "You don't understand. It doesn't work that way." That was a very frustrating experience, and I am not exaggerating.

Now I am at the World Bank, and the man who was the Secretary of Defense is now the president of the World Bank. Among the many problems he has, he has that same one. We have about 200 economists in the World Bank, we have lots of specialists in various sectors—agriculture, transportation and power, et cetera—and we have many, many good people, and I am not knocking the organization, but there is not a single anthropologist in the Bank, to my knowledge; there are no other social scientists at all. There is one alleged philosopher, and there are a lot of economists and a lot of specialists.

Mr. MacNamara made a speech a couple of months ago here in New York in which he appealed to the development community at large, and that means everybody who is interested in development, the university, bilateral government aid organizations—everybody—and he said, in effect, Look, we're starting the second development decade, and we want to reach the new targets of GNP growth that have been suggested by very distinguished men like Professor Tinbergen. But we want to do something more than that in the Bank. We want to do something about the quality of life; we want to do something about social uplifting in these countries, and we need social indicators and we don't have them. He said that we have a tremendous number of questions and we don't have the answers. The bank has a lot of leverage. It lends $2 billion a year. And the Pentagon has a lot of leverage. It has a lot more than $2 billion.

The point I am trying to make is: We dismiss these agencies, whether they are government or nongovernment, and we criticize them for doing a lot of harm. Perhaps they do, but my experience at the World Bank and in government is that we cannot find people who will come to us with workable ideas that our kind of bureaucracy can understand and put into operation.

I completely agree that we ought not to prostitute basic research. But if we want to change the world, the World Bank's whole purpose is to try to change the world for the better: it wants to change it in accord with the way the world wants to be changed. There are 112 member governments, so I end this with a plea: If you don't believe what I say, please write to me and I will send you Mr. MacNamara's speech, in which he says very clearly that we want research organizations, we want scholars, we want people to tell us the answers to some of these questions from all over the world. Furthermore, we are the only institution in the city of Washington that I know anything about that is hiring people; everybody else is firing people. We are doubling the staff of the bank in two years.

Consider me a mail box. You all have a standing invitation to come to lunch with me at the bank's expense any day of any week in Washington if you can tell me how we can proceed more intelligently. It is very, very difficult to get the kinds of information that we need to progress rationally in the kind of operation that the World Bank is engaged in. And I am putting the blame on you people; I put the blame on us. We don't know how to get in touch with you.

MEAD: I am glad this plea came from the World Bank, and not from the Pentagon, because we are looking forward in this whole meeting—as I understand it—to the creation of the worldwide instruments that will be able to muster all this expertise and create this kind of cooperation.

NG: I am glad Mr. Maddux spoke up, because I was thoroughly enjoying the intellectual exercise and I was wondering whether I was in danger of sitting on the other side of the fence. We think that we are separate, but I would like to point out that we are all in this together. First of all, I would like to point to the title of this Conference, "Environment and Society in Transition." Man is conspicuously absent in this, and I hope it is not inauspicious, meaning to say that man has disappeared in the process of transition. What particularly bothers me as a member of the younger generation is that the world we have been discussing, and the whole concern of this conference, is that this is a world that the younger generation is going to inherit.

I would like to point out also that in the current environmental-crisis interest, again man is conspicuously left out. It appears that man in this whole process is becoming a slave asked to clean up the environment, yet man is a very integral part of the total environment. And while we are cleaning up the mess, we also have to begin to clean up the mess that is man. Let us look at the world through the eyes of a child. To a child there is no environmental crisis. This is a world consisting of toys. Environmental crisis exists as it is perceived in the minds of adults. I am not denying the existence of a problem to be dealt with, but, unfortunately, in the process of growing up we substitute atomic reactors and supersonic jets for toys.

It seems at this time that environmental crisis is an artifact created by our artifacts, and unless we get dragged into creating more artifacts to counteract these artifacts, we had better begin to ponder what we are doing because, as Robert Oppenheimer once said, in the process of trying to save our own skins we may, by that very process, accelerate our own destruction by building more and more powerful weapons to counteract the existing weapons.

I am a neurologist, and I would like to share with you some thoughts that have been bothering me as to the state of man. The hard, cold pavements underneath us, as well as the existence of the Conference, are derivatives of a pulsating deep within us. I have no answers, but I want to pose some questions. Why is it that man, apart from several other species—namely, the red ant and the brown rat—is the only species that, as far as we know, organizes willful killing to any significant extent? Could it be that the external inconsistencies we are seeing in society today are a reflection of some basic "inconsistencies" that exist within us?

I want to toss out a hypothesis that has been proposed before: perhaps if we look at this from an evolutionary perspective, man is a relatively weak creature; he depended more on flight than fight for survival. Perhaps at that distant point defenses against killing were superfluous, yet when we began to invent weapons, first, the jawbone of a dead animal, and then sophisticated machinery, we took evolution by surprise. Now we have become a very powerful creature, and perhaps at that juncture we begin to invent defense mechanisms necessary for our survival. I want to ask, could it be that conscience, guilt, and all these responses are an invention that man invented as man evolved? We have taken evolution by a second surprise; we now have weapons that kill at a distance, kill so rapidly in a millionth of a second that our traditional defenses are useless. These defenses will perhaps evolve in face-to-face combat situations, and I feel a lot of pity when I stab with a knife. I have a lot of pity when I see a dog run over by a car. I cannot sleep for days. Yet we talk about the pulverization of millions with detachment.

This is one of the problems I think we have to deal with: how do we go about inventing new mechanisms that will be effective? This is a task of cultural social engineering, and I would like to pose this as a question that hopefully the Conference will consider. I would like to bring up two points with reference to the generation gap. This is one way to look at it, from the quantification aspect: For a man living 6000 years BC, the doubling time of population at the time took, probably, 6,000 years. Today it takes roughly about 35 years, let us grant 100 years, to double. Divided 6,000 by 100, and you get 60. What this means is a man living 6000 years BC had to cope with a rate of change once in his lifetime. Let us equate rate of change with doubling time. Today, we see a doubling time in 25 years as the rate of change increases— and this is my point—to the very point where we begin to reproduce ourselves at puberty. In other words, the doubling time is 13 years. That perhaps may be a critical point, because before we are able to reproduce, we do not even recognize ourselves. And I leave this as a question.

ERAJ: Dr. Mudd has very eloquently dealt with the problem of population and family planning, but there are a number of points that should be considered. She has mentioned the pitiable plight of a woman having unwanted children and the effect it has on her health and the health of the baby. I will tell you that medically the unplanned pregnancy is the biggest source of mortality and morbidity in women between the ages of 16 and 44. It is not only that they have to undergo abortions; we know full well that when they keep on having pregnancies they are much more susceptible to many diseases. This has led to a simplification of the solution. We have accepted that anything that affects the social, physical, and mental well-being of man is really a disease. That is why we call alcoholism a disease. That is why we call delinquency a disease. I do not understand why we cannot accept the unplanned pregnancy as a disease. This is what we have done in Kenya; clinics in Kenya are open to any woman to go and ask for guidance and help or assistance in family planning.

Now, another point was that reference was made to Japan, Taiwan, South Korea, and that family planning has been very successful there. I must state that family planning in itself was not very successful in those countries. The income per capita had increased, the living standard had become very much better in recent years, and family planning came along as a process of that evolution. It really did not start those trends. We have actually no example of any traditional society where family planning today has made any successful dent in the population.

You have asked the question, Dr. Mudd, will family planning ever achieve any control over population? I will tell you that it will never do so. We have never, scientifically, sat down and tried to discuss it, but where it has been practiced we know it does not reduce population growth. The population in the United States is going to increase by 16 million over the 1964 level. The population in the United Kingdom is going to increase, and in the countries where family planning has been practiced, it is going to increase. We have closed our eyes, trying to think that family planning will control the population and that population control will control the other evils that are prevalent today: hunger and stravation. This is not true. We in this scientific organization have a duty to limit it, but I don't think we will do it

Now, we were discussing what could be done. It is known that if *some* money is spent on family planning, if *some* research is done, or if a quarter of the money that is spent on cancer, which does not affect as many people as are affected by unplanned pregnancy were spent on that, tomorrow we could devise a test paper so that every women would be able to tell by using it if she is or is not going to ovulate tomorrow.

Here again the question of religion would perhaps be circumvented.

LOWENTHAL: Dr. Eraj is right about the environmental crisis and children. Some of you may have noticed that in the kindergartens in New York City children were playing "garbage," and some of them would lie on the floor and get swept into trash bins by others, and there was no role that was preferred.

I would like to put a serious question to the working groups that will meet later, and this assumes that many of the questions that were asked this morning are soluble. I am speaking now of the questions of deciding how you get interdisciplinary work going in a profitable manner, of how we resolve the question about what is relevant and what is not relevant, of how we resolve decisions between what seems to be pure and what seems to be applied research? Efforts can be made through a great deal of such research to solve many of the hard problems of world resources and of environmental control, although I think not all of them are soluble.

What I would like to put in this question is: how can you persuade the new generation that it is reasonable to wait to solve some of the questions in the only ways that, presumably, they can be solved? I have in mind two special points here; one is that practically all the questions involving environment and society are questions that require time to work out. The whole science of ecology is one involving problems that we cannot begin to understand until there has been time to see what the problems are. We do not know the answers in advance and it takes years, and in many cases generations, to know what problems are going to emerge.

In order to understand this, one must therefore have a sense of what ecology is about and one must have a sense of what history is about, and yet we see a demand everywhere not only for relevance and for activism, but for immediate solutions. I would like to have a consideration of how do we—assuming we think we know how to go about answering some of these questions—then proceed to convince young people and the world in general that it can afford to wait and undertake this kind of thing in the kinds of institutions that we have?

How do you avoid burning the Institute for Advanced Studies? How do you avoid having professors beleaguered in pursuit of what may be inappropriate methods of solution?

MEAD: I will make one closing comment from this panel in reply to David Lowenthal's question: How can we get people to wait so they won't burn things down? I would suggest, by doing the things that need to be done; there is a systematic relationship between people who wait too long and people who then won't wait long enough. One of our jobs is to figure out how we can work out a balance between the things that must be done today, and if they are not done today, that will make a lot of people burn before tomorrow.

ON THE STATE OF PSYCHOLOGY

Gardner Murphy

George Washington University
Washington, D.C

If from a high tower above the world you could view in perspective the state of psychology as it takes shape day by day, what would you see? First of all, you would look at the universities. There, in a "life-science building" or in a building shared with other experimental sciences, you would see a busy hive of psychologists and their students, experimenting on human or animal functions with their new electronic gear, probing in skin and brain of mouse, monkey, and man. There you would also see the Psychological Clinic, with its interviewing and testing, sometimes utilizing the same gadgetry as the experimental portion of the department of psychology, but sometimes oriented more to the "talking world," especially the talking world of the life problems of child or adult. Throughout everything that is considered psychological there is emphasis on change, growth, development: how things got to be what they are.

In many of these university life-science buildings or other buildings associated functionally with them, there is also concern for the activities of men, women, and children, in their homes, neighborhoods, communities. One may wonder whether there is an effort on the part of psychologists—or of the university administration—to restore the balance between the "biological" and the "social" aspects of human life, although in terms of personnel, man-hours, space, and budgets, the balance would seem difficult to restore. What seems to be happening is the creation within the laboratory of a biological or, we might say, behavioral world. Because most biological laboratories are concerned with man in a planned laboratory setting rather than in the everyday "naturalistic" patterns of his life, his broad ecology may rarely be mentioned. The word *environment* and specifically the word *situation* are held to be appropriate in describing what makes him behave, but there is very little orientation to the sociocultural world as a whole. One does indeed discover an occasional "social psychology laboratory" rich in maps, photographs, and charts of urban trends, with real concern for the larger environment; but on the whole, this is the preoccupation of sociologists, or anthropologists, or political scientists, or economists. Vividly and pervasively one would therefore register the impression that a new biological science is being created, with intense seriousness and high ambition, and that the experimental methods of science are everywhere dominant.

The picture thus drawn emphasizes American university life; other research centers, such as Federal and state-supported hospitals and research institutes, would confirm this picture. In general, this identification of psychology with modern laboratory science would apply, on a somewhat reduced scale to other English-speaking lands, to Western Europe, to the Soviet Union, and in some degree to Japan. But for economic and other reasons these other lands show relatively more interest in the great historical questions ("the persistent problems") of psychology, represented in the United States by courses in "history and systems," and a less concentrated concern with experimental methods, especially those that require expensive equipment. The trend described is, however, world-wide; even in India, Pakistan, and certain portions of Latin America, the new trend towards "hard science" is visible.

What are the *ideas* that go with this package? What is it that keeps the psychologists so busily and eagerly preoccupied? First of all, it is the behavior of organisms. Organisms are viewed in terms of their evolutionary history, their own individual heredity and life history, their functional capacities for receiving and elaborating messages from outside, their capacity to symbolize and represent within themselves the manifold of their immediate environing situations—with special emphasis on their capacity to benefit by experience, in order to adapt to situational requirements. The psychology of sensation and perception has a prominent place, but an even more prominent place is held by the psychology of learning. "Learning theory," abundantly documented by experimental findings and constantly changing as ingenious experiments alter the balance of evidence in one direction or another, takes the bulk of man-hours in many an American laboratory.

Behind this daily panorama of research on the behavior of living things, there is everywhere an identification with the general scientific point of view as represented, for example, in the journal *Science,* official house organ of the American Association for the Advancement of Science. Psychologists everywhere profess their intellectual and moral adherence to the goals and methods of science. The prevailing scientific world view as it has taken shape since the time of Galileo and Newton is imbibed by the young psychologist as he moves from the elementary course into his "psychology major" and graduate-student years. He may, as a student of social psychology or of personality or of clinical practice, belong to the "soft" pole of the modern psychological world rather than the "hard" pole represented in pure experimentalism, but almost as ardently he will pledge his allegiance to the output and the methods of science.

But in science it is legitimate and necessary, when you can't get into the subject matter directly with your hands and feet, as in building equipment, or running animals, or plotting curves, to *reflect* and *discuss;* and in the psychological laboratory and classroom you also *discuss* all sorts of issues, such issues as the life histories, or attitudes, or personalities of those who took part in some recent experiment. A "talking science," at a lower prestige level than a "doing science," has a place of secondary, yet real, honor in the total adventure.

If now we ask how such a state of psychology could possibly have come into existence, the answer lies in certain considerations relative to the general history of science since the 16th century. The revival of Greek and, especially, of Alexandrian science, along with gains traceable to the growth of mathematics and to incidental new knowledge from explorers, military, and naval men, threw into sharp relief the mathematical order of the world and the elements of mechanics as utilized by Copernicus, Kepler, and Galileo. The mathematical genius of Sir Isaac Newton made possible the development of astronomy and of a universal rationale in the form of physics. Then, late in the 18th century, came Lavoisier and a quantitative chemistry. Science was seized upon by the French, and a little later by the Germans, as a central part— and an ennobling part—of a rational and enlightened civilization.

The biological sciences took shape in their turn, especially in the German universities during the early 19th century, followed, in the middle of the century, by the evolutionary tide of thought made dominant by Charles Darwin. The biological sciences were not as exact as the physical, but their proponents and adherents strove for exactness. The career of Helmholtz illustrates the movement from the exact observational and mathematical methods of the physical sciences, to the study of optics and acoustics as expressions of physiological function, rather than simply the study of lenses and tuning forks as physical objects.

There was a possibility for psychology, through this developmental era, to become a science in the serious sense; that was the possibility of dealing with the living organism in essentially physical-chemical terms, using the methods, especially the experimental methods, of physics and chemistry. As soon as a biological platform was established on which to stand, one had to develop a "physiological psychology" of the processes of *perceiving, learning,* and a little later even *imagining* and *thinking,* in which the methods of the laboratory would be given maximal play.

All this happened exactly as one could predict, and, late in the 19th century, the German universities began their march towards a quantitative and systematic experimental physiological psychology. The social sciences were in the meantime borrowing as much as they could from the physical and biological sciences, especially from psychology; but there was not enough of it, nor was it scientifically good enough, to enable the social sciences of that period to compete. So psychology took its shape without benefit of any serious contributions of scientific methods from the social sciences to influence the laboratory-minded experimental scientists. Sociology and anthropology, political science and economics, though dealing imaginatively with vast quantities of complex observational material, could not claim the status of science, and before long psychology became crystallized as a biological science with only incidental utilization of social science methods and concepts. We see how the stage was set historically for the one-sided psychology which we described above. The physical sciences had essentially established both the problems and the methods for the biological sciences, and later for psychology and the social sciences.

As Kurt Lewin has pointed out, there were some unfortunate, even destructive, consequences. For one thing, much of the world of physics and chemistry relates to reversible operations in nature. There are reversible reactions in chemistry. In physics objects may go back and forth, up and down, with energy conserved but patterns in time-space conceivable as ultimately bidirectional. For the developmental sciences, notably for evolutionary biology and for individual growth, this would be a fantastic assumption. Hand in hand with this is the fact of *uniqueness* in individual development—a grossly misleading assumption when one speaks of atoms or molecules, but a patent fact of primary reality in the life sciences, specifically in psychology.

This is my attempt to answer truthfully the question regarding the state of psychology—as seen from a psychologist's point of view. But there are other points of view. A literate nonpsychologist would probably suggest that the richest source of information about individual personal development available among modern psychological methods is the yield from the work of Sigmund Freud. Whatever the controversies regarding his work, Freud did much to define the intricacy of those aspects of personal life that operate outside of personal awareness and that make for its happiness or sorrow, its effectiveness or failure. It was Freud who, through a theory of the "dynamic unconscious," showed how the biological and sociocultural forces that make their impact on early childhood growth predetermine the individual towards a particular vision or blindness, a specific enthusiasm or revulsion. The clinical psychology that emerged began, in the early years of this century, to glimpse a richer and more complex world from which dreams and fantasies, ideals, and ambitions took shape in the family and community interactions of normal development. The contribution to psychopathology was conspicuous, but this was a minor point compared with Freud's massive impact upon the understanding of individual growth as a function of biosocial forces: love, fear, hate, injured self-esteem, as all of these made their sorrowful way through the record of the early years. Very much in the same manner in which physical disease has forced medical attention to basic issues of life, growth, and health, aside from all specific disease processes, psychoanalysis and the related dynamic psychiatries of the last seventy-five years have drawn attention to biological and biosocial issues that were scantily considered until the new clinical context and clinical observations were available.

Cross-Cultural Psychology

There are, however, still other ways of gleaning psychologic perspectives. The cross-cultural methods and findings, both in their own right and in association with dynamic psychiatry, have forced the development of new perspectives all along the line. It is not enough to be told what everybody knows: that the social environment "shapes personality." The issue is different when we confront the fact that the oedipal situation may take forms different from our own when the child's hostility to a frustrating disciplinary agent is focused on the mother's brother rather than the father, or when we read that a certain schizoid "awayness" in the adult may reflect the mother's trenchant teasing habits of the small infant who has learned that all which is warm and good must be given up forever, or when one learns that the quality of parent-child relationships may alter the ideal ethical and religious system to a point where the cosmos as a whole and all its laws reflect the pattycake world of infancy. We may brush aside all such suggestions but we come guiltily back to the question: "Why have we not thought of the possibility that the emotional struggles in the growth period cast long shadows over the deepest issues of adult living." Or, as James Harvey Robinson said, "We are children at our most impressionable age."

The social science view, often in conjunction with depth psychology, has created a massive impact on our civilization, and it is clear that whatever the well-controlled and accurately described results of animal or human experimentation may yield, in the long run the results will have to be seen in the perspective of the historical and cultural science modes of observing and thinking. It is one humanity, not a diversified ragbag of scattered human images, that serious science will have to confront.

But perhaps we have not named all the resources for a study of man that could help in planning for the decades ahead. We have mentioned experimental, biologically oriented psychology, sociocultural psychology, and clinical psychology. But what shall we say of the status today of the huge fund of human observation, reflection, and wise utterance that comes to us from folklore and history and from the biographies, the letters, the skillful interpretative studies of unusual men and women—all standing in their magnificence, apart from the texture of modern science. Must all these sources of psychological insight and reasonable hypothesis be swept away today by the more systematically oriented psychological scientist? G. W. Allport speaks of ways in which "personal documents" may offer at least hypotheses and, perhaps in many cases, good generalizations regarding the attributes of boys and girls that may presage the attributes they manifest as men and women. The psychologist may take Euripides or Shakespeare, Pushkin or Ibsen, as his source for a delineation of human nature or for the definition of unique attributes of specific individuals. A "psychology of personality" suitable for the study of Goethe or Darwin has made use of social, historical, personal sources of all these types and has, for nearly a century, received and turned back to the literary world conceptions of individuality effectively used by the modern psychologist.

Ecology

From the viewpoint of all the psychologies considered thus far, psychology is concerned with the behavior of organism. In view of the fact that the life sciences are concerned with the interaction of organisms and their environment, one might think that the environments in which psychological events occur would be studied in a very sophisticated fashion. In a sense this is true. But it is soon learned that it is the *immediate* environment, the laboratory room, the stimulator, color-mixer—or tone, or food pellet, or electric shock that is regarded as the energy package that is impinging upon the living subject.

Occasionally the time of day will be of interest, and one may think of the earth's magnetic field or of broad weather and climate factors as making a difference in a subject's behavior. There are superb modern research rooms in which light and shadow, temperature changes, air currents are nicely controlled and measured with excellent photography and cinema. One may attempt to bring the "macrostructure" of the social environment into the laboratory setting. It is generally agreed, however, that to *simplify* the stimulating conditions is a scientific virtue, and it is generally held that there is no feasible way to study the social life of man by the more hard-nosed procedures that hold the greatest prestige.

What can one do about this? At least three things: First, by the procedures just mentioned, one may simplify the macrostructure. One may, as Sherif has done, create a laboratory situation that produces essentially the more naturalistic human situation response pattern. Noting that men in society, from their different vantage points, make different interpretations of events going on before their eyes, he has standardized a variety of devices by which their convergence of judgments, when exposed to one another's pressure, can be observed and measured; i.e., men in the laboratory who at first disagree move from great disagreement to less disagreement—or may, indeed, move towards identity of report in what they observe. Large issues remain in regard to the persuasibility, suggestibility, negativism, obstinacy, or the various idiosyncrasies of men in society as they make everyday social judgments, as has been observed in these and many other studies. Just as the industrial psychologist creates a miniature man-machine system that duplicates the larger man-machine system of assembly line or cockpit, so the social psychologist attempts to capture all the essentials of the free-flowing social life of everyday.

A second path open to the social psychologist is the establishment of principles or "laws." By gathering large quantities of representative data in which samples of situation-response interactions can be observed, and, noting the dependence of the responses upon the attributes of the situations, he builds a primitive *ecological psychology,* in which emphasis is placed on the demand properties of the situation. Outstanding here is the work of the Midwest Research Station, Oskaloosa, Kans., under Roger Barker, and Herbert Wright and their colleagues. In this approach the experimental method is wanting; it is the controlling role of the situation that inevitably receives primary emphasis. One records and classifies the types of behavior that characteristically occur in each clearly defined "behavior setting," such as church, school, or street corner.

Third, one may place one's emphasis on *individual differences* in response to behavior settings, as has been done by Gordon Allport, Kurt Lewin, and their students, or one may actually succeed in considering both the special properties of the situation and the special properties of the individual person and attempt to see both in interaction.

But had not the public opinion and market research specialists already taken these various steps? Had they not, even in the days of Stuart A. Rice, studied behavior quantitatively as an expression of a given region—comparing the farmers of fertile and of barren land, for example, with reference to their views on simple economic and political issues, and then gone on to compare the response of these farmers or other citizens by the use of "cross-tabulated" data concerning the age, religion, and other attributes of these individuals? Yes, the economists, political scientists, sociologists, and anthropologists had indeed already made these studies of ecology and of individuality, and of these two factors in interaction; and in spite of the identification of the individual research men with the specialty to which each belongs, there has been in recent years a good deal of intercommunication between psychologists and these other students of social behavior. Everyone wants to be "cross-disciplinary." Occasionally there will even be a *clinically* and *personally* oriented study of public opinion or of social attitudes, in which the massive pressures of daily life, region by region, town by town, are brought into relation to the individual-response tendencies of specific persons. But from the social-science point of view, the tail wags the dog; the studies of individuals again overwhelm the studies of social settings. The clinical information available about an individual observed in a modern clinical setting (life history, health history, autobiography, psychometric and projective tests) so far outweighs in mass and in relevance the best we can gain in ecological terms that the imbalance appears absurd. It is as if we had the most marvelous information about a tropical humming bird and all we could say about its ecology is that it lives in a large cage.

Why, then, is the confrontation not seriously, intensively, deeply considered and the integration of all these materials not seriously attempted? The question brings its own answer in the form of another question. *Who* could effect this integration? One may say: the psychologist, of course; he who has given his professional life to deep probing into these issues. But for the most part, the psychologist has been absent when the roll is called. His professional image of himself, indeed the deep channeling or cathexis of his scientific love, has drawn him towards the specialized biological world of science, Science with a capital S. Even in the liberal-arts preparation he received, he had begun to eschew the mushy issues of social science, and his measure of maturity has for him progressively fixed on the degree of his adherence to a hard-nosed ideal. If he belongs to a clinical, social, or other "applied" branch of psychology, he will not know enough about biological methods and results to be taken seriously when he discusses them. And indeed it will be evident that his aim is poor, his touch gauche, when he attempts that breadth of integration in which all the views of human kind would be made to coalesce.

No, the psychologists who can effect this larger integration are few; indeed, they are hardly to be found. The William James kind of psychology, with its medical and its philosophical integrations, has long since been an idol of the past to which we turn back, like Lot's wife, to be turned, if not to salt, at least to a stony helplessness. Here and there flashes of insight into the modalities of such an integration may occur to us. We have already noted that Kurt Lewin and his followers made noble sallies into this domain. Erik Erikson, with his clear delineation of biological growth stages and his vast preoccupation with preliterate and advanced societies as expressions of psychoanalytic dynamics, has kept the flickering light alive. But among the men who have matured and undergone their specialization within the last 20 years, it would be hard to find any who actually lived through the days of their lives in such preoccupations as are demanded today. If humankind is to have such wisdom, it will be a bit easier to formulate and express if the young psychologist has training in at least the elementary methods and facts that go with participation in today's scientific and philosophical undertakings.

But we do not really have even the methods and results to offer to an earnest, brilliant young integrator if he should come along. We have a lop-sided psychology, a psychology of the organism, not a psychology of men, women, children, and society. The inexactness and fuzziness of much social psychology is shared by sociology, anthropology, political science, even most of economics and human geography, and our young man would not only have to be offered a vision of integration but some extensive and serious scientific understanding of human ecology.

But what are the problems in creating such a scientific ecology? Such a scientific ecology is taking shape, of course, around the more organized factual domains, the physics and chemistry of pollution, poisoning, and malnutrition by inadequate, faultily balanced or faultily processed foodstuffs. When people themselves, as the environment of a given individual are considered, elementary measurable units, as in "population density," come into the reckoning. A well-known experiment by Calhoun describes the pathology that develops among rodents under conditions of extreme overcrowding, and human ecologists dream of a time when similar exact documentation of the effects of overcrowding could be made by the urban sociologist. But it is entirely possible that a complex of psychological factors having to do with loneliness on the one hand and lack of privacy on the other, with pressures toward competition, toward forced cooperation, toward pathological loss of self-awareness or inflation of self-awareness, and a host of complex psychological responses to a manifold of types of interstimulation, may within two or three decades give us a human ecology. It would be concerned much more with human *interpersonal* pressures than with response to industrial or human waste or to the sheer geometry, mechanics, and biochemistry of changing personal contacts. Ecology as a psychological science would then be different indeed from the ecology of today. Development in the direction of the newer ecology will come from the creation of new models of interpersonal relations and their testing through observation and experiment and, above all, through extrapolation and the discovery of weaknesses and necessary corrections in models. On a rolling and pitching ship a gunner must aim not at the invisible target ship controlled by an unknown distant admiral but at the probable position the zigzagging admiral may have assigned to that target ship at that *future* moment. One-dimensional predictions based on one-dimensional extrapolations are probably worse than useless. The serious science of society organized in time-space units, of course, but organized also in higher units of individual and group purpose, will have to see where we will be when we need to use the specific new methods now in the process of being invented. Relevant science will call for the fantastic multidimensional thinking that alone can cope with such multidimensional problems.

Our argument regarding status (*qua* scientist) as a major factor in the determination of the world outlook for the young psychologist must, however, evoke one very large exception. It is not experimentation as such but rigor, clarity, order, the inescapability of the documentation of a fact or principle that constitutes the prestige-bearing aspects of science, and the world of *mathematics* possesses this on the same footing as does the world of experimental documentation. It would be ridiculous to maintain that the theoretical physicist, with his complex capacity to use mathematical reasoning, is automatically inferior to the experimental physicist. No such invidious distinction can be made. The time for a similar development appears to be ahead in psychology. In fact, Cronbach has already signaled this event, for mathematical psychology—or what are called today mathematical models of psychological events—is coming to have the same status-bearing qualities as experimental method. It is likely then, as mathematical sophistication increases and as the capacity for stating an issue in mathematical terms becomes more and more understood and respected, that the new pschology we are trying to describe will be both a mathematical psychology and an experimentally validated psychology. When experimental verification is impossible, the mathematical method will prevail. This means that the development of higher-level mathematical training will be more and more important for the psychologist and that types of social reality that admit such quantitative method will begin to play a larger and larger part in the ecological psychology we have been describing. It is to be feared that this will usher in another type of lopsidedness, if those future ecological relationships that can be well predicted prove more and more an inducement to able young psychologists to become specialists in ecological problems. But as they do so, they will choose those particular ecological problems that are most readily converted into the kind of thinking respected by the men of the mathematical models. As long as our prestige needs are fixed at the level of exactness, it will be the principles that are "clean" and bare, rather than those that are most comprehensive, which will become the center of our science.

EDUCATION: BALANCING CONDITIONED RESPONSE AND RESPONSIBLE CLAIMS

Ralph W. Tyler

*Director Emeritus, Center for Advanced
Study in the Behavioral Sciences
Stanford University
Palo Alto, Calif.*

If modern man is to survive and is to realize his potential for intellectual achievement, for human fellowship, and for deeply satisfying esthetic experiences, he must be capable of coping with change, be skillful in attacking new problems, and be· open to and seek new experiences, including new human relationships. Education that inculcates "final answers" to questions, that forms rigid habits of response to symbols and other stimuli, and that develops a perception of the world as fixed and ethnocentric is maladaptive for the species and, as well, denies the individual the kind of life that is most fully human. Yet man has made much greater progress in perfecting techniques of conditioning human behavior than in developing learning situations that are conducive to acquiring the attitudes, understanding, and skills essential for responsible human participation in this rapidly changing postindustrial society.

Behavioral science fiction, such as Aldous Huxley's Brave New World and Orwell's 1984, furnishes a vivid warning of the perils of regimenting people by the use of systems of immediate reward, hypnosis, and drugs. Fortunately, the experimental evidence affords much less support for the belief that men can be as fully conditioned as these books imply. Nevertheless, most of us have been and are being conditioned to a larger degree than is necessary or desirable for a responsible citizen in a world society. What is a proper balance, and how may it be attained? These are questions for which no generation will have final answers, but each generation must seek a balance that appears to provide for both stability and change, recognizing that this balance is a dynamic and shifting one.

As a background for this discussion, a general conception of human learning that includes conditioning as one type is useful. Human learning is the process by which one acquires new patterns of behavior through experience. When one carries on a new behavior pattern and finds it satisfying, he continues it until it becomes part of his available repertoire. Then it is said that the behavior has been learned. The term behavior is used here in the inclusive sense to refer to all kinds of human reactions—thinking, feeling, and acting. Thus, one can learn a new way of attacking problems, he can memorize a statement, he can perceive persons, events, or situations in a different way, he can respond to esthetic elements to which he did not respond before, or he can respond to them in a different mode, and he can learn to swim or to drive a truck. In short, any kind of behavior a human being is capable of carrying on can be learned, maladaptive behavior as well as behavior that has great survival value.

The term "conditioning" is commonly used to refer to the learning of a behavior which is initiated by a clear stimulus and consists of an automatic, fixed response. Most of the behavior of a driver of a car represents conditioned responses to traffic lights, to the approach of other cars and pedestrians, and to the sensations that he receives from the car's movements. To be a good driver, he must respond swiftly to stimuli that present themselves suddenly, and he has no time to view the traffic scene from various perspectives and to analyze the several traffic problems that might be identified. For most people, habits of cleanliness, of eating, of punctuality, are conditioned responses. The way we respond to authority, to brothers and sisters, to strangers, includes a large component of conditioning. The demands on a man for reactions in modern society are so great that he would soon perish if each one had to be examined, analyzed, and dealt with in a problem-solving way. Hence, conditioning is a necessary and important type of learning. An attempt to eliminate it altogether in an educational system would be disastrous. It fails, however, when it furnishes an automatic response where such a reaction is inappropriate. The problem is to identify the situations where conditioning is essential or at least helpful and the other situations, where a fixed response is not only not essential but where it would lead to the destruction of the species or the denial of significant opportunities for man's fuller development.

How can one make the distinction? It can be made only as an approximate adaptation to the present and foreseeable situations. Habits of eating, sleeping, exercise, speaking, obedience to accepted rules, coding and decoding stylized symbols are likely to be seen as requiring automatic responses, so that conditioning is a proper means of learning. However, we recognize that our society is undergoing continuing change, and we can conceive of the possibility that some of these types of behavior will require re-examination and the formation of new patterns. That is, human education seeks to help the student understand human behavior, particularly his own, and to be able to choose new learning objectives and to work on their attainment. In this way, each generation has a means for re-examination and self-renewal of even basic habitual reactions.

The inadequacy of conditioned responses arises from the changing environment, which requires new human behavior patterns for coping with these changes, the increasing understanding of the world and of man, which opens new possibilities for men to achieve their aspirations by effective utilization of the new knowledge, and greater acceptance of the ideal of the brotherhood of man and a world of greater equality of opportunity, the attainment of which requires new attitudes, skills, and deeper understanding. Conditioned response learning does not furnish a model to guide education that enables men to deal with a changing environment, to gain and use new knowledge, and to form and strengthen new relationships of man and society. These more general and dynamic goals are attainable through a more complex model.

One that is compatible with the development of responsible men in a changing society conceives the learning situation as one in which the learner himself seeks to acquire new behavior and the rewards of learning include the satisfaction of using the new behavior successfully as he copes with the problems he encounters. In animal conditioning, the trainer seizes on incidental reactions of the animal that represent a piece of the total behavior the animal is to be trained to carry on. He seizes on this animal reaction by reinforcing it, that is, by giving the animal a reward of some sort. This reinforcement is continued when the animal repeats the behavior until it becomes an habitual response to a stimulus that the trainer has chosen to use as the trigger. Similarly, other pieces of the complete behavior pattern are reinforced when the animal carries them on. Incidental reactions rather than intentional ones are the common bases for conditioning. Consciousness and understanding are not necessary and may interfere.

On the other hand, the development of the behavior required for human responsibility implies consciousness on the part of the learner and an increasing understanding of the goals of his learning and their means. Hence, an initial step is for the learner to perceive learning goals that he will put forth effort to attain. We often wonder why a child will spend so much time and effort to learn to speak, or to play baseball, or to acquire a new dance step and yet seems to tire very quickly when working on school assignments. In the former cases, he sees others speaking, or playing baseball, or dancing and he can see that they are getting satisfaction from it. Furthermore, he can perceive at least roughly what the behavior is he is trying to learn, so that he can emulate those who seem to be carrying on successfully. It is all too rare for teachers to demonstrate in their normal actions much of the behavior they would like to help children learn, and even more rare for them to express in ways children can perceive how meaningful and satisfying this behavior is. Hence, a considerable part of the motivation for learning in school is based on the child's desire to make good grades in order to get along or to avoid reprimands or failures rather than the desire to learn what the school attempts to teach. Furthermore, the students' perceptions of what they are trying to learn are inaccurate and frequently in conflict with the desired objectives. It is important in this connection to recognize that for education to go beyond conditioning, the learner must perceive something to be learned that is attractive to him, or, to use the current phrase, it must be relevant and meaningful. He must also see clearly enough what he needs to learn so that he can take the initial steps in emulating this behavior.

This model of learning also includes rewards for the student as he successfully carries on the behavior he is seeking to learn, but the nature of the reward system itself must be consistent with the role of a self-directive, responsible person. Students should be helped to discover the satisfaction that comes from having acquired and used new understanding, new interests, new attitudes, new skills, rather than depending largely on the rewards that are extraneous to the learning itself. As a student develops character structure and conscience or, to use another current phrase, a stable self-image, rewards that arise from learning what he believes is in harmony with this self-image are to be preferred to rewards that depend on the favor of others. The techniques of conditioning commonly place great reliance on rewards that gratify appetites. These are played down in other learning models, because habitual responses to physical gratification makes a human being more dependent on those who can use force and material power than on those whose importance derives from intellectual or social influence.

Another feature of this model of learning is the availability of opportunities for practicing this new behavior until it becomes part of his usual repertoire. Availability of opportunities means that there are many chances to carry on the behavior and also that the student has time for the necessary practice. Too often, students spend most of their time in school passively while the teacher performs rather than actively engages in the thinking, feeling, and acting that they are expected to learn. Daily, weekly, and yearly school schedules need thorough reconstruction to furnish time for complex learning required for responsible human living.

Another aspect of opportunities for practice is that they should be sequential.

Sequential practice means that each subsequent practice goes more broadly or more deeply than the previous one. Sheer repetition is quickly boring to the learner and has little or no further effect. Only as each new practice requires him to give attention to it because of new elements in it does it serve adequately as a basis for effective learning. That is important for the student in gaining understanding, because it means that concepts and principles are brought in again and again, but each time in new and more complex illustrations, so that the student continually has to think through the way in which these concepts or principles help to explain or to analyze the situation. In developing a skill, it is important to see to it that each new practice of the skill provides opportunities for greater variety or complexity in its use. Sequence is also important in the development of appreciation, for it means that each new work of art should be demanding something more of perception and be providing opportunity for a greater variety and depth of emotional response.

This model of learning also includes some kind of feedback by which the learner not only is informed that he is successful or unsuccessful in his attempt to carry on the behavior but also is furnished some analysis of his unsuccessful efforts that can serve to guide him in his next attempts. Many forms of feedback are possible, varying with the type of behavior being practiced, the conditions under which the practice takes place, and the availability of human aids and technological devices. For example, videotapes of a human situation furnish feedback for a participant, but if such tapes are not available, the student's own effort to recall and reconstruct the situation and review his behavior in memory can be a crude but useful form of feedback. A colleague's notes, samples of one's products, comments of friends, criticisms by teachers when based on dependable observation, can also be used. Feedback is important because a learner who attempts to change inadequate behavior into successful practice simply by trial and error finds learning very slow. He is often discouraged and gives up. Some means of indicating to him more promising reactions serve to guide him and usually improve his learning.

The purpose of outlining this model in some detail is to show that there is an alternative to conditioning as a conception of learning that can be used in the planning and conduct of education designed to develop persons who are socially responsible, humane, and self-renewing. Too often, those who question the use of conditioning offer no definite alternative that can guide practice. More commonly, critics of conditioning expect that the spirit of the school or the undefined interpersonal relations, particularly the relations between pupils and teachers, will develop children into "good specimens of humanity." Some even profess that the content inherent in certain school subjects liberate and develop the responsible man. Evidence does not support this claim. If schools generally are to become effective in helping to educate responsible human beings, there must be available a conception of the educational tasks and the means of accomplishing them that furnish a constructive basis for the work of the school. The model just presented is one such possibility. It is not a vague, global conception but delineates features that can be defined, principles that can be followed, criteria that can be used to test the effectiveness of the model in action. Systematic efforts to develop responsible men can be made.

Thus far, the discussion has centered on the nature of conditioned response learning and of purposeful learning and the need for both. We now turn to an examination of the interrelations and balance between these two ways of stimulating and guiding learning appropriate for the education of man in the contemporary world.

First, it should be clear that there are ambiguities in both conceptions of learning that make it difficult or impossible to predict in the case of an individual person whether a situation constructed for the learning of a particular behavior pattern will be effective and what the individual will learn. This ambiguity arises from the complexity of human beings. Even a small child has a range of drives and motives that result in differential response to situations offering particular rewards. Although most children respond positively to the offer of food, an individual child may not be hungry or he may be more concerned for the approval of a friend, or his image of himself as a stoic may result in a negative reaction to a food reward. As a child grows up, his hierarchy of drives, motives, and satisfactions develops into an individual pattern that helps to characterize him as an individual. This pattern can be one in which sensual gratification and social approval are high, in which case he is quite easily manipulated by the power and weight of others, or the pattern can be one in which hard work and helpfulness to others have become major features of his ideal self-image, in which case it is difficult or impossible to get him to learn behavior that he perceives as in conflict with these characteristics.

A second kind of complexity in the human motivational system is in the extent to which the individual postpones immediate gratifications for more distant rewards that he perceives to be greater. Some children have discovered that greater satisfaction is obtained by picking up their toys and washing their hands before running to eat because they obtain a warm response from Mother as well as getting a meal, whereas when they dash in to eat without these preliminaries, they may not get the meal and they do not get Mother's warm approval. The vividness of the human memory makes it possible for one to keep in mind and to anticipate the joys of an achievement that requires a long time to complete, so that he will forego many immediate satisfactions in order to gain the greater but more distant ones. The fact that a person can, in effect, live in the future makes it difficult to predict in an individual case whether the learning situation is one that will motivate and be rewarding, since he may be reacting to an anticipated future situation rather than to the present one.

Another factor that makes individual prediction hazardous is the differential nature of perception and attention. Real situations have many facets, so that an individual cannot perceive all that the situation, in fact, includes. He cannot see all, he cannot hear all, he cannot sense all that could activate sense organs. In addition to the selective sensations, his past experience influences his interpretation of the sensations so that one cannot be sure, in an individual case, what he makes of the situation—that is, how he defines it. Combined with this necessary limitation to any individual's perception of complexity is the influence of selective attention. One's orientation, expectations, and purpose strongly influence what one attends to in a complex situation. This focus of attention not only makes it possible to bring to the foreground elements that are objectively weak, such as whispers in a noisy place, or objects in shadow, or a faint aroma in a garlic-scented café but also to shut out sensations that are clear and obvious to others. The importance of selective perception and attention to the learning process lies in the fact that in conscious learning the student is responding to perceived stimuli. If his perception of the situation is different from that of the teacher, he is not in a position to learn what the teacher intended.

These ambiguities are mentioned here not merely to suggest that the notions sometimes presented of brain-washing or total conditioning are not likely to corrupt the whole species because of individuality, but primarily to suggest the importance of early childhood years in developing human beings adequate for the conditions and opportunities of tomorrow. Individual hierarchies of motives, the wide variations among individuals in the length of time they will postpone immediate gratifications for more distant rewards, individual differences in perceptions of complex phenomena and in focus of attention are influenced markedly by early childhood experience, much of it in the form of conditioned learning. Hence, the nature of early conditioning is important for the education of contemporary man.

For this task, one of the first objectives is to develop confidence in the child that he can explore new places and new behavior without losing the love and nurture of his parents. Modern man must be able to explore his environment, investigate new ideas, and gain wider experiences with men of different backgrounds. This requires confidence rather than fear, sensitivity to the positive and negative aspects of new experience to which he can respond rather than behavior guided by tradition. Too often a child is cut off from parental love when he explores the wider environment, and he then senses that he can enjoy the security of a loving home only by repressing his exploratory behavior and accepting his parent's definition of what the outer world is like and how to behave in it. With young children, avenues for exploration need to be opened up in a way that furnishes new experience where the child must cope with some disagreeable features as well as enjoy the pleasant ones. As the range and difficulty of the places he explores are increased, he learns to approach new experiences with confidence and to be guided by his own observations and interpretations rather than to accept traditional attitudes without question. This kind of early conditioning is very important for the development of a genuine world community.

A second emphasis in the early conditioning of a child is in developing faith or confidence in the capacity of each generation of men to solve their problems rather than to place this faith in the answers that have been given in the past. This confidence in working out new solutions to problems is important because the problems and the criteria for effective solutions are in continuous flux. Now, for example, many people recognize pollution as a serious problem. This was not the case 50 years ago. Furthermore, an acceptable solution to the problem of pollution today would be significantly different from what would have been acceptable in 1920.

To develop this confidence in the power of collective intelligence to deal with difficult problems, the child needs to have many examples within his observation and participation that show the world as something men are always learning more about—not one in which everything is known. He needs to see people at work on the analysis and solving of problems rather than waiting for someone else to tell them what to do. He needs to discover that knowledge is made by man and is growing and changing as man finds out new things from his experiences. Education for contemporary man is not finding ready-made answers from books but learning how to inquire, how to formulate ideas and test them, how to use his intelligence in understanding the world and himself, and in solving his problems.

The development of faith and confidence in our own human intelligence conflicts with faith in tribal gods and belief in the teachings of tradition. These latter, although often of value in giving security and reassurance to the members of the tribe that they will succeed and prevail in spite of obstacles, have a serious limitation in their emphasizing the inherent superiority of the chosen tribe against all others. They also give children a conception of a fixed world of wise men who know all the answers, thus failing to emphasize the necessity for world-wide cooperation in working out new ways of dealing constructively with world problems.

Fortunately, curiosity is characteristic of children, and their drive to find out and to explain things can be a powerful motive for conditioning them to seek to understand, to discover that answers to questions can be worked out by men, and that they themselves can develop better and better explanations as they approach problems from different perspectives and increase their own experiences with the phenomena about which they are curious. The positive reinforcement of children's efforts to understand, combined with guidance in ways of learning, can contribute strongly to building this faith. The difficulties lie in the easy way weary adults have of shutting off curiosity and the desire of parents to furnish answers to children rather than help them to push on with their inquiry.

A third center of early conditioning is concerned with the interpretation and enjoyment of human relationships. Most children are negatively conditioned, so that they expect and seek a very few elements in their relations with other human beings. They do, of course, learn to expect from parents and friends help in gratifying physical needs and opportunities to participate with limited roles in social life. But, on the negative side, they expect aggression, rebuffs, or aloofness from most of the others with whom they come in contact. By the time they are young adults, their social relationships are greatly circumscribed both in the variety of people with whom they have meaningful relations and in the range of human interaction in which they participate. Along with exploring the wider world, children need to be positively conditioned to seek out new persons from different backgrounds and to be involved in an increasingly wider range of relationships. They can participate in work as well as play in helping older people, as well as being helped by them, in listening to stories of experiences of persons from different backgrounds as well as in recounting anecdotes they think interesting. Too often, children gain from adults conceptions of what is proper, what is enjoyable, and what are the values in human relations that are entirely too limiting—especially when they are conditioned to this view throughout life. A world society requires citizens who participate with others in a wide range of situations, not because they feel it a duty but because they find satisfaction in a common life shared with others throughout the world. But to appreciate a multicultured and a multivalued society, children need to have experiences with persons who come from other backgrounds and whose hierarchies of values are different. They need help in playing various roles in their relations with others; they need help in understanding other customs and practices, other beliefs and values. They need to see parents and friends enjoying interactions with others without losing their own identities in these experiences.

The purpose of early conditioning in the area of human relationships is not to give children the belief that the people of other groups and nations are model angels wholly to be loved but rather to help them perceive as accurately as they can what human beings from many backgrounds are like. They have strengths and weaknesses, characteristics to be admired and characteristics to be deplored. Most are interesting and friendly and can be helpful. But they are not all alike, and, in their variety, they have to be understood as individuals, not as stereotypes. Realism regarding men and women from different parts of the world is essential in establishing a workable world order. Early exploration of human relationships is the base on which realism can develop.

A fourth center of conditioning deals with learning to postpone immediate gratifications in order to enjoy greater but more distant rewards. Most of the problems of modern society take time to work out. Most of the skills, techniques, attitudes, understanding, and appreciations that contemporary man must acquire take a relatively long time to develop. Individuals who are unable to discipline themselves to a long span of time before their effort pays off are ill-adapted to the conditions of modern life. Hence, the development in children of the ability to postpone immediate gratifications in order to reap great rewards later is quite necessary. A child can be started with a short span of working for a desired reward, such as devoting two weeks to earn two quarters to obtain something of greater value than one quarter would have purchased had it been spent when it was earned the first week. Situations should then be developed in which there are increasingly longer spans of time during which some rewards are foregone to gain a greater one later. In this conditioning, a child is aided by periodically talking about and imagining the satisfactions he will have later and, by receiving a variety of reinforcements when he has carried through to the end, including the reward he sought, the approval of parents and friends, and additional comments and praise for the achievement.

Of course, the purpose of this program of conditioning is not to eliminate all immediate gratifications but to enable the child to choose for himself whether he will take immediate satisfactions or work for a larger and more distant reward. There are occasions when spontaneity and gaiety are enhanced by seizing the opportunities of the moment, while there are other times when such a choice would make it impossible for the individual to gain something he really values much more. He must be able to choose and to follow through on whatever choice he makes. Without early conditioning, postponing gratifications is difficult, often impossible, for him.

As parents and teachers use conditioning to help the child gain the perspectives, the attitudes, the habits, and the practices on which more flexible behavior can be developed, more and more of the purposes and techniques of conditioning should be made apparent to the child and become a matter of conscious discussion. This is necessary in order that conditioning can be controlled as far as possible by the individual and used as a technique for his own purposes rather than being a hidden weapon utilized by others to manipulate him. The objective of the conditioning that has been recommended, like the aim of all education, is to increase the range of life choices available to the individual in modern society. The four conditioning programs are suggested because they do not narrow the outlook, restrict the exploration, reduce the boundaries of social experience, or limit the individual's habits of learning to those that furnish immediate rewards. They are not designed to seal individuals into niches but to enhance their autonomy and increase their capabilities for further learning and development. Parents and teachers should point out and demonstrate that each learning program really adds to the individual's range of choice and his ability to follow the lines chosen. It opens new doors rather than closing any. In seeking learning opportunities, children and youth can be helped by understanding this basic educational purpose.

When the child understands what conditioning is, how it works, and for what kinds of learning objectives it is effective, he is able to employ conditioning for his learning goals that can be achieved in this way and to avoid conditioning himself or being conditioned when this would produce an undesirable result. For example, if he wants to develop skill in a new sport, conditioning is likely to be effective and efficient. If he desires to develop proficiency in hearing and speaking another language, he can use conditioning. On the other hand, he may note that he is being conditioned, without realizing it, into a narrowly circumscribed social and civic life, and he may then decide to develop a consciously planned program to extend his understanding of and participation in civic affairs.

Thus far, the discussion of early childhood conditioning that can provide a sound base for education emphasizing responsible choice has not raised the question of the likelihood of such a development of child rearing on a wide scale. This is very hard to predict. In the past, children have been taught to believe and act very differently. Group loyalties have been inculcated, ethnocentricism has been developed both consciously and unconsciously, children's curiosity has been stifled by many parents and teachers. Children have been taught answers rather than encouraged to carry on lifelong inquiry. Security has come from acquiescence and from remaining at home rather than from exploring and questioning. Yet, in spite of this situation, the widespread evidence of world change, social crises, and the development of international communication are making more difficult the efforts of those who would condition the young to respond to the old ways. The initial revolt is from the old to a new, rigidly conceived panacea. This, however, entails no great change in child-rearing practices. The usual political, economic, or social panacea substitutes one dream for another rather than helping children to explore experiences and to trust the evidence of experience. But even in countries where new doctrines are substituting for old ones, the awareness of the people of the relevance of experience is growing. Increases in the practice of the kind of conditioning outlined above can be anticipated because they are in harmony with the growing awareness of people in many parts of the world. But several decades seem likely to elapse before most men can feel secure without the convictions that are built around narrow group loyalties that are often divisive. The experience of minority groups in the United States as more avenues of participation are opening to them is that their sense of identity and respect can easily be lost unless undergirded by narrow loyalties and threats of aggression. Especially when the real position of a group is an intolerable one does it seek the assurance of ultimate success and ultimate victory over its real or imagined oppressors. These narrow loyalties and divisive beliefs are not likely to wither away until tribes and other groups are conscious that they are achieving and can find ample evidence for identity and respect in the real world.

The prospectus for this Conference states: "Among the options for critical examination are, for example, the development of planet-wide programs of basic education in the prehistory, history, and potentiality of human society, supplemented by exchange of students on a massive scale." This suggestion offers great possibilities as an educational program to follow earlier stages of conditioning. At present, the schools in all nations lack authentic materials regarding the development of human society and the potential evidenced by this development. Most of the material used with children and youth is either grossly distorted myth or a narrow selection of the field of study based on ethnocentric perspectives or limited notions about the important areas of man's achievements. Furthermore, most such material fails to associate the student in seeking to understand man's condition, his aspirations, his striving, and his achievements. It leaves him untouched without having learned anything significant.

The exchange of students on a massive scale is likely to have a positive influence in developing world citizens, but the relatively brief time that a student spends abroad has, in the past, been largely occupied in activities entailing limited cooperation. Instead of the usual exchange program, a more promising option is a carefully designed work-study program in which students from various nations serve together on task forces responsible for achieving a goal recognized as important by all. This is one of the values of the Bristol Channel Rescue Program, for which the students at Atlantic College are responsible.

The youth there are from many nations. When a ship is wrecked near the college and people are thrown into the heavy seas, the work of the students in rescuing them is serious and effective cooperation is essential. In such associations, students get to understand themselves and others much more than in the usual school situation. Of course, not all international student groups can have rescue experience, but it should be possible to find hundreds of types of important tasks that can be undertaken by teams of youth in ways that furnish the invaluable learning that comes from working together under stress.

In the university years, international teams of students engaged in inquiry and problem solving in areas of great social or intellectual significance can contribute greatly to world understanding on the part of young men and women as well as accomplishing the purpose indicated in the prospectus: "No doubt a principal theme should be the growth of knowledge and the arts through freedom both to differentiate and to integrate."

The prospectus raises two other important questions: What specialized education, if any, beginning at what age, is desirable for leaders in government and politics? By what strategies can the rising generation be smoothly incorporated into adult roles without sacrificing youthful drive and inventiveness?

I shall comment on the second question first, because my response to the first builds on the second. In most highly industrialized countries, the demand for a large percentage of persons with 10 to 16 years of education has been met by supporting the continuous schooling of youth from five or six years of age to about 20 years. Typically, they are kept out of the labor force and other opportunities to participate in the adult world until their formal schooling is completed. Yet adolescents are conscious of their physical and intellectual maturity. They want tangible evidence that they are able to fill adult roles, but they are isolated in an adolescent island without any sense of significant responsible participation in the adult world.

What is needed is to open the doors of secondary school and college to the wider community and to arrange for students to spend part of their time in responsible activities, in work, in civic affairs, and in social service. The kind of cooperative education in which college students alternate quarters at work and quarters in residence has been shown to result in increased motivation to study and learn, increased personal and social maturity, and easier transition to full-time adult roles than is characteristic of typical college programs. The college employs coordinators to work with employers and students to provide for increasing complexity of work responsibilities and significant connections between the job experience and the college curriculum. Some programs of cooperative education are found in high schools, but this does not involve a large proportion of students. A wide expansion of responsible experience, like the suggestions for student exchanges, would add greatly to the positive induction of young people into the larger society.

Given such a background of second-school and college education, the specialized education for leaders in government and politics starts when the individual seriously aspires to such a role and begins to explore the possibilities through reading, observation, and interview and by utilizing the choices available to him in cooperative work, social service, and civic participation in school and college. With this background, his university specialization can be greatly strengthened through intern experiences in government and politics—locally, nationally, cross-nationally, and in international organizations like the United Nations. Assuming that he has had an international exchange experience in his youth, his plan of internship should associate him in alternating experiences in political or governmental organization within his own country and those without. The kind of planet-wide perspective required of a political leader for the future can better be attained by working in situations of more than one national perspective and also in international organizations in which no single national view is adequate.

The intern experience should be more than simply working in an organization. It should also include periodic seminar meetings in which a group of interns are able to report what they are doing, what they have observed, and what they see as problems and can discuss possible interpretations in order better to understand what is going on and to gain a sounder basis for working effectively in such agencies.

In summary, the position presented in this paper is that conditioned-response learning can be employed in early childhood training to build a disciplined base for voluntary, consciously directed learning. A combination of conditioning and a system of consciously directed learning is necessary to develop individuals adequate to meet the complex requirements of a world society and able to develop their own unique human potential.

ANTHROPOLOGY TODAY

Margaret Mead

Curator Emeritus of Ethnology
The American Museum of Natural History

I

Anthropology as the science of man has manifested an extraordinary number of changes of direction and emphasis since it began to emerge as a unified science in the early quarter of the 20th century—in the U.S. in the work of Boas, in the United Kingdom under the leadership of Rivers and Haddon, in France in the work of Durkheim and Mauss, and in Germany and Austria with the work of Graebner, Ratzel Frobenius and Father Schmidt. The number of types of enquiry included differed from one country to another in the extent to which paleontology, prehistory, archaeology, the comparative anatomy and physiology of living man, comparative study of primate or mammalian behavior, and theories of evolution were blended or combined uneasily with studies of the forms of social, political and economic organization. The study of earlier historical periods combined with that of isolated folk cultures was important on the continent, while in the United States anthropology became primarily the study of the language, archaeology and contemporary behavior of the surviving nonliterate peoples of the world. Analysis of the customs and languages of peoples for whom we had documentary materials of any kind was considered the concern of other disciplines. Thus, the study of the great civilizations of China, India and the Middle East were treated as the purview of specialists, and the economic and social behavior of members of contemporary European or Asian cultures was regarded as the subject matter of sociologists, economists, social psychologists, students of human growth and development, specialists in contemporary linguistic theory, or students of technical development.

Crossover points came in the field of applied anthropology, when it was necessary for purposes of government, war or directed social change to consider both primitive and modern economic or educational systems, or systems of contrasting values, or when questions of race and disease brought the characteristic disease picture of a modern population into line with an understanding of the evolutionary value of such a condition as sickle cell anemia—protective in its heterozygotic form for populations exposed to malaria, dysfunctional when malaria is brought under control.

Nevertheless, in spite of such crossovers between disciplines, there has been a strong tendency for anthropology to do intensive work with many preliterate peoples and to leave those cultures in which writing is an outstanding characteristic to another set of specialists—historians, economists, sociologists, demographers, etc. This has been particularly true when it came to specialized methods of research. Although, for example, the historian was trained in the use of documents and inscriptions, the archaeologist in the combination of archaeological remains and inscriptions and documents, the sociologist and economist in the analysis of complex written or tabulated data, the anthropologist maintained and maintains a closer relationship to all the materials that provide him with the basis of his knowledge and theories. He studies those cultures that can be approached only through a first-hand study of living members, specializes in languages that he is the first to reduce to written form, excavates in fields in which there are essentially no decipherable records, measures and somatotypes living men, and includes, today, the study of primates in the wild. His techniques remain essentially direct firsthand observation, whether recorded with pencil and paper, tape recorders, cameras, or the apparatus of videotaping. When he explores modern societies, although he may invoke documentary materials, statistical tables, and laboratory experiments, his methods remain essentially the same—observation of the living scene, the treatment of written materials such as historical records of party congresses or movie scripts as if they were records of living behavior, so that he can analyze them with the help of native informants as he would had he been present as an observer. He does not, as a rule, make any attempt to master the techniques of the human sciences that rely on documentary or laboratory techniuqes. If he wishes to use them, in working on modern problems, he must either work in teams with numbers of other disciplines or rely on consultation and secondary sources.

There has been a continuing interest in dealing with wholes: with mankind as a species, the single hominid species now existing on this planet, with many variations in climatic and breeding conditions, but still essentially one species, exhibiting complete intraspecies fertility and hybridization between varieties as a source of strength; with human history and prehistory as one adventure extending over hundreds of thousands of years as early hominids spread out and human cultures evolved by virtue of a series of biological changes—the opposable thumb, erect posture, changes in the brain, neotony—and by virtue of cultural inventions—language, tool using, use of fire, forms of social organization, horticulture—so that man's earliest beginnings can be systematically related to his contemporary history. Language also is treated as having arisen in the course of human history. When comparative studies of primates are admitted, it is again to see development of species-characteristic behavior within groups of related forms. Any contemporary observation, any recovery of the past through the paleontological or archaeological record, is fitted implicitly into this entire scheme, man developing on this planet, spreading out, inventing, borrowing, earlier forms being superseded by more elaborate forms of social organization, religion and technology developing in parallel with the growth of cities and an increasing population size. Essentially there is no observation of any kind of behavior or record of behavior, from the shape of the human jawbone and the contour of a hand axe to the form of a gesture or the way of holding a knife or a pen, that cannot be fitted into this whole.

In his search for data, for bodies of materials that can provide a growing corpus on which theory and understanding can be based, the anthropologist has, therefore, sought for wholes that were most manageable and that could be subjected to the most complete analysis—small tribes, unwritten languages, small communities, the archaeological sequences of small isolated peoples, such as the Eskimo or the American Indians of the Southwest—where he has sufficient command of his material to see the interrelationships of all the parts. In the objection to fragmentation of his subject matter, he stands with the vanishing school of historians and classical scholars who attempted to master all of the existing material on a period in history, with those naturalists who were specialists in all of the fauna and flora of small islands, and with those students of personality who are willing to give hundreds of hours to the exploration of the psychological experiences of a few patients or subjects.

These emphases are particularly apparent in the work of Franz Boas, who recurrently tackled new aspects of the diverse subject matters of anthropology, either himself or by encouraging his students to make in-depth probes into new places. Each probe, whether it was study of the design of Eskimo needle cases or the shift in idiolects among different households within the language spoken by a small Northwest Coast Indian community, was as thorough as possible. In the pursuit of such detailed, controlled, particular data, the whole was not lost sight of, although further data that would be needed to connect the results of these single in-depth studies would not be collected for many years. Comparably, physical anthropologists interested in early man have worked with single fragments—a jawbone or a few teeth—using them as clues to a whole skeleton, a whole that could be conceptualized before it was known. New data would alter many of the factual conclusions, change the temporal horizons, alter our ideas as to where man originated, how he wandered and differentiated, where an invention was made, a song first sung. But the grand design remained to give form to the persistent search.

In the face of the growing fragmentation of both the sciences and the humanities, and as more materials became available for the exercise of scholarship, it has been increasingly difficult to maintain this integrated approach to man, to include man's biological history as well as his cultural and social history within one area of study. Periodically, we attempt to draw together, to reassert, in conference or curriculum, the inclusiveness of our subject matter, and to define the way in which our discipline's specialized conclusions relate to the whole of mankind. Such integration takes different forms: a real, as in the series of International Congresses of the Americanists drawing together all those who studied the early cultures of the Americas; international, as in the International Congersses of Anthropological and Ethnological Sciences, where European folklorists and etymologists find themselves jostling elbows with English and American students of aboriginal Australians and New Guinea cultures and languages. Thus, the difference in emphasis in different countries provides new levels of inclusiveness out of new forms of specialization peculiar to different areas of the world. More recently Japanese, Chinese, and Indian anthropologists have been making their specific contributions to such specializations, as are the scientists and scholars of the U.S.S.R. and the other countries of Eastern Europe. Scandinavia has maintained a consistent tradition of specialized interest in early European culture and in the Eskimo. Countries have differed both in their own traditional area of concentration and in the subject matter, partly because of the archaeological sites, primitive or tribal peoples, or exotic peoples to which they have the readiest access. Thus, it is as a former colonial power with a rich background in Indonesian culture that Netherlanders contribute to studies of jurisprudence, or the Japanese to the early history of their island possessions. Integration has also been provided by such attempts as the 1952 Wenner-Gren Foundation International Symposium on Anthropology, which reaffirmed the unity of the historical subject matter and introduced new areas of interest and specialization.

So also the Human Relations Area Files and the system of classification of cultures developed in Paris provide new ways in which diverse subject matters, many geographical areas of the world, and new methods of research—all essential to the wholeness of the subject—can be brought into new relationships. Anthropology, more than any other discipline except geography itself, is informed and extended by materials drawn from different geographical settings, by data on man's body and man's culture that are closely related to the climate in which he has lived, the nature of the soil he has tilled, the forests in which he has hunted, the materials and terrain that have guided and shaped his houses and towns and cities, and the landscape and weather that have given form to his cosmologies. As each new nation enters the modern world and as its members engage in the study of the human sciences, each group brings with it a special, hstorically given ingredient, without which the subject as a whole would be different and less complete. Mexico, through the work of both Mexican anthropologists and anthropologists from other countries, is an outstanding example. Each new country, like each new archaeological find that opens up yet another civilization, or each study of a different primate group that opens up new psychological possibilities, changes the shape of the developing whole, a science of man that parallels, reflects on and reflexively influences man's fate on a planet that is rapidly becoming known in all its furthest reaches.

At any given moment in history, the relevance of a particular anthropologist's work, or that of a particular anthropological school to the concerns of the other sciences or of the informed and questing laymen, varies greatly. One may encounter in the course of an afternoon at a large conference specialists on the diffusion of some design or minute object of decoration, on the problems of linguistic form, on the archaeology of a continent, the present social condition of a new nation, or the future of this planet. Governmental agencies, intent on finding advisors on large and small policies, often seek out anthropologists and find themselves inexplicably saddled with an anthropologist whose level of interest is quite different from their needs. We are bound as anthropologists to each other by a common set of assumptions about the unity of mankind, the nature of man's history, the interrelatedness of human culture, the nature of learning, of cultural transmission and cultural borrowing, and the relevance of the smallest detail to the largest whole. But this is not always immediately apparent to those who integrate ideas in quite other ways.

II

In this paper I shall address myself to developments in anthropology between the Wenner-Gren Symposium in 1952, which was international and inclusive, and the present time. The end of World War II saw a very rapid growth of interest in anthropology. Academic departments proliferated, research funds became available, purposeful technological change posed new practical problems, just as the war had posed the problems of studies of culture at a distance, and there was also recognition of the need to analyze the cultures of modern industrialized societies as well as primitive and exotic ones. But the 1952 conference, although it was called international and included a few Asian delegates, was still, in its published form, primarily Euro-American in representation.

So a first significant development has been the inclusion of Asian anthropologists and renewed communication with the Russian anthropologists, particularly at the Fifth International Ethnological Congress in Philadelphia, 1956, the meeting of the same Congress in Moscow in 1964 and a later Wenner-Gren conference in 1968, which attempted to bring together representatives of different countries to complete a statement of what each national school could contribute.

Parallel to these conferences were the Pacific Science Conferences, which brought together anthropologists with circum-Pacific interests, and a wider interdisciplinary representation from the biological sciences, especially those interested in ecology.

The International Biological Program, focusing on parallel enquiry into issues that needed to be studied simultaneously and on the concentration on vanishing natural situations, has also brought biologists and anthropologists close together; the UNESCO Conference on Biology as History, held in Chichen Itzá in January, 1969, and the search for optimal solutions rather than for the maximizing of single variables, which is a major contribution of biology to contemporary thought, and combined the anthropological emphasis on functional wholes and cultural relativity with complex, multidimensional models of biology.

It is possible, then, to say that anthropology can make significant contributions to the unification of thought as the specific contributions of different major cultures are made conscious through anthropological forms of cultural awareness, to the unification of the sciences and humanities and policy sciences by providing a basis from which the more specialized disciplines can take off, and to the development of a planetary ecological model that includes man by extrapolation from an approach that includes every aspect of small and self-contained societies. The use of macro-approaches, in which former work on preliterate peoples, unwritten languages, and small inbreeding groups, can be used as models for meeting planetary problems, complements the contemporary search among anthropologists for small units that will facilitate crosscultural comparison in which mathematical methods and computers can be used. Anthropologists have been preoccupied for the last 70 years with the problem of adapting a science based on observations *in vivo* and dependent on historical circumstance rather than experiments *in vitro*, adapted to the demands and methods of experimental science. There have been a variety of premature attempts at codification of data too various to sustain the coding, and at gross statistical methods that have yielded only gross results, often followed by reactive returns to more humanistic, less precise, methods of description. The advent of the new technologies of recording—tape, film, videotape, and oscillographic and sonographic techniques, etc.—has gradually made it possible to record human behavior in ways that are relatively independent of individual intervention. And the appreciation of the relationship between the individual researcher's integration, the enormous quantities of minute observations, and the capacities of computers has made possible new approaches to the problems of quantification of such extremely complex data. Prior to the development of recording and computers, we were dependent on the education of individuals who could deal with these minutiae and who, if exposed to different cultures or periods, sequentially, could produce viable hypotheses. The size of the communities with which anthropologists dealt, the fragility of systems that had already begun to disintegrate under contact, the brief duration of the periods of observation, and the coincidence in life span of the anthropologist and that of his human subjects, all militated against the assemblage of data with sufficiently long runs and large enough samples for traditional types of quantitative analysis.

Two developments, in addition to the development of adequate forms of instrumentation, have made enormous advances possible. One is the conceptualization of two methods of analysis: one *etic*, that is the use of crossculturally valid units—as phonetics—in which the particularities of any cultural system can be analyzed for descriptive and comparative purposes; and the other *emic*, in which the distinctive historical style of a culture can be defined in its uniqueness, by using etic units for translation into crosscultural comparabilities. This overall methodology has been supplemented by a number of other methods developed for the intensive microanalysis of small groups and complex materials, like cluster analysis profiles and factor analysis, which have yielded impressive results in the study of cantometrics and choreometrics and made it possible to establish a world picture of man's behavior in greater depth than we have ever been able to do before. The concept of the *idioverse*, each precisely specified individual version of a set of cultural constructs, can be expanded into a large number of sets to deal with the hitherto perplexing problem of how to relate the individual to society through the medium of differentially shared cultural constructs.

Systematic analysis of aspects of human cultural behavior—paralinguistic, kinesic, proxemic, interaction as measured by time—relationships between actors and something acted upon—is beginning to yield useful units, *kines, proxemes, actons,* or computerized methods of parallel analysis, as in the machine analysis of the Chapple Interaction Chronograph. As more of these units are identified, we come closer to methods that will be suitable for the rapid analysis of small samples of complex behavior, as well as for handling the enormous amounts of information made available by modern methods of recording and made necessary by the size and complexity of the world's population.

It is also significant that the integrating tasks performed by single historians, single scientific innovators and single anthropologists in the past are now becoming impossible in today's world for two reasons: one, today's young people, reared within a set of fragmented, internally inconsistent, multimedia bombardments from every part of the globe and every school of thought, are currently imperfect human instruments to perform such integration; and two, the corpus of observations itself is too massive and complex to be handled by any single human brain. Only methods of observation, codification and analysis that call on modern recording and machine analysis have any hope of making a massive contribution to the solution of today's problems.

III

Anthropology has passed through successive periods of cooperation with the other human sciences. In its earliest period it provided materials for speculative constructs about early man and early cultures. This was followed by a period of purposive correction of culturally limited hypotheses about such matters as stages in human development, the nature of human perception, inevitable sequences of economic or political behavior, and linkages between different aspects of culture, the laws of memory, or the nature of thought. This corrective phase, which lasted substantially until World War II, was followed by a third phase, in which anthropological research began to generate hypotheses about human behavior that were not primarily reactive to earlier, more culturally limited theories. A fourth phase is now developing, in which it is possible to draw on anthropological research or to construct new research that will generate not merely hypotheses about human behavior but also new models for culture building in the future, new political, economic, and social forms.

The addition of each new function has not, however, meant that former functions were superseded. Speculation—on the basis of new knowledge of early precursors of man and early forms of man, and contemporary studies of primates—still will make a substantial contribution to our handling of contemporary problems. The study of various contemporary primitive forms of self-identification, warfare, and territoriality provides a necessary and useful corrective to hypotheses about the innate nature of human aggression, or the limits to human crowding, or the limits to the exercises of human intelligence. Experience of different forms of perception and cognition widen our view of the different kinds of human intelligence available and unused or distorted or destroyed within our contemporary forms of civilization. All of these contribute to the design of a world in which the most accurate and extensive estimate of human capacities that we can produce is essential to the success of cultural building.

Unlike the present style within which the natural sciences are organized, where the past is treated as a series of discarded and irrelevant paradigms, anthropology, dealing as it does with precious records, the partial, unrepeatable fragments of man's behavior, must continue to include and use the earliest findings as well as the most recent and to return over and over again to the same materials as new methods of analysis and new scientific formulations become available. Because anthropology deals articulately with cultural change, it should be, and sometimes is, easier to include in new structures the relationships between past theory and past data collected within the theoretical structure than it is in sciences where culture change is not a substantive part of the subject.

Such relationships are illustrated by the extent to which anthropological materials—by their very nature unreplicable because any culture event simple enough to be replicated exactly yields too little information—can nevertheless function as experiments. If adequately recorded by the new technologies, the same event can be reanalyzed many times in many different frames of reference. Such carefully recorded complex events not only function as experiments but, just because they were collected without any infusion of a current paradigm, provide a continuous form of correction of the extravagances of any fashionable point of view. We have found that it is useful to take photographs in depth, so that the behavior of individuals in the background on which the photographer did not focus can be used as a corrective for the interpretation of the behavior on which he did focus, with increasing degrees of hypothesis formulation during the course of a long field trip. In the same way, earlier work can be used as background correctives for current, focused research.

IV

The uses of anthropology as one of the disciplines that, in combination, must provide us with solutions of worldwide problems are, at present, seriously compromised by circumstances that formerly constituted one of its major strengths. In the face of chauvinistic theories of national, racial or religious superiorities or inferiorities, anthropology has depended for the very heart of its principle paradigm—the concept of culture—on recognition of the psychic unity of mankind and the right of respect to all human cultures. This has meant that the anthropologist has based his research on collaboration; informants have been trusted collaborators, not subjects or objects. Subterfuge and the use of deception, of stooges, of falsification has been rigorously eschewed for both practical and methodological reasons. Although there has been a widespread and generous solicitude for the peoples he studies, the anthropologist has attempted to avoid the kind of political partisanship that makes it necessary to embrace totally the cause of any people, no matter how oppressed, in disregard of the welfare of the larger society.

While a mere recitation of these standards of behavior can be presented in a favorable light when compared with the procedures of many of the other human sciences—which, while deprecating the need for human experimentation have also treated human subjects like guinea pigs—there is a growing confusion in the minds of leaders of emerging nations and in the minds of young anthropologists themselves between the ethics of the past and the ethics required in a world in which the search for self-identification competes with accusations of racism against all other groups with self-identifications. Standards of absolute and uncompromising moralities in human relations call for an uncompromising espousal, for example, of the cause of black South Africans, even though it might lead to the massacre of white South Africans, and to the assumption that work done under colonial auspices in the past was necessarily contaminated and exploitive of the peoples under colonial rule who were studied by anthropologists. Records of earlier and more primitive phases of cultures now reaching nationhood, which the anthropologist made as an obligation to the people themselves and to his science, are often found incompatible with the current versions of a new nation's past. As the insistence on the ethical position that all racial differences should be ignored has been succeeded by an insistence that racial differences should be reified, emphasized and given specific recognition, work done within the older anthropological conventions become suspect. Other complications in the last 30 years (Nazism, Soviet dogma about sequences of human evolution, opposition to Lysenkoism, the Civil Rights movement in the United States, anthropological interpretations of poverty) have all conspired to confuse the understanding of the anthropologist's role by others and by himself. Responsible objectivity and commitment to the longest time perspective and the largest whole that he can envisage, responsibility for all "foreseeable effects," are difficult to maintain within this rapidly changing world scene. Certainly, in any planning for the participation of anthropologists in world policy planning, these changing ethics must be be taken into account.

LEGAL BASES FOR SECURING THE INTEGRITY OF THE EARTH-SPACE ENVIRONMENT

Myres S. McDougal

Yale University, New Haven, Conn.

Like Wordsworth's world, the facts about man's contemporary damage—and threats of even more perilous future damage—to his environment are almost too much with us. They scream in horrifying detail, not merely from the face of nature but from every medium of communication. In urgent summary, United Nations Secretary-General U Thant finds a mounting "crisis of worldwide proportions," with portents long apparent "in the explosive growth of human populations, in the poor integration of powerful and efficient technology with environmental requirements, in the deterioration of agricultural lands, in the unplanned extension of urban areas, in the decrease of available space and the growing danger of extinction of many forms of plant and animal life." [a] When one recalls also the accelerating damage to the oceans, it is not surprising that the Secretary-General should conclude that "if current trends continue, the future of life on earth could be endangered." [b]

Similarly, although our need for new and more precise information is enormous, our knowledge about the causes of all this damage, both actual and potential, appears to be increasing. Ecologists have come to emphasize what community planners have long known, that there is a maze of complex and intimate interdependences—an "indivisible web" of interrelationships—both among the features of the natural environment such as air, climate, topography, soil, geologic structure, minerals, water resources and access to waters, natural vegetation, and animal life and between such features and the institutions and practices by which man seeks to satisfy all his many social and psychological needs and demands, as well as basic bodily needs for nutrition, procreation, shelter, safety, movement, and so on.[c] These interdependences extend through many different interpenetrating communities, from local or minute to global or earth-space in range. It is in the violation of these interdependences—in the transgression of many different resources, technological, and utilization unities —that the root causes of damage to the environment are beginning to be revealed.

There would appear, further, to be a growing consensus among the peoples of the world about the appropriate overriding goals for general community action in lessening the near-disastrous damage with which we all threaten each other. The emerging aspiration of mankind is not so much for some simple conservation of resources or environment in a pristine, untouched state of nature as for an appropriately conserving, economic, and constructive employment of resources in the greater production and wider distribution of all basic human dignity values. In many contemporary conceptions, the resources of the globe are increasingly regarded as the common patrimony of the whole of mankind; practices in the exploitation of resources are being assayed in terms of their aggregate consequences for all who are affected, and costs-benefits analyses are being extended beyond mere quantitative calculations about wealth to qualitative assessment of impacts on the shaping and sharing of other representative values, such as power, enlightenment, respect, health, skill, rectitude, and affection. This would appear to be the empirical reference of the somewhat amorphous demand, so often expressed in popular discussion, that resources and environment be protected and maintained for improving "the quality of life."

The invention of effective remedial measures—of appropriate technical solutions—for the better securing of these overriding goals of the general community would not, once again, appear to be beyond the reach of man's creativity. The damage is not yet universal or irreversible, and the same science and technology that contribute to the difficulties also immensely enhance the potentialities for alleviation. In our national communities, many of us have had a rich and enlightening experience in the allocation, planning, development, and control of resource use for multiple-value goals. If, however, appropriate account is to be taken of the interdependences that pervade both the natural environment and man's practices in the shaping and sharing of values, any remedial measures or technical solutions that would be effective through time and for consequential geographic areas must eventually extend to planning, development, and controls that are comprehensive, integrated, and rational for the whole global community, as well as for its many internal communities: no national community today can be an island in a universe of interdetermination.[d] In addition, it is obvious that if such comprehensive planning, development, and controls are to be achieved and made to serve the overriding goals of both maintaining a secure environmental base and promoting and augmenting human dignity values, many delicate and continuing adjustments will be required in the management of processes of authority and effective power at all levels of government, from local through national and regional to global.

For specialists in law, the important question posed, in this gradual identification of the problem, is: What are the creative potentials in the management of processes of authoritative decision for contributing to the clarification and implementation of the common interests of the peoples of the world in protecting, and securing the more constructive enjoyment of, their shared environmental base?

It is the thesis of this paper that these potentials are sufficiently high to afford ground, not for complacency, but for at least a modest optimism.

It is not, of course, our suggestion that existing processes of authoritative decision, or the governmental structures in which such processes occur, are adequate to secure mankind's basic goals in relation to its environment. What we would suggest is rather that we do today have sufficient general knowledge about, and skills in, the management of processes of authoritative decision and sufficient assets in our inherited, and rapidly improving, global constitutive process and presently established public order policies about the control of resources greatly to encourage the expectation that, if the perspectives of the effective élites of the world can be appropriately shaped, much more rational and economic processes of authoritative decision can quickly be constituted and managed for better securing basic goals.

The full documentation of this thesis would require performance of a sequence of intellectual tasks, including:

(1) the detailed specification, in their context of causes and consequences, of the more important problems arising from man's contemporary interaction with and exploitation of his environment;

(2) the clarification in detail, from the perspective of an observer identifying

with the whole of mankind, of basic general community policies in relation to each of these particular problems;

(3) a survey of past experience, of prior trends in decision, at all levels of government, from local to global, in terms of approximation to clarified policies;

(4) an investigation of the factors that have affected past decisions on particular problems;

(5) the projection of probable future decisions and conditioning factors in relation to particular problems; and

(6) the recommendation of new alternatives in constitutive process and public order prescriptions for the better securing of clarified policies

The most that can be attempted here is an impressionistic indication of the basic legal assets at our disposal for undertaking inquiry and, ultimately, action. These assets include, as suggested above, our general knowledge about, and skills in, the management of legal processes, the existing global constitutive processes of authoritative decision, and the presently established public order policies relating to the allocation, planning, development, and exploitation of resources. We will glance briefly at each of these different types of assets and note certain possible directions toward improvement.

PERSPECTIVES ABOUT, AND SKILLS IN, LAW

The increasingly predominant theory ("jurisprudence" or "philosophy") about law today is explicitly sociological or policy-oriented in emphasis. In this conception, law is not some frozen set of pre-existing rules or arrangements that inhibits constructive action about environmental and other problems but, rather, a dynamic and continuous process of authoritative decision through which the members of a community clarify and implement their common interests. Rejecting the mysticism of natural law emphases, the fatalism of historical emphases, and the arid technicality of analytical or positivistic emphases, this contemporary sociological or policy-oriented conception seeks to facilitate the bringing to bear upon community problems, within the constitutive processes through which common interests are clarified and implemented, all the findings and techniques for modern social and physical science. The major features of this emerging emphasis may be briefly indicated.[e]

Observational Standpoint

In contemporary inquiry about law we have learned to distinguish the standpoint of the scholarly observer, primarily concerned with enlightenment and skill, from that of the decision maker, primarily concerned with power—that is, for participation in the making of effective decision. The scholar, detaching himself and his procedures from the events (including the purposes and procedures of the participants in processes of decision) that he has under observation, seeks to create and employ a *functional* theory both to establish a comprehensive and relevant focus of attention and to facilitate performance of certain indispensable intellectual tasks in reference to the flow of authoritative decision and of the *conventional* theories employed to explain and justify such decision. The identifications the scholar seeks are not merely with some single parochial community but, rather, with the whole of man's many different—often concentric, and always interpenetrating—communities and the enlightenment he seeks is that relevant to clarifying and implementing the common interests of all.

Focus of Inquiry

Eschewing definitional exercises, contemporary jurisprudence aspires to a focus of inquiry that is both comprehensive and selective, effectively relating authoritative decision to the larger social and community processes by which it is affected and which it in turn affects. This comprehensiveness and selectivity are sought by certain subordinate, interrelated emphases.

(1) A balanced emphasis upon perspectives and operations: The central focus is squarely on *decision*, as effective choice, composed of both perspectives and operations. Exaggerated emphasis is avoided either on technical rules of law, too often assumed to be an accurate expression of community perspectives, or on bare, behavioristic operations, the choices in fact made and enforced by threats of severe deprivations or promises of extreme indulgences. In balanced inquiry about patterns in both perspectives and operations, the manifest content of conventional rules of law is pierced for examination of the choices in fact made, while perspectives are still subjected to systematic and realistic study as among the factors importantly affecting decision.

(2) Clarity in conception of both authority and control: Authoritative decision is distinguished from naked power, and law is regarded not merely as decision, but as *authoritative* decision, in which elements of both authority and control are combined.

Authority is found, not in theological or metaphysical or allegedly autonomous abstractions, but rather—in a conception at least as old as the pre-Socratic Greeks—in the empirical perspectives of community members about who is to make what decisions, in respect to whom, in accordance with what criteria, and by what procedures.

By control is meant participation in effective choices, in choices that are in significant degree put into practice; control, in this sense, may obviously be based on any of the wide range of community values.

(3) Comprehensiveness in conception of processes of authoritative decision: Relevant inquiry extends beyond occasional, isolated choices, to the whole, continuous, ever-changing *process* of decision by which a community shapes and shares its values. In any community, the process of authoritative and controlling decision, which is the appropriate reference of "law," is seen to be composed of two different kinds of decisions: the "constitutive" decisions that establish and maintain the most comprehensive process of authoritative and controlling decision; and the "public order" decisions that emerge from the process so established for the regulation of all other community value processes.

The constitutive process of a community comprises the decisions that characterize and identify the appropriate decision makers, specify and clarify oasic community policies (which may be demanded in varying degrees of intensity), establish necessary structures of authority, allocate bases of power for sanctioning purposes, authorize procedures for making the different kinds of decisions, and secure the performance of all the different kinds of decision functions (intelligence, promotion, prescription, invoking, applying, terminating, and appraising) necessary to making and administering general community policy.

The public order decisions of a community are those that shape and maintain the protected features of its different value processes. These include the decisions by which resources are allocated, planned, developed, and exploited; by which populations are protected, regulated, and controlled; by which health is fostered or neglected; by which human rights are protected or deprived; by which enlightenment is encouraged or retarded; and so on.

(4) Explicit relation of law to social process: Since authoritative decision is a response to events in social process, is affected by such events, and, in turn, has effects on the future distribution of values, a comprehensive set of value-institutional categories—making detailed, empirical reference to interrelations among people—is employed to locate authoritative decisions in the larger social processes that envelop them. By such categorizations the different types of claims people make to authoritative decision and the varying responses of established decision makers may be compared through time and across territorial boundaries in study of the factors that affect decision and of the public order consequences of decision.

(5) Location of law in its larger community context: In contemporary conception it is recognized that mankind today interacts on a global and even earth-space scale, not merely in the sense of a universal science and technology, but of an interdetermination with respect to all values. One component of this larger, if still primitive community, is seen to be a process of *effective* power, in that decisions are in fact taken and enforced, by severe deprivations or high indulgences, which are inclusive in their effects. Among these effective decisions it is observed that, although many continue to be made by naked power or sheer calculations of expediency, some are taken from perspectives of authority and achieve enough control to be of high consequence. Like the all-inclusive transnational social processes, this latter transnational process of authoritative decision—sometimes ill-described as "international law"—is seen to be maintained at many different community levels and in many different interpenetrating patterns of authority and control, in affecting and being affected by the value processes of all the component communities of the larger earth-space community. A global public order is, thus, seen to affect the internal public order of all particular communities, and the public order of each particular community to affect, in turn, the global public order.

Relevant Intellectual Tasks

In lieu of traditional exercises in derivational logic and concern for limited conceptions of "science," contemporary theory about law emphasizes the deliberate, systematic, and differentiated performance of a comprehensive set of intellectual tasks, each of which is relevant to problem solving about the interrelations of law and social process. The economic and effective performance of any one task requires its relation, configuratively, to the formulations and findings achieved by each of the other tasks.[f]

(1) Clarification of community policies: The most relevant clarification, avoiding circular derivations, deliberately seeks the detailed specification of

postulated goals, whatever the level of abstraction in their initial formulation, in terms that make clear empirical reference to preferred events in social process. The findings and techniques of each of the other intellectual tasks are employed to estimate the aggregate consequences of alternatives in choice and to relate such consequences to common interest.

(2) Description of past trends in decision: In supplement to conventional summaries of complementary rules and concepts, the description of past trends in decision is related to specific, detailed types of claims to authority, and trends are appraised in terms of degrees of approximation to clarified policies for constitutive process and public order. For the more effective comparison of decisions and their consequences through time and across community boundaries, the events that precipitate claims to authoritative decision, the decisions actually taken, and both the immediate and longer-term consequences of decision for the claimants and others are all categorized "factually," in terms of value-institutional references to social process.

(3) Analysis of factors affecting decision: In performance of the scientific task, hypotheses are inspired by the "maximization postulate" that responses are, within the limits of capabilities, a function of net value expectation, and emphasis in inquiry is placed on both predispositional and environmental variables. The techniques and findings of modern science are brought to bear in appraising the significance of multiple factors from culture, class, interest, personality, and previous exposure to crisis.

(4) Projection of future trends in decision: Expectations about the future are made as conscious, explicit, comprehensive, and realistic as possible. Developmental constructs embodying alternative formulations of the future are deliberately formulated and tested in the light of all available information. The simple linear or chronological extrapolations of conventional legal rules are subjected to the discipline of knowledge about conditioning factors and past changes in the composition of trends.

(5) Invention and evaluation of policy alternatives: All the other intellectual tasks are synthesized and brought to bear on the deliberate invention and assessment of new alternatives in policy, institutional structures, and procedures. Every phase of decision process, whether of constitutive process or relating to public order, and every facet of the conditioning context, are examined for opportunities in innovation that may influence decision toward greater conformity with clarified goals. Assessment of particular alternatives is made in terms of gains and losses with respect to all values and disciplined by the knowledge acquired of trends, conditioning factors, and future probabilities.

Explicit Postulation of Basic Community Goals

In recognition that policy choices are ineradicable components of any process of authoritative decision, a policy-oriented jurisprudence recommends that scholars explicitly postulate, and commit themselves to, a comprehensive set of goal values for the guidance of inquiry and decision. Postulation and clarification, rather than derivation from the premises of some particular faith, are recommended both for economy and for increasing the number of potential co-workers. The basic goal values postulated for preferred world public order cannot, of course, be expressive only of the exclusive, parochial demands of some particular segment of mankind, but when overriding goals are shared, particular values can admit of a very great diversity or functional equivalence in the institutional practices by which they are sought and secured.

The World Constitutive Process of Authoritative Decision

It is sometimes lamented that, while environmental problems are global in their reach, the processes of law are not. This is a profound misconception. The contemporary world arena does exhibit a constitutive process that, though it has not yet achieved that high stability in expectations about authority and degree of effective control over constitutent members that characterize the internal processes of some mature national communities, still affords, in at least rudimentary form, all the basic features essential to the effective making and application of law on a global scale. In recent decades, this emerging transnational constitutive process of authoritative decision has been expanding and improving itself at an accelerating rate, and it would not appear that vast, and possibly grandiose, structural alterations are any more necessary for coping with environmental than for other problems. Conversely, environmental problems would indeed appear so global in their reach and so immense in proportion that a whole global process for the continuous clarification and implementation of common interest, and not merely some new specialized organization or cluster of organizations, is required for their management and amelioration. The basic features of existing constitutive process, as they suggest potentiality for the more effective management of environmental problems, may be briefly indicated.[g]

Participation

In recent decades, participation in world constitutive process, as in the embracing process of effective power, has been tremendously democratized— with not merely nation-states but also international governmental organizations, political parties, pressure groups, private associations, and individual human beings playing important roles. With this increase in the range of effective participants has come also a rapid proliferation in the number of territorial and functional entities demanding and being given voice. Similarly, a multiplying host of private associations, operating within the larger constitutive process, are increasingly international in membership, goals, and arenas of activity. Groups and individuals especially concerned with environmental problems have abundant opportunity to participate in all aspects of making and applying law.

Perspectives

The perspectives of the effective élites of the world, on which processes of authoritative and controlling decision must depend, would appear to exhibit both an increasing stability and a turgid, but perceptible, movement toward the demands, identifications, and expectations appropriate to public order of human dignity.

The demands of peoples upon constitutive process increasingly emphasize the necessity for protecting common interests, with the rejection of all claims of special interest against community. The provisions of the United Nations Charter, including Art. 2(4) and various ancillary articles, have made a tremendous contribution to the clarification, if not consistent implementation, of the common interest in minimum order (the minimization of coercion and the protection of reasonable expectations created by agreement and customary behavior). Other provisions of the Charter, the Universal Declaration of Human Rights, and the multitude of more specialized covenants about human rights have, similarly, done much to clarify the details of common interest in optimum order (the greater production and wider sharing of all values).

The rich experience of mankind—as expressed in customary development, United Nations resolutions, and specialized agreements—in clarifying and implementing common interests in the enjoyment of such great sharable resources as the oceans, airspace over the oceans, international rivers, polar regions, and outer space could fortify efforts to achieve a comparable, more generalized clarification for environmental problems, which embrace all resources, including even the land masses.

Increasing interactions on an earth-space scale would appear to be fostering expanding identifications among all peoples with the whole of mankind, as well as occasional, parochial, defensive reactions. Authoritative decision makers increasingly achieve an appropriate balance between inclusive and exclusive identifications.

The ineluctable spread of a civilization of science and technology carries with it, as one important component, as increasingly common map of reality and expectation about social process and environment for all men. With respect to environment, if not social process, expectations could become more contextual, realistic, and rational, and all men might come to see that they share a common fate.

Arenas

The structures of authority and other situations in which the participants in world constitutive process interact have exhibited in recent years both an enormous expansion and a modest movement toward organized, inclusive form. A principal contribution of the United Nations and of the great host of specialized agencies and regional organizations has been in the supply of a new abundance of diplomatic, parliamentary, mixed diplomatic and parliamentary, adjudicative, and executive arenas in which the other effective participants in world power process can interact. A comparable increase has occurred in the patterns of interaction established by burgeoning private, nongovernmental associations primarily dedicated to values other than power. In consequence, the interactions of the decision makers whose choices in sum create global policy have become more timely and continuous—i.e., less episodic and more alert and responsive to crisis.

Similarly, though some official arenas remain closed to some effective participants, there has been a general trend toward openness in arenas and a parallel movement toward making appearance compulsory for participants whose choices in fact affect community policy. Both openness and compulsoriness are promoted by increasing interdependences in effective power.

Bases of Power

While many of the more important bases for influencing decision remain under the relatively exclusive control of nation-states, there appears to be a modest trend toward allocating to representatives of the inclusive community the authority and other assets required for the better securing of both minimum and optimum order.

With respect to authority, one encouraging development is the attenuation of the concept of "domestic jurisdiction" (the exclusive competence of states) and the expansion of that of "international concern" (the inclusive competence of the general community). The distinction is now made in terms of the relative impact of activities on exclusive and inclusive interests, and the competence of the general community is being extended, through varying decision functions, to all matters of transnational impact. This extension of the competence of the general community is, of course, of particular relevance for the regulation of environmental modifications and practices in resource exploitation whose transnational impacts are obvious.

Similarly, though the myth abides that very little authority is being conferred on specific international organizations, the facts about peoples' expectations appear to be quite different. The actual authority of the United Nations, the specialized agencies, and the regional organizations would appear to be enormous, in the sense of the support they receive from the demands, identifications, and expectations of the peoples of the world.

The more traditional principles of jurisdiction, which allocate an exclusive competence for self-help among nation-states, are also coming to be interpreted in terms of the relative impact of activities on different exclusive interests. Under contemporary interpretations, states are authorized to protect themselves, by necessary and proportionate measures, against activities from the outside that substantially affect their internal community processes. This competence, too, would appear to be of direct relevance to environmental problems.

In the allocation of the effective control necessary to sustain authority, there is, similarly, a slow movement toward greater inclusivity in fact. It may be noted that an allocation of control designed to secure an appropriate balance between inclusive and exclusive decision making need not imply the complete centralization of control; a pluralistic distribution of values among the peoples of the world may better maintain an appropriate balance.

The worldwide spread of enlightenment and skills, facilitating perception of interdependences and common interests, and modern institutions of instantaneous communication have, of course, concentrated immense assets in the hands of inclusive decision makers. The expanding identities and loyalties of peoples, as transnational interactions accelerate, and the increasing internationalization of standards of rectitude and of respect for human rights add to these assets. More tangible assets may, perhaps, yet be found in the great sharable resources—such as the oceans, the airspace over the oceans, the polar areas, and outer space—which, despite recent inroads on behalf of special interests, are still largely under inclusive competence and control.

Strategies

Improvements in communication, expanding scientific knowledge and skills in observation, and accumulating experience in large-scale administration have all combined in recent years to facilitate a gradual rationalization of the procedures by which participants manage base values in performance of the different policy functions necessary to the making and application of law. The exploration of potential facts and potential community policies is being made more dependable, contextual, selective, and creative; the final characterization of facts and policies in prescriptive or applicative decision is being made more deliberate, rational, and nonprovocative; and the communication of the shared subjectivities indispensable to legal process is being made more effective.

(1) The diplomatic instrument: With the multiplication of new, more stable, and more continuous arenas, the old diplomacy of occasional official or élite communication is being transformed into a kind of parliamentary representation and activity, as in the United Nations. Important new multilateral conventions have been formulated and accepted for guiding and assisting diplomatic interactions and the making and performance of agreements. New procedures are constantly being devised and tested for executive and adjudicative arenas.

(2) The ideological instrument: The potentialities of instanteous communication about the globe, to mass audiences beyond élite groups, promise both greatly to increase participation in global constitutive process and profoundly to affect performance of many policy functions, such as—especially—promotion, prescription, and application.

(3) The economic instrument: Though direct control over resources and the management of wealth processes are still largely reserved to the exclusive competence of nation-states, the organized general community is acquiring an increasing experience in the promotion of economic development and in the management of credit and monetary policies. Some expansion of this role could greatly enhance the potential of the economic instrument in promoting desired world public order.

(4) The military instrument: Despite the failure of the original plan to establish within the United Nations a permanent military force for the maintenance of minimum order, the United Nations has had some minor successes in different parts of the world in assembling and employing small forces. Somewhat greater cooperation has been achieved at regional levels. Perhaps these successes could again stimulate plans for a more ambitious inclusive employment of the military instrument in support of general community policies.

The promise of the integrated employment by the organized general community of all four instruments of policy, in the management of all relevant base values, for the improvement of sanctioning process in the establishment and maintenance of desired world public order remains, of course, largely for future exploration. In the meantime, the more important sanction for transnational law, as for most national law, resides in the perception by community members of their interdependences and common interests and in their expectations about reciprocal, unilateral indulgences and retaliations in relation to such interdependences and interests.

OUTCOMES

In consequence of the gradual modification and improvement in all these varying phases of constitutive process there has been a parallel improvement in the culminating outcomes of the process in the different types of decisions taken. The relevant decisions appear to be becoming more comprehensive, in the sense of embracing all necessary policy functions; more inclusive, in the sense of extension toward participants and interactions affecting common interests; more rational, in the sense of conformity to the basic public order demands and expectations of the peoples of the world; and more integrative, in the sense of molding the potentially divisive claims of peoples into the perception and fact of common interest.

(1) Intelligence: The recent proliferation of international governmental organizations and the enhanced participation in constitutive process of political parties, pressure groups, and private associations have immensely increased facilities for the gathering, processing, and dissemination on a global scale of the intelligence necessary to rational decision. The emerging technology of observation and communication by outer-space instrumentalities offers still further augmented potentialities. The massive and complex intelligence required for effective management of environmental and resource problems appears to be well within reach but will require cooperation on a larger scale than heretofore throughout the entire constitutive process.

(2) Promotion: The increasing democratization of participation in world processes of effective power, the availability and openness of the new structures of authority, and the contemporary instrumentalities for communication have, similarly, brought a new comprehensiveness and intensity to the active advocacy of policy alternatives before authoritative decision makers. The ease with which demands can be formulated and propagated, and support mobilized, for the enactment and application of new authoritative prescriptions is already being demonstrated on a global scale in relation to environmental problems.

(3) Prescription: Historically, the making of transnational law has gone forward by way of articulated multilateral agreement and of unarticulated, habitual, cooperative behavior, from which expectations about authority and control are derived. The practices of the United Nations have both given a tremendous boost to these traditional modes of lawmaking and added a new dimension that reflects a closer approximation to parliamentary enactment. The activities of the International Law Commission and of the General Assembly, through its committees, have greatly rationalized prescription by multilateral agreement, as witness the important new conventions about the oceans and outer space. The opportunities afforded in the General Assembly for the representatives of many different communities to state their conceptions of prevailing law and to articulate these conceptions in formal resolutions have, further, greatly eased the historic burden of identifying customary law and clarifying its content. It is this latter modality of General Assembly resolution, greatly foreshortening the time necessary for establishing customary law and affording an economic mode for articulating consensus about common interest, that increasingly bears the hallmarks of parliamentary enactment. Clearly, if inherited prescriptions about the protection of the environment are inadequate, the prescribing process offers few impediments to their being made adequate.

(4) Invocation: Though some arenas remain closed to some participants, as the International Court of Justice at present is to individuals and nonstate entities, most participants today either have many arenas open to them or else can easily find a surrogate or champion who does have access, for stimulating the application of community prescriptions. The community member who would complain about the violation of prescriptions for environmental protection has abundant opportunities for a hearing.

(5) Application: Historically, the great bulk of the applications designed

to put general community prescriptions into controlling effect in particular instances have been made in interactions between foreign office and foreign office or in national courts. The fact that the same participants have had to be, alternatively, both claimants and appliers has been not so much a source of bias as a guarantee of aggregate decision in terms of common interest. In recent decades there has been a modest movement toward third-party decision—through international courts, arbitral commissions, and various structures of authority within international organizations—and toward compulsory attendance by participants whose activities are alleged to transgress community prescriptions. Important recent illustrations of this trend appear in conventions relating to the conservation of fisheries and to the law of treaties. If compulsory jurisdiction becomes acceptable in relation to environmental problems, there is no want of models for making it effective.

(6) Termination: The same developments in United Nations practice that have brought a new economy and flexibility to performance of the prescribing function have resulted in comparable improvements in procedures for putting an end to outmoded prescriptions and arrangements, with appropriate measures for the compensation of those who suffer disproportionate loss. In lieu of traditional assertions of unilateral naked power or reliance on a mutual consensus difficult to achieve, parties contending about termination today more frequently resort to organized authoritative arenas, such as in the United Nations or *ad hoc* conferences, and to collective determinations.

(7) Appraisal: The same factors that contribute to improved performance of the more general intelligence function serve equally to facilitate specialized inquiry about the adequacy of past decision processes to secure postulated goals. The expanding contemporary world focus of attention could encourage all participants, official and nonofficial, to undertake more systematic and intensive appraisal of the success of past process in coping with environmental problems.

An examination of the internal constitutive processes of the different particular territorial communities contained within the larger global process would of course reveal that most of these communities possess much more mature and fully developed processes—amply endowed with supervisory, regulatory, entrepreneurial and corrective competences—for making and applying the law necessary to the more effective management of environmental problems within their exclusive territorial boundaries.[h]

INHERITED PUBLIC ORDER IN RELATION TO RESOURCES

Our inherited prescriptions about the allocation and exploitation of resources, comprising in sum our environment, are not as archaic, irrelevant, and inadequate as is sometimes suggested. The exigencies inherent in the cooperative, interdependent exploitation of the world's resources have imposed severe restraints on the assertion of special interests, even in relation to the land masses, and have encouraged the mutual recognition and reciprocal protection of common interests, both inclusive and exclusive, especially in relation to sharable resources. The general community, acting through the constitutive process described above, already allocates resources between inclusive and exclusive uses in a way designed to maximize inclusive competence, seeks to regulate and limit the injurious employment of resources by communities in relation to each other, makes abundant provision for facilitating the productive and harmonious employment of resources, exhibits the beginnings of an appropriate network of planning and development institutions, and even aspires, through an increasing concern for human rights, to achieve a more rational relation of peoples to resources. The potentialities afforded by our inherited prescriptions for assisting movement toward improved environmental protection may be economically indicated in relation to certain basic, perennial problems.

(1) The allocation of resources: The principal achievement of past constitutive process in relation to resources has been in establishing that the great sharable resources of the globe—that is, those admitting of a high degree of shared use by reasonable, mutual accommodation, such as the oceans, the airspace over the oceans, international rivers, the void of outer space, and the celestial bodies—remain subject to relatively inclusive competence, open for inclusive use and not subject to exclusive appropriation by particular states.[1] This outcome, though increasingly threatened today by accumulating demands in terms of special interest, is fortified by the clear experience of mankind that it is inclusive competence and use that most often promote the greatest production and widest distribution of goods and services for the benefit of all. Only with respect to the land masses and closely proximate waters and airspace, which admit of the least degree of shared use, have states reciprocally honored each other's claims to comprehensive, continuing, and exclusive competence, and with the contemporary accelerating interdependences in the use of land masses, even this exclusive competence is becoming attenuated and being made a matter of degree.

Similarly, with respect to those resources that it has permitted to be subjected to exclusive appropriation, the general community has imposed quite severe conditions and limitations on such appropriation, including a genuine "occupation," in the sense of a comprehensive and continuous process of enjoyment and utilization, made known to all the world; limitations on the quantity of the resource subject to appropriation; and requirements for development within a reasonable time.

From these perspectives of authority, the general community clearly retains inclusive competence over many of the resources importantly affecting the quality of the environment, it could preclude the exclusive appropriation by states of any new resources made available by advancing science and technology for control of climate and environment, and it could condition the continuing control even of resources subject to exclusive appropriation to conformity with the dictates of common interest in the maintenance of a necessarily shared environment.

(2) The regulation of injurious employment in use: The general community, through its present constitutive process, seeks to minimize the losses caused both by major, deliberate attacks by states on each other's territorial integrity and by the less comprehensive, often unintended deprivations that inevitably attend transnational interaction.

The protection that the general community establishes against major, intended deprivations that threaten territorial integrity and independence today derives principally from the United Nations Charter and the ancillary procedures established thereunder. The basic prescription is that of Article 2, paragraph 4, which provides:

> All Members shall refrain in their international relations from the threat or use of force against the territorial integrity or political independence of any state, or in any other manner inconsistent with the Purpose of the United Nations.

It would appear the common expectation of most of mankind that this policy of minimum order, indispensable to law in any community, applies not merely to activities on earth and to traditional exercises with the military instrument, but also to man's activities anywhere, as in outer space, and to any new techniques of coercion and deprivation made possible by manipulation of environmental variables.[j]

The protection established by contemporary constitutive process against less comprehensive, relatively nondeliberate, more ordinary acts of deprivation builds largely on customary international law, as occasionally reinforced by explicit agreement. A continuous flow of authoritative community decisions and of unilateral acknowledgements for many decades appears to be establishing that states regard themselves as reciprocally responsible for such deprivations, even to a degree approaching absolute liability. Perhaps the most famous and influential case in this line of decision is the *Trail Smelter Arbitration*[k] between the United States and Canada. In this arbitration, the tribunal found Canada responsible to the United States for damage caused by sulfuric fumes emitted by a smelter in British Columbia and affecting large areas in the state of Washington. Certain language from the opinion has been often quoted as a concise formulation of the international prescription:

> . . . under the principles of international law . . . no state has the right to use or permit the use of its territory in such a manner as to cause injury by fumes in or to the territory of another or the properties or persons therein, when the case is of serious consequence and the injury is established by clear and convincing evidence.

> . . . the Dominion of Canada is responsible in international law for the conduct of the Trail Smelter and, apart from the undertakings in the Convention, it is the duty of the Government of the Dominion of Canada to see to it that this conduct is in conformity with the obligation of the Dominion under international law as herein determined.

> . . . The Trail Smelter shall be required to refrain in the future from causing any damage through fumes in the State of Washington. To avoid such damage the operations of the Smelter shall be subject to a regime or measure of control as provided in the present decision. Should such damage occur, indemnity to the United States shall be fixed in such manner as the governments acting under the convention may agree upon.

The same authoritative policy that infuses this decision could be documented in many other specific decisions and in a great variety of state and private practice. Among the more famous cases are the *Corfu Channel* decision in the International Court of Justice and the *Lake Lanoux Arbitration*.[l] The most relevant practice, exhibiting expectations about responsibility, would include the recognition as lawful of "contiguous zones"—beyond the territorial sea—that are reasonably designed to protect against external injury; the mutual tolerance of "self-defense" or "self-help" anywhere on the high seas that is necessary and proportionate for dealing with threatened deprivation; and the emerging, generally accepted customary regime with respect to international rivers, which establishes "reasonableness" as the touchstone for adjusting the equities in use among riparians who cannot avoid affecting each other. The easy acceptance in multilateral convention and otherwise of strict liability for space activities and nuclear damage reflects comparable expectations.[m] The contribution of private, nonofficial practice to authoritative expectation is dramatically demonstrated in the recent voluntary settlement of the *Torrey Canyon* case and in the establishment of a continuing private association to deal with comparable future disasters.

(3) Facilitating the productive and harmonious employment of resources: The best exemplification of the high potentials of inclusive competence for effecting the productive and harmonious exploitation of sharable resources is the historic international law of the sea, which for some centuries has served the function of clarifying and securing, by shared reciprocity and mutual restraint, the common interests of all peoples in the greatest possible production and widest possible distribution of values from the oceans.[n] The basic framework of this law—established in the more comprehensive world constitutive process by habitual, cooperative behavior—has been, in a largely decentralized and unorganized arena, a few simple customary prescriptions, applied with an extraordinary economy and effectiveness: that all peoples enjoy equal rights of access to the oceans and to the appropriation of resources, such as fish, with respect to which exclusive appropriation is permitted; that each state makes and applies law to its own national ships and that no state may assert its exclusive, unilateral competence over the ships of other states save for violations of international law; and that no state may question the competence of another state to confer its nationality on a ship. These few simple prescriptions have, of course, been supported by a whole host of ancillary, implementing rules, established both by custom and by multilateral agreement, for fixing rules of the road, securing safe navigation, repressing piracy, promoting conservation, restraining pollution, accommodating and integrating different interests, and so on. The evidence would seem clear that over the centuries, despite the difficulties of recent days, this regime of inclusive competence, with a minimum monopolization of either authority or use, has served the whole of mankind well in creating the greatest net gains both in the indivisible value of general security and the divisible values of wealth, enlightenment, well-being, and so on.

A comparable regime is in process of being extended, both by customary prescription and by explicit agreement, to the sharable resources of outer space and the celestial bodies but, unhappily, has as yet been only partially extended to airspace over the land masses.

Even with respect to the land masses, commonly regarded as admitting of the least degree of shared use, a regime of modestly inclusive competence may be observed to serve the interests of a world economy and society. The prescriptions most importantly comprising this regime are those popularly described as "private international law" and as relating to the "responsibility of states." [o] The first set of these prescriptions are principles of jurisdiction— the principles of territoriality, nationality, protection of interests, impact territoriality, passive personality, the "proper" law of "contracts" and "torts," and so forth—which confer a competence on states to protect their internal community processes and their nationals from injury, even that originating beyond their boundaries. The second set of principles constitute an international "bill of rights," antedating the contemporary prescriptions about human rights, which impose certain limits on the competences of states, with respect to events otherwise within their "jurisdiction," for the protection of aliens and their property. In recent decades this aspiration in customary prescription toward a world economy and society has been significantly implemented by explicit agreements and new specialized organizations, as in the International Monetary Fund, the International Bank for Reconstruction and Development, and the General Agreement on Tariffs and Trade.

(4) The performance of planning and development functions: The inclusive performance of planning and development functions has received great impetus from the recent proliferation of international organization. The number of organizations—governmental and nongovernmental, specialized and nonspecialized, general and regional—engaging in such functions in relation to activities with a direct bearing on environmental control is so vast as to defy even listing.[p]

The most comprehensive and important inquiries and activities obviously stem from the United Nations and its affiliated enterprises, but regional organizations, such as the NATO, and private associations are beginning to play increasingly significant roles.

For some indication of the range and complexity of contemporary planning and development activities, we note in relation to each of certain major resources a few of the organizations engaging in such functions.[q]

Energy Resources:

> The International Bank for Reconstruction and
> Development
> The International Atomic Energy Agency
> The World Meteorological Organization

Natural Environments:

> UNESCO
> The World Wildlife Fund

Animal Life:

> The Food and Agriculture Organization
> The International Whaling Commission

Vegetation and Soils:

> FAO
> UNESCO
> International Society for Plant Geography and Ecology
> Inter-American Institute of Agricultural Sciences

Water:

> FAO
> WHO
> UNESCO
> WMO
> IAEA
> IMCO
> The International Association on Water Pollution Research
> The Council of Europe

Air:

IAEA
IMCO
WHO
The Council of Europe
The International Association Against Noise

Minerals:

Various United Nations agencies
Many specialized private associations

Weather and Climate:

WMO, including World Weather Watch Program
International Civil Aviation Agency.

(5) The relation of peoples to resources: The most obvious inadequacy in inherited public order concerns the relation of people to resources. The access of peoples to communities (and hence to the resources exclusively appropriated by such communities) and movement between communities is made dependent largely on the "nationality" of individuals, and states are permitted to prescribe and apply highly restrictive policies in the granting and denying of nationality.[r] In a global arena in which expectations of violence are high, states recognize, and reciprocally honor, a common interest in the exclusive protection of their bases of power and take the most severe measures to make certain that their members remain available to themselves and their allies and do not undermine internal power structures. They impose, and acquiesce in, highly arbitrary restrictions on freedom of circulation and freedom to change memberships, and employ stringent negative sanctions to secure conforming conduct.

Some amelioration of historic attitudes has begun to appear in the contemporary human rights program, which projects prescriptions favoring freedom of movement between communities and freedom to change community memberships; in the increased honoring in the United Nations of the right of peoples to "self-determination" when based on appropriate conditions of security and economy in the shaping and sharing of values; and in the accelerating movement toward a more rational regional organization of the world arena and world society. The preferred policy, most compatible both with human rights and the productive employment of resources, appears to favor a high voluntarism in travel, affiliation, and participation. For the rational shaping and sharing of most values, the present boundary lines between nation-states are as outmoded as the comparable demarcations between provinces within nation-states.

Transnational efforts to control population growth raise even more difficult, and almost unprobed, issues of policy and legal technique.

ALTERNATIVES FOR IMPROVEMENT

One of the premises with which we began our discussion was that changes in the environment, whether beneficial or injurious, are an inextricable component of the continuous process of social interaction by which resources are employed in the shaping and sharing of values: injury to the environment may not be inevitable in such process, but the practices by which injury is inflicted are indistinguishable from the ordinary practices of production and distribution and are omnipresent. Another premise was that emerging general community goals in relation to protection of the environment extend, beyond the mere minimization of particular losses, to the maximization of gains in the production and distribution of all human dignity values.

From these perspectives it is apparent that the demanded quality of environment is, and must in a world of ever-increasing interdependences continue to be, affected by every feature both of global constitutive process and of the internal constitutive processes of its lesser constituent communities. It is sometimes proposed, as by George Kennan,[s] that there immediately be created a new specialized international governmental agency charged with the performance of intelligence, promotion, invocation, and appraisal functions with respect to environmental problems. Conceivably, such an organization— if accorded the necessary bases in authority, manpower, and finances—could do much to improve the functioning of global constitutive process in relation to its purposes. Clearly, however, it would not be enough: there would remain the functions of prescribing new and more appropriate standards as general community policies and of applying these policies in particular instances to put them into controlling practice; the new agency could not be given authority even in relation to the intelligence, promotion, invoking, and appraising functions over all the activities in production and distribution that affect environment without constituting in effect a comprehensive world government in lieu of present constitutive process; and the internal constitutive processes of the different territorial communities making up the world arena would still require renovation.

The observer who would be effective in recommendation about environmental protection need not decry, and may on the contrary rationally participate in, efforts to create new and more appropriate institutions for the performance of policy functions in the world arena. Such efforts, however, if they are to be most effective, should be made but integral parts of a more comprehensive, incremental endeavor to reshape the whole of the features of global constitutive process, including the processes of the constituent communities, for the improved performance of the necessary functions. The problem of establishing and maintaining an appropriate global environment is just as indivisible and all-pervasive as is the problem of maintaining the general security against military attack. Neither problem is likely to be solved either by partial, halfway measures or by utopian schemes that would shatter existing processes and begin anew.

The most fundamental approach to new alternatives will seek, at both global and lesser levels, systematically and deliberately to improve performance, in every available structure of interaction, of each of the seven policy functions described above. When basic community goals are clarified and kept constant, many different structures of authority may serve with equivalent effects in facilitating performance of the necessary functions. Some of the implementing, ancillary policies that might be sought in the improvement of each function may be indicated:[t]

Intelligence: dependable, comprehensive, selective,
creative, available.

Promotion: rational, integrative, comprehensive,
effective.

Prescription: effective, rational, inclusive.

Invocation: timely, dependable, rational,
nonprovocative.

Application: rational, contextual, uniform, effective,
constructive.

Termination: prompt, balanced, effective,
ameliorative.

Appraisal: dependable, continuing, independent,
contextual.

When consideration is given in detail to improvement of the application function for minimizing losses from particular injury, attention might extend beyond judicial applications for fixing responsibility for injury already inflicted to a whole range of potential sanctions for serving various particular subgoals embraced within the larger policy of minimizing losses. These more particular subgoals include prevention, as long-term efforts to minimize the occasions for injury; deterrence, as precluding injury immediately threatened, restoration, as putting an end to injuries already in process; rehabilitation, as the short-term binding up of wounds; and reconstruction, as the longer-term redesign of the situation to preclude future injury. The range of potential sanctioning practices for serving these subgoals obviously infuses the whole of the larger community social process.

The task of highest priority for all genuinely committed to the goal values of human dignity is, of course, that of creating in the peoples of the world the perspectives necessary both to their more realistic understanding of their common interests in relation to the environment and to their invention, acceptance, and initiation of some of the many equivalent measures in constitutive process that might better secure such common interests. It is the message of the maximization postulate that men act, within their capabilities, to enhance their values. If this Conference makes its appropriate contribution, that contribution will be in increasing the capabilities of men for preserving and enjoying their world.

ACKNOWLEDGMENT

The author expresses his appreciation to John A. McCullough, of the third-year class in the Yale Law School, for assistance in the preparation of this paper.

NOTES AND REFERENCES

[a] Report of the Secretary-General, Problems of the Human Environment, United Nations Economic and Social Council, E/4667, 26 May 1969, at p. 4.
[b] Ibid.
[c] These interdependences are developed in McHale, The Future of the Future (1968) 66 et seq. and McDougal and Rotival, The Case for Regional Planning, With Special Reference to New England (1947). See also Murphy, Governing Nature (1967); Dimbleby, Restoring the Ecological Balance in Keeton and Schwarzenberger

(Eds.), The Yearbook of World Affairs (1969); Mayda, Conservation, "New Conservation," and Ecomanagement, 1969 Wisconsin L. Rev. 288.

[d] Global perspectives are appropriately stressed in Kennan, To Prevent a World Wasteland: A Proposal, 48 Foreign Affairs 401 (1970).

[e] These perspectives are amplified in McDougal, Lasswell, and Reisman, Theories About International Law: Prologue to a Configurative Jurisprudence, 8 Va. J.I.L. 188 (1968). For historical background see Friedrich, The Philosophy of Law in Historical Perspective (2d ed., 1963); Brecht, Political Theory (1959).

[f] Lerner and Lasswell, The Policy Sciences (1951); Dror, Public Policymaking Reexamined (1968).

[g] A more detailed statement is offered in McDougal, Lasswell, and Reisman, The World Constitutive Process of Authoritative Decision, 19 J. of Legal Ed. 253, 403 (1967); Falk and Black, The Future of the International Legal Order (1969) Ch. 3.

See also Jenks, The World Beyond the Charter (1969); Corbett, From International to World Law (1969); Jessup, Transnational Law (1956); Schachter, The Relation of Law, Politics, and Action in the United Nations, Hague Academy, 1 Rec. des Cours 167 (1967); and Scientific Advances and International Lawmaking, 55 Calif. L. Rev. 423 (1967).

[h] This theme is documented with respect to the United States in McDougal and Haber, Property, Wealth, Land: Allocation, Planning, and Development (1947). See also Oscar S. Gray, Cases and Materials on Environmental Law (1970), and Malcolm F. Baldwin and James K. Page, Jr. (Ed.) Law and the Environment (1970).

The essay in the latter volume (at p. 297) by Professor A. Dan Tarlock, "Current Trends in the Development of an Environmental Curriculum," is an excellent history of scholarship and teaching in the United States.

A vast recent literature offers detail with respect to United States law. For examples, see: Symposium, Law and the Environment, 55 Corn. L. Q. 663 (1970); Symposium, Control of Environmental Hazards, 68 Mich. L. Rev. 1073 (1970); Comment, Toward A Constitutionally Protected Environment, 56 Va. L. Rev. 458 (1970); Comment, The Air Quality Act of 1967, 54 Iowa L. Rev. 115 (1968).

[i] For detail, see McDougal, Lasswell, and Vlasic, Law and Public Order in Space (1963) Ch. 7.

[j] These expectations are described in McDougal and Feliciano, Law and Minimum World Public Order (1962).

[k] Trail Smelter Arbitration Tribunal Decision, March 11, 1941. Text in 35 Am. J. Int. Law, 684 (1941).

[l] Citations and discussion appear in International Law Association, Helsinki Rules on the Uses of the Waters of International Rivers (1966) 20 et seq.; Taubenfeld, Weather Modification and Control: Some International Legal Implications, 55 Calif. L. Rev. 493 (1967); and Some International Implications of Weather Modification Activities, 23 Int. Org. 808 (1969).

[m] Nanda, Liability for Space Activities, 41 U. Colo. L. Rev. (1969); Vlasic, The Space Treaty: A Preliminary Evaluation, 55 Calif. 2 Rev. 507 (1967); Dembling and Arons, The Evolution of the Outer Space Treaty, 33 J. of Air Law and Commerce 419 (1967).

For more genera' relevant discussion see Utton, Protective Measures and the "Torrey Canyon," 9 Boston Coll. I. of A. L. Rev. 613 (1968); Jordan, Recent Developments in International Environmental Pollution Control, 15 McGill L. J. 279 (1969); Healy, The International Convention on Civil Liability for Oil Pollution Damage, 1969, 1 J. of Maritime Law and Commerce 317 (1970); Mendelsohn, Maritime Liability for Oil Pollution: Domestic and International, 38 Geo. W. L. Rev. 1 (1969).

[n] For detailed exposition see McDougal and Burke, The Public Order of the Oceans (1962); Burke, Toward a Better Use of the Ocean (1969); Johnstone, Law, Technology and the Sea, 55 Calif. L. Rev. 449 (1967).

[o] Yntema, The Historic Bases of Private International Law, 2 Am. J. Comp. Law 297 (1953); Dunn, The Legal Protection of Nationals (1933).

[p] A modest survey is offered in Report of the Secretary-General, see note 1.
[q] This itemization of resources is borrowed from Crane Miller.
[r] Weis, Nationality and Statelessness (1956).
[s] Kennan, note 4.
[t] This presentation draws, as does other segments of this paper, on unpublished collaborative studies with Harold Lasswell and Michael Reisman. For a somewhat less sanguine view than that here presented, insisting that "systemic" as well as incremental changes are required, see Falk, Toward Equilibrium in the World Order System, 64 Am. J. I. L. 217 (1970).

WORLD PRODUCTION AND ITS DISTRIBUTION

Josué de Castro

University of Paris
Vincennes, France

Because the progress of economic science has permitted better definition of the concept of gross national product and measurement of its annual growth—that is, practically speaking, since the Second World War—the wealth of nations is no longer considered as a gift of nature, climate, race, soil, or subsoil, but the result of certain historic conditions that have led all the nations to the organization of their productive forces at various levels.

Certain countries have profited from historic circumstances in developing and enriching themselves, while others have been subjected to restraints that have prevented them from attaining this level of development.

Without going into the historic roots that have created this gap in the course of development of nations, which has split them into two groups—that of the rich and well-developed nations and that of the poor and underdeveloped—without, furthermore, discussing the ethics of the rich, industrial societies, their human and moral results, and their purposes, one must recognize that this society, called "technological," which dominates the world, must be better managed, in the interest of each and every man on our planet. Some statistics show very well the absurd concentration of wealth in certain limited zones of the world while most of the people remain peripheral or marginal to the dominant economic system.

A small number of countries, where 20 percent of the world's population live, possess 85 percent of the world's wealth. The average income of the inhabitants of these countries is above $1,000 annually. These countries are concentrated around the North Atlantic, the Soviet Union, South Africa, Australia, New Zealand, and Japan. Their total production in goods and services is more than $2,000 billions, of which one country alone—the richest, the United States of America—accounts for 40 percent, or $800 billions.

All the rest of the world, represented by the underdeveloped countries, where 80 percent of the world's population live, possesses only 15 percent of the world's wealth—that is, $300 billion—and the mean income per inhabitant is below $250.

This concentration of wealth on one side and poverty on the other reflects an enormous distortion in the map of the world's production (FIGURE 1). It has thus become commonplace to speak of the tremendous gap that separates the rich industrial countries from the poor and underdeveloped ones. One no longer even speaks of eliminating this gap because it seems a utopian task. Rather, one speaks today of throwing some bridges over the abyss that separates the two groups situated on either side of this abyss, in addition separates these two worlds in every aspect, economic, social, and political, creating an antagonism so deep that it is putting the peace and security of the entire world in peril. What is more serious is that this situation, far from easing, is becoming more tragic because of the disparity in the speed of growth of these two worlds, which is constantly widening the gap. With the Conference of Bandung, which took place in 1955, and at which the economically dependent countries became conscious of the reality of their socioeconomic situation, was born a new entity called "The Third World." The Third World is now in its greatest economic, social, and political crisis since the Second World War.

Faced with this dangerous situation, the United Nations established, beginning in 1960, a plan called "The Decade of Development," in which was noted the need to multiply every effort, on a national and international scale, in the struggle against worldwide economic imbalance in order to promote the underdeveloped regions. But, after ten years and in spite of all the efforts that have been made, the first "Decade of Development" may be counted a "Decade of Disappointment." It is true that a certain number of well-provided countries have collaborated to speed the dangerously slow development of these

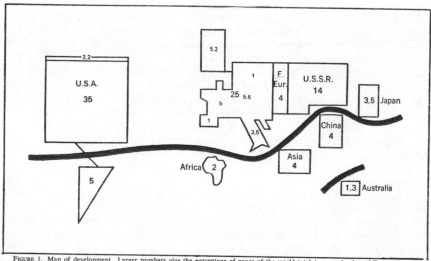

FIGURE 1. Map of development. Larger numbers give the percentage of zones of the world total (e.g. production of Europe represents 25 per cent of the total). Smaller numbers give the share of certain countries (e.g. Italy: 3.5 per cent). Gross national product of the industrial countries: 85 per cent; of the Third World: 15 per cent.

rebellious peoples: this was the beginning of international cooperation and of technical assistance for development. Unfortunately, all these efforts were made without enough coordination and, above all, in the absence of a real *policy* of development. As the former Secretary General of the Conference on Trade and Development, Raul Prebisch, has written, "The Decade of Development is being made without a policy of development." It cannot be denied that the crisis of the Third World and of its frightening economic lag is, above all, the product of a mistaken attitude of not wanting to recognize that true development is an essentially political phenomenon. The very notion of development always implies a certain conception of man and society and leads to a political choice. Without a worthwhile politics of development that encompasses a whole ensemble of measures in every sector of life and at every level of responsibility, from the individual to the universal community of all nations, we will never have world-wide economic equilibrium for resolving the problem of underdevelopment.

This characteristic of a notion of complexity and interdependence, capable of unifying and concerting efforts for the general organization of society on a political level, is the basis for what one may call "global strategy for development."

In support of this thesis—the ineffectiveness of aid or, indeed, its veritable failure in development—it is enough to present a picture of the socioeconomic reality of the present world from various angles.

The disclosure of the work of the last Conference on Trade and Development at New Delhi has made it clear how critical the situation of the Third World is, how serious the drama of underdevelopment is. It is no longer possible to hide the fact that, despite efforts made through aid and international cooperation since the last World War, the level of life in the Third World is, each time viewed, farther from that of the well-developed and industrial countries.

All the world seems to agree on the failure of the programs of development employed in the less developed regions. It has become evident that this strategy was built on principles and systems of thought that were far from able to lead to effectiveness. Efforts are now being made in research and in sustained consideration to discover the weak points in this unsuccessful program and to conceive a new strategy of development, able to save the world from the grave dangers represented by the alarming disparity in the two worlds' pace of growth, juxtaposed as they are, but unrelated—development and abundance, on one hand, and underdevelopment and misery, on the other.

Profound analysis of this resounding setback, which turned the decade of development into one of disappointment, reveals certain very serious errors that were committed and that have made all efforts for development to amount to little.

The greatest of all these errors was to consider the developmental processes of all the world similar to those of the rich countries of the West. A kind of ethnocentrism has led the theoreticians of development to base their ideas and to build their systems of thought on conceptions of classical economics that almost completely ignored the socioeconomic reality of the economically dependent regions. They forgot that there did not exist, and does not exist today, an integrated world economy but only a Western, capitalist economy, full of contradictions, plus a socialist economy in accelerating development, alongside a network of raw material and markets in the *rest of the world*. They did not bother with the economic structure of this *rest of the world*, abandoned to the sociologists or, rather, students of folklore. But, above all, they completely forgot the inhabitants of these regions, with a traditionalist culture and so distant and so different from those who are products of Western civilization. Besides, in its frenzy of productivity, the West everywhere forgot man. The theoreticians had the notion that well-regulated doses of capital and technology—the product of Western discoveries, inventions, and innovations—would change the general picture of the structure of traditionalist cultures of the non-Western world, automatically launching all kinds of development.

The miracle did not occur, and the illusion of aid, the glib solution of the first years, was followed by disappointment and pessimism, which led to the idea that the lagging behind of the Third World is an almost insoluble problem. These countries are underdeveloped, say Western pessimists, by the force of things—biologic fatalism or geographic determinism, by natural and human conditions that prevent access to true autonomous development.

The two attitudes, the great postwar illusion and the current defeatism, strengthened by a false comparison between the rapid reconstruction of the European economy by the Marshall Plan and the stagnation of the economy of the Third World are false positions, attitudes more emotional than rational, built on no scientific foundation.

The countries of the Third World are underdeveloped not as a result of natural causes—the force of things—but as a result of historical causes—the force of circumstance. Unfavorable historical conditions, mainly political and economic colonialism, has kept these areas outside the world's economic process, which is in rapid evolution.

The secret of development resides, above all, in changing the out-of-date structures of these societies created by colonialist overlords; outside their world, these structures appear in the form called by the American sociologist Lewis "cultures of poverty," in which factors of animation and dynamics are never found but, rather, a powerful system of forces that block development and maintain the status quo, forces of economic and cultural nature. The phenomenon of development in these regions appears as a complex, at the same time economic and cultural, but, unfortunately, it was not customary to bring these two ideas together—economy and culture—as the French economist André Piatier has wisely stressed.

In truth, economic growth cannot be launched except on the basis of a certain level of cultural development: 18th century Europe, which witnessed the birth of industrial civilization, had already reached this threshold, but the Third World had not. In other words, it cannot be denied that social conditions affect economic conditions. Ignorance of these ideas, today widespread, was the cause of underestimation of social conditions on the part of classical economics; such conditions were considered a matter of sentimental feelings or claims. In the establishment of criteria for priorities in investment, already so slim and insufficient in the Third World, there was always an invincible drive to invest solely in "profitable" sectors—which is to say, in order to furnish productive machinery—in which man was always excluded from consideration. This aberration frustrated all attempts to accelerate industrialization of the countries where the weakness of the human bases and the absence of masses of consumers did not permit such economic adventure. Also forgotten was the fact that precapitalist societies are ritualistic and each action is a rite whose conduct cannot be changed by the mere introduction of a technique divorced from its ideology or of a new system of thinking. Fortunately, we are beginning to recognize that there are not two different kinds of investment, profitable and nonprofitable—those made to better human conditions by education and health.

A new science has been born, economic psychology of somewhat wider vision than that of classical economics, and through which is being formed, for example, an economics of education, which considers education an essential instrument to the formation and growth of human capital. It is by this means that the international policy of development should begin, because only the labor of men makes wealth. Economic dynamics occurs in direct ratio with the quantity and quality of this factor of production—man, the motor of the economic machine. Economic factors are put into motion only by the brains of men and the will to organize a more advanced economic world, as Jacques Dibs has written in *Le Monde*. But brains, those factories sparked by creative energy and imagination, were almost always unemployed in underdeveloped countries. International plans for development had in mind the equipping of other types of factories, which tended to further the economic interests of the giver rather than that of the country receiving aid.

The nations that have for a long time been politically dominated by foreign powers always mistrust certain types of aid, in which they suspect hidden intentions to dominate. The West must give ample proofs that it has changed its colonial mentality in order for its cooperation to be accepted without reserve by the starving masses of the Third World. It is not enough to *have* more in order to aid others; one must *be* more—that is, it takes greatness to know how to give. The West, which—unfortunately—holds its civilization the only worthwhile one and tries hard to impose it on the whole world, setting the example of the wealth of its accumulated innovations produced by the material development of what is called the "great postindustrial society"—has not such fine things to show in terms of moral or spiritual development, things that people in search of a new philosophy of life could admire and adopt in the process of making dynamic social progress. A thoroughly reformed world leadership, as that philosopher of economics, François Perroux, points out, implies that the leaders have become capable not only of changing the world materially but also of offering a model of new men, or free men. And, paying very dearly for a still precarious political liberty, less developed nations are afraid they will fall back into a state of total servitude confronted with technocratic societies that appear, as a general rule, too domineering. One cannot make real social progress if the people are not liberated from moral and spiritual underdevelopment. This leads us to the statement, fraught with consequences, that certain well-developed countries are, in fact, as Louis O'Neil says, "rich underdeveloped countries." And from this fact it may be easily deduced that the condition for balanced development of the world is the development of man—the formation of men capable of and responsible for establishing a real dialogue between the two worlds— that of abundance and that of poverty—which, today, do not listen to each other and understand each other less and less.

In spite of the extraordinary progress of the techniques that have created polar masses in the center of wealth in the rich and well-developed countries, this prosperity, concentrated in the poles, has not radiated to the periphery of the world's economy, where two-thirds of mankind live, excluded and deprived of the benefit of this technology, of which it has learned through the means of communication and information so widespread throughout the world.

To understand what a global strategy of development must consist of, we must first of all define, with precision and clarity, the notion of development. The world has always lived under the force of myths or governing ideas, and the great myth of the 20th century is "development," as "liberty" was that of the 18th. How did we arrive at that notion of development? It is linked to the idea of progress, a new idea, because until the end of the Middle Ages it had no part in the repertoire of current thought. Man then lived in a world considered finished, therefore unchangeable. Sociology of the time was concerned with immobility. The idea of progress, that is, of changing the world, was born in the modern world with the philosophy of Spencer and Hegel. Spencer called progress "the passage from the homogeneous to the heterogeneous," and Hegel defined it as "the dialectic process of history."

This progress is made by continual evolution or by breaks with the past or transformations that provoke radical change. It is the passing from the quantitative to the qualitative that can serve as a scientific definition of the word *revolution*. The first industrial revolution followed the myth of progress. Next was formed the myth of wealth or, better, of the growth of wealth, based on the economic conceptions of Lord Keynes, centering around the idea that, in maximizing the productivity of labor and capital, we would achieve the promotion of progress and wealth everywhere in the world. This distribution of wealth would be made automatically by that force called by Adam Smith "the invisible hand," which, in the liberal economy of *laissez-faire*, would be occupied with promoting the economic balance of the world. The Manchester school of economics assured us that the economic specialization of classes and countries would benefit all mankind with an increase in wealth and result in a realm of universal abundance.

This myth, unfortunately, proved false. The invisible hand has never acted in the interest of humanity, and the visible hand of dominant groups and the privileged has always reserved to them the rewards, leaving in misery and nakedness the great marginal masses who today constitute what is called "the populations of the underdeveloped countries." Underdevelopment, however, is not the lack of development but the product of a universal type of ill-conducted development. It is the abusive concentration of wealth—especially in this historical period dominated by capitalistic neocolonialism, which has been the determining factor in underdevelopment of a great part of the world, the regions dominated precisely as political colonies or economic colonies.

In view of the limitations of this notion of the growth of wealth as a solution to the problems of the world inherent in the idea of maximization of production, another myth was born—that of development based above all on the utilization of natural and human resources. The superiority of the notion of development over that of the growth of wealth and over that of "progress," is that *development* implies both the growth of wealth and social change, both in the service of man. Development is the development of man, which is why the Secretary-General of the U.N., U Thant, is correct when he condemns the distinction made between economic and social development. There is only one type of real development, development of man—man, the creator of development and the beneficiary of development. The workshop of development is the brain of man. It is the brain of man that must flower in the use of the products put at his disposal by development.

It must be recognized, however, that the idea of development, although more complete than earlier notions of progress, is not yet entirely detached from emotional prejudices and is not yet scientifically precise. If development is the passing from a lower to a higher human level, we are not yet unanimous on the criterion of value for determining these different levels. The notion of development does not imply inevitable improvement in the human condition or greater happiness for man.

There is, in truth, progressive and regressive development—or, rather, both positive and negative aspects are included in the complex of the idea of development. For example, if man frees himself from natural constraints through technological development, by machines constantly more perfect, he is, on the other hand, shackled by other types of constraints. Development, therefore, does not mean liberty. Man chained in a mechanical network is indeed the symbol of industrial or postindustrial civilization where there is no longer freedom of choice because he is always governed by the machine.

There is no longer the right of voicing hopes, because hopes are inculculated from without by the subtle system of publicity, which is part of the consumer civilization in the so-called well-developed countries. An example that sadly reveals the situation of the well-developed man is the advertisment of a great store in a great metropolis that says: "If you don't know what you want, it doesn't matter—enter; we have it."

The contradictions of development are several. Development means both change and discipline. But discipline sometimes prevents change. The conservatism of societies that have achieved a high degree of development makes them consider themselves the ideal model of society to the point of combating the desire for change. Unequal progress in the various sectors of society is one of the causes of the grave crisis in the contemporary world. Dissemination of information on a huge scale allows the less-developed peoples to be acquainted with the images of prosperity, indeed of abundance of the great, dominating centers of the world, with their modes of life and their habits of consumption, and they despair of being able to reach that level; hence, dissatisfaction and revolt.

The old conflict between the aspirations of man and the means of satisfying them intensifies constantly when the image of the great centers is projected in the poor and underdeveloped countries. The contradiction that one augments wealth by raising production is not less serious, because this increase in production generally ends in a fall in prices, and, therefore, checks efforts for development.

Technological development is full of these contradictions, for, in reality, technology has no ethics; it is neither good nor bad, and it can therefore be used for good or for ill—to create progress or to prohibit progress. Until today, technology has worked more against the development of the Third World than in its favor.

To complete the econometric analysis of the process of development of the Third World, we must stress that the increase of five percent in universal development, decided on as a minimum objective for the decade of development, has been attained in only a quite restricted number of countries of the Third World. In fact, the average increase achieved in 54 countries— where 87 percent of the population of the Third World lives—has been only 4.5 percent. It must not be forgotten that the increase of population in these countries is the highest in the world—around three percent—which leaves a gain of only 1.5 percent a year, a ridiculous amount.

Among the factors capable of explaining the failure of plans for development of the Third World, one must stress the following points:

(1) What has been called "international aid" has been insufficient, badly distributed, and badly applied. The U.N. recommendation according to which the rich countries devote one percent of their gross national product to international aid has never been followed. The total of all forms of aid— including gifts, bilateral and multilateral loans, and private investments in the Third World, which has never been enough—far from increasing, is in fact decreasing. Since 1961, transfer of capital from the rich countries to poor countries (which once grew to 15 percent yearly) has leveled off and, in comparison with the income of the rich countries, has probably decreased.

(2) The various structures of the underdeveloped countries bar the necessary changes for the true process of development. It is, above all, traditional distribution, based on an unequal system of ownership of the earth and

on other built-up privileges, that rules out the rational use of the forces of production.

(3) Political instability, the result of tensions and disturbances caused by governments that do not truly represent the national interests of the great marginal masses of the Third World, spur neither foreign investments nor the reinvestment of profits realized in these countries.

(4) The absence of teams of qualified and capable workers constitutes another factor limiting economic expansion and utilization of modern techniques of development.

The absence of plans for development that are not just a kind of visiting card of a government that has no real rapport with the disparate measures and thinly spread initiatives of that same government acts as a negative factor in the economic development of the underdeveloped countries.

(5) Partial development, by application of capital to the more profitable sectors of the economy—above all the obsession with industrialization without the parallel development of agriculture—is also one of the most serious factors blocking the process of development, as has occurred in certain countries of Latin America, for example.

(6) The tendency to exterior imbalance, caused by an always unfavorable balance of trade (the familiar problem of the relationship of the prices of raw materials and those of finished goods) a tendency that we have not been able to lessen despite concerted efforts of countries producing basic raw materials during the two Conferences on Trade and Development that took place in Geneva and New Delhi—continues to weigh crushingly on the countries on the economic periphery and those of economic dependence.

There are many other factors produced by an absence of rational criteria, of categorical options to choose among or absence of actions to undertake, that have kept the introduction of the Third World into the economy of the world at large far from accomplishment.

The problem of underdevelopment is not a problem of these countries alone, but a universal problem that has no solutions except on an equally universal scale. To live in luxury in a world where two-thirds of mankind huddle in misery is not only dangerous, it is criminal. The social tension in which we live today is most often the product of this well-known social injustice—well known to the submerged nations once they have learned the socioeconomic reality of the world. In the phase of the history of man during which we live, a phase of transformation—essentially characterized by various explosions—the psychological explosion of the exploited nations is no less dangerous than the atomic explosion that opened a new era on our planet.

To struggle against this discrimination, which divides the world into two—a minority of nations in dynamic expansion and a majority in stagnation—we must conceive a true policy of development, an effective strategy for global development, with the goal of giving value to mankind throughout the world.

The task is immense, but it is urgent. In the years to come we will see either the economic integration of the world or the physical disintegration of the planet. Peace depends more than ever on the economic equilibrium of the world. The security of man is more important than national security based on arms. This idea, unfortunately, has not passed from theory to practice, and the great powers always depend on arms and spend $160 billion for what they call their "national security" and devote only $8 billion (5 percent of the earlier figure) for universal security through the development of man.

To be realistic, global strategy for development must be based on a premise. We must make every effort to promote the conversion of the world's war economy to peace economy. We must, however, recognize that the current political situation of the world, with the reigning ideological antagonisms and the mutual distrust among the great powers, does not allow us, in short, the establishment of this economy of peace. We must content ourselves with obtaining the maximum of expendable funds for peaceful investments, remaining aware that a great part of disposable income will assuredly be—and be for some time—wasted in the armaments race.

The struggle against underdevelopment will for a long time be carried out under a system of penury; we must, therefore, plan to organize this penury to achieve the maximum of benefits in terms of development. This implies, first of all, a new world-wide political consciousness of the fact that the problem of underdevelopment is common to all the countries of the world— to the rich and well-developed countries and to those that are peripherally developed. The responsibilities in this struggle must, therefore, be shared by the two groups in the framework of a global strategy. To the underdeveloped countries themselves belongs the greater responsibility and the greater effort but these countries cannot get out of the vicious cycle of underdevelopment without international cooperation on the part of the more advanced countries, which have a surplus of capital and techniques and knowledge. International cooperation is thus indispensable to the effectiveness of this effort of development. The strategy also must be global in respect to the various measures that must be taken and the several sectors that must be touched on. We cannot limit ourselves to developing one sector while we ignore measures vital to the parallel development of others, and risk the distortions and restrictions that would compromise the totality of the process of development. Such is the case when nations aim at industry at the expense of agriculture; we soon see that development is blocked.

A global strategy of development must respond essentially, as Raùl Prebisch stressed, to the three great problems for which a rapid solution must be found:

The struggle against the persistent tendency to exterior imbalance.

The struggle against the lack of saving.

The struggle against exterior vulnerability of peripheral economies.

We must conceive an entirely new plan for international trade that would permit the underdeveloped countries to intensify their exports at competitive prices on the international scale. It is clear that if exporting raw materials and basic products is daily increasingly difficult because of technological progress that causes a relative diminution in the consumption of these raw materials, there remains the possibility that the new lands could export finished products. Unfortunately, the industrial countries have so far resisted the yielding of the smallest place to the peripheral countries in the world market for industrial products. It is true that the rich world is beginning to have some misgivings in the face of protests and threats from the poor. At a recent congress in San Francisco, the President of the Chase Manhattan Bank, David Rockefeller, declared: "If we do not succeed in arresting inflation without at the same time reducing trade and investment, we will run the risk of provoking such bitterness among two-thirds of humanity that it will take several generations to restore contact. Meanwhile, we may find ourselves forced to live in isolated citadels." He has also predicted "the most terrible disasters in the world if nations do not open their frontiers wider to the products of poor lands." What Mr. Rockefeller said has been stated and restated by us for the last 20 years in the following formulation: "We cannot live in a world divided between, on one hand, two-thirds who do not eat and, knowing the causes of their hunger, revolt, and, on the other, one-third who eat well—sometimes too much—but who can sleep no longer for fear of revolt on the part of the two-thirds who do not eat."

During the Second Conference on Trade and Development, held in New Delhi, the industrial countries nonetheless were deaf to the claims of the poor countries in regard to international trade. We must also have a new view concerning capital investment in the Third World. The underdeveloped countries much distrust speculative capital not directed to the interest of global development but, rather, to the particular interests of certain products and certain groups that will garner the benefits. Investments must be made in obedience to real plans of economic development, conceived in the interest of the people who are to participate in the realization of these plans. It is true, and it is fair, that the benefactors or the investors know these plans and begin, in part, to make studies capable of judging their viability and productivity. But we must not confuse these studies with unwarranted interference in the political life of countries considered the beneficiaries of these investments. The study of a country's policy of development is already being practiced by such financial institutions as the World Bank, but, considering the delicate responsibility that lies in examining a plan with contradictory interests at work, it is preferable that this judgment be made not by the investors themselves nor by the beneficiary country but either by groups of independent experts or by a consortium of investors from the various countries participating in the project or the plan of development in any country.

Another important aspect of a global strategy is that of fixing limits between the obligations of recipient countries to reinvest in development the capital given and the obligation to raise the level of consumption of their people. Each nation must have the right to study the problem in the light of its own social situation and to establish norms compatible with the biological and psychological tolerance of its masses. It is in this way that criteria or directives for plans of development are established. Furthermore, he who speaks of "plans" says also "discipline in development," and this discipline imposes rules that both parties must obey.

In the strategy of development we cannot limit ourselves to the conception of short-term plans but must also consider perspectives, that is, long-term planning. This does not mean that we are obliged to make prophecies or foretell results, but we make prospective analyses that will permit the formulation of possible and desirable outcomes and allow for working toward them. There too, it is up to the most competent men of every country to know what their aspirations are and how to labor to satisfy them. Global strategy implies action at all levels and in all directions, but it is plain that a criterion of priority must be set up in accordance with the disposable means and the most urgent needs.

In my opinion, one of the highest priorities for the entire Third World is that of *human development,* that is, the creation of men responsible for and capable of starting global strategy on its way. We must integrate into one world the side-by-side bits of contradictory economies; this is the business of tomorrow's man. To bring this task to a successful conclusion we must educate this type of man.

The problem of developing the Third World, and even the entire world, which in some aspects is still underdeveloped, is above all a problem of developing men. If the industrial revolution dominated the 19th century, it is cultural revolution that must dominate the 20th; we must strive for creation of a culture capable of finding real solutions to the great problems of mankind. In this new culture, science and technology will certainly have a great role to play, but they cannot comprise the whole image of that culture. There are many other, equally important, values. We must not forget that science is not wisdom. Science is knowledge. Wisdom implies knowledge *and* judgment. And on this point—on the judgment of values—we are far from having a clear idea of the hierarchies of factors to be brought into play in order to build a global strategy for development that does not separate the economic from the human but, on the contrary, considers man—groups of men, all of humanity—the final goal of development. Father Lebret spoke of the urgent necessity for proceeding to the conversion of man, that is, on one side to change the mentality of power and domination of some and, on the other side, to change the mentality of fatalism and conformity of others, by a mentality nourished with the taste and desire for progress and with the will to acquire the benefits of true development, as well. In this new view, it is instruction, education, and human development that must constitute the requisite investment and, surely, the most profitable investment, for unleashing the rush of development. It is evident that, in order to measure the profitability of investments of a cultural nature, we must use a norm of economic calculation entirely different from that we now employ; thus, in the light of efforts made in certain countries, we may realize, as the economist André Piatier has declared, that the profitability from intellectual investment is infinitely superior to that obtained from the equipping of machines, tools, or works. In recent years, efforts towards development have begun to be obtained in this direction.

In the working budget of the Third World, 20 to 25 percent of funds are now devoted to education. Further, the agreed-on effort by the industrial countries in the policy of international aid—after appropriations for the purchase of armaments—is the mobilization of human resources; education, food, and health, which have the highest priority. In spite of this relative effort, we have progressed little in this domain, and the more serious problems of the underdeveloped countries remains today that of the educational level of the inhabitants, considered as factors in production. Besides, credits intended for this sector are in reality quite limited in relation to the huge task of developing men at all levels, from the education of the masses to the formation of élites—specialists and experts in adapting foreign technology to domestic needs and realities. Nor must we forget that, in underdeveloped, and hence under-capitalized, countries, planning—as an Indian economist Ika Paul-Pont has emphasized—has for its primary objective the organizing of poverty to spur development. It is, thus, with the budgets of poverty that the rescue from ignorance and illiteracy of these great masses of mankind—the damned of the earth—has been attempted. It is a difficult task, and is made much more difficult because we do not yet know the means appropriate to such a complex undertaking. We realize again that it is not enough to assign credit, large or small, in order to have worthwhile training, the type of education adequate to the real needs of these countries. The methods of teaching, the educational programs put into practice have always been ill-adapted to the social context of the economy of poverty in which traditionalist societies are mired, with their archaic social structures and their forces of production strangled by the domination of alienated élites whose interests do not correspond with those of the disinherited masses. The mere transfer of culture—that is, the exported utopias, in terms of education—can never provide a means of development of the kinds of men the Third World needs to expand its economy in a human sense, one that respects the cultural roots of the people. It is being asked more insistently whether development is, indeed, dehumanization, in this frenzied search for wealth in accord with the formula instituted by the West to maximize revenues instead of maximizing the mental energy that more quickly enriches the lives of men and can give them much more happiness. The only way of making the world develop in balance is to look for the means of integrating scientific and technological values into the heritage of values representative of other, non-Occidental, civilizations—but not under the dangerous banner of domination, which everywhere provokes revolt. An education for free men; that is what the peoples of the Third World aspire to. And this presupposes a pedagogy of freedom, one that liberates them from the constraints of nature and also from the constraints of other groups of men—of all types of domination. This means that they must be educated to free themselves economically, politically, and spiritually. What has been done in the Third World until now has been far from attaining that objective. Until today, culture represented in these countries a privilege, an extension of the other privileges of the dominant minority. In most of the countries on the way to development, not one percent of the youth can go to the university, and the university in these countries is, in general, merely an apparatus for the adoption of a false culture, an importation, where élites are formed. The universities are only centers of professional training in the model of the halcyon days of the *Belle Époque*, where a sterile culture is served up, one incapable of creative and renewing stimulus. Students leave these universities with a mentality prefabricated to accept and defend the status quo that has given birth to these privileges, privileges that must be defended at all costs, even that of underdevelopment and national servitude. It is, thus, a type of particularly antidemocratic culture. To make a popular type of education available would be to unleash an irresistable movement toward social change to which the dominant minority, hostile to the idea of worthwhile educational reforms, is opposed. The true reformers of teaching methods in many underdeveloped countries are regarded as dangerous elements, subversive to established order, dangerous to the maintenance of these *democracies without men,* where a handful must know everything and rule everything and the masses must know nothing and must always obey.

Confronted with the sharp crisis that the West is undergoing, always based on an economy of war, we have been led to believe that we must conceive a human development for the new countries based instead on an economy of peace. But to achieve it, we must mobilize mass public opinion with modern methods of communication, public opinion on the part of masses capable of creating a mass culture sparked by an egalitarian ideology. This mobilization of public opinion can be created only by an intercommunication quite different from the unilateral methods of information by which the instructors permitted only the learning and the ideas that interested the dominant group to filter down to the masses. The director of UNESCO, René Maheu, told me recently that he wanted with all his might to expedite the world-wide revolution by education. This is the great objective of education in the countries of the Third World, where already the insight has occurred that only revolution can achieve real solutions to the problem of the marginalization of these people. But this revolution must be prepared for, or, better, men must be prepared to make it. One may improvise a revolution that, in general, will be abortive, but one cannot improvise a revolution that would be an act of thoughtful creation, as Louis Armand has pointed out in his "Plea for the Future."

To conclude the "scandal of underdevelopment," let us cite the expression of Jacques Austruy: the only road onto which one must rush is that towards validating and mobilizing one's human capital by revolutionary methods. Unless this spiritual revolution is made, all international aid—gifts, loans, whether bilateral or multilateral—private investments as well as the alteration in the terms of international trade and the stabilization of the prices of basic materials —all these trifles will not end in plucking the proletarian countries from their chronic misery, the hideous inheritance of colonialism.

Underdevelopment is a kind of undereducation, undereducation not only of the Third World but of the whole world. We must educate and develop the spirit of men, which has everywhere been deformed. Only a new type of man, daring to think, daring to reflect, and daring to act, can realize a true economy, based on human and balanced development.

I would emphasize that a global strategy of development *must not* be considered a semantic formula, a play on words that could cover the ineffectiveness of an action until now badly performed. The strategy must correspond to a new conception of the world's economy, which must be totally reviewed. Putting it into effect will require a world-wide movement to convince the economically dominant countries of all the profit they will get if they direct their efforts toward a plan of true international solidarity and to convince the countries in need of aid of the necessity of proving their firm wish to develop their nation and to become equals in the economic community of the peoples of that world.

FROM ECONOMIC TO SOCIOECONOMIC DEVELOPMENT

Jan Tinbergen

Netherlands School of Economics, the Hague, the Netherlands

THE "ECONOMICIST" CONCEPT OF DEVELOPMENT

There is no doubt that human life in the more prosperous countries has changed very rapidly during the last few centuries. The availability of many forms of comfort has increased at a high rate. To a considerable extent, the forces behind this change are increased scientific and technological knowledge manifesting itself in the large numbers of new goods, in improvement in their qualities, and in a continuous change of production processes using increasingly ingenious and increasingly complicated means. A very considerable portion of these innovations have been created by individual minds and by individual acts, in which the individual was guided by personal interest and personal gain. Inventors, engineers, managers, and owners of means of production were moved largely by such personal motives. Scientists' and technicians' work was for quite some time one-man work, and so was employers' activity. To be sure, they cooperated in groups of increasing size, but for a long time this cooperation was based on contracts that could be easily discontinued. And even though groupings of individuals of increasing size came to play their role, for the period under review the process of our society's development was described as a process in which each person pursued his own interest. Attempts were even made to prove that such an attitude was conducive to the maximum of satisfaction for all and was creating the "best of all conceivable worlds." This was typically the attitude of economic science, represented by its "father," Adam Smith, and continued to be the approach adhered to by economists for a considerable portion of the 20th century. The political representatives of the élite in power in most modern countries continued even longer to believe that individualism had to be the basis of economic development, stressing the role of individual responsibility as an invigorating force fostering further development of human society. A strong belief in private enterprise and economic freedom prevailed, which were thought to take care automatically of everybody's interest, providing employment and a living to all, with incomes reflecting social justice: everyone got from society the equivalent of what he was giving to it. We will call this view the "economicist" concept of development. For a long time, this view was held especially in the United States.

REACTIONS IN EUROPE

The economicist view was held in Europe also, but reactions to it came earlier than in the United States. The first reactions came as early as 1850, especially on the part of Karl Marx and his followers. For a considerable period their influence was limited, however, to the workers' world, where they gave rise to the creation—or the reorientation—of trade unions and of political parties with only slight influence. Some influence penetrated into parliaments, however, and in the second half of the 19th century a beginning was made with social legislation, tending to interfere with complete economic freedom in favor of some of the very weakest social groups, such as children, and some affected by accidents or professional illnesses. In the more advanced countries, the income tax was introduced. Socialist ideas penetrated into the working class to the extent that the Pope was induced to launch the encyclical *De rerum novarum* (1891). One of the features of the Roman Catholic approach of these days was to recognize the right of every family to a minimum of decent living.

More important changes took place during or immediately after World War I, which led to the two Russian revolutions, of which the second brought the nondemocratic wing of the socialist movement to power in Russia and at the same time accentuated the split in that movement between the democratic and the nondemocratic wings, later also known as the *social democrats* and the *communists*. The Soviet Union established a society that did away almost completely with economic freedom and to a large extent with private enterprise. In the most advanced Western European countries, the public sector, social legislation, tax legislation, and some other types of restriction on private enterprise and economic freedom were pushed, leading to the form of integrated socioeconomic development later known as the *welfare state*. One of the limitations observed was that most workers, or even trade union leaders, were not particularly interested in what is nowadays called "participation" (in those days "industrial democracy" or "codetermination").

REACTIONS IN THE UNITED STATES

As already stated, in the United States the philosophy of free enterprise was adhered to for a longer period. It was only in the Great Depression, after 1929, that the belief in the automatic cure of all evils by free enterprise received a bad shock, simultaneously with another shock in Europe. In the United States, the New Deal brought a number of limitations to freedom and introduced social legislation and other forms of government intervention at a scale until then unknown. In some respects the United States went further even than a number of European nations. This was partly due to its higher average level of income, which permits some types of interference to be stronger; thus, the ratio of direct to indirect taxes is higher in the United States than in most European countries, and capital gains taxes do not exist in many European countries.

The Great Depression also brought the recognition of the necessity to regulate untable markets, especially numerous in agriculture, and the need of an anticyclical fiscal policy. Elements of planning were introduced, and not only in the public sector. On the contrary, planning had its greatest victories in big business.

The Second World War reinforced the understanding for the necessity of government guidance, at least during emergencies, and developed several techniques of intervention. Some type of government planning, whether of the rather penetrating French type or the much looser Anglo-Saxon type (Council of Economic Advisers in the United States, Cabinet Offices and, later, "Neddy" in Britain) entered the scene in almost all Western countries.

BACK TO ECONOMICISM

The end of World War II brought some reinforcement of economicism, however. In the United States, the untimely dissolution of the Office of Price Administration—leading to world inflation—was one example; in Western Germany, as a reaction to Hitlerism, a revival of slightly old-fashioned ideas under the beautiful name of "social market economy" brought back some economicist attitudes, which, to be sure, helped to rebuild the German economy (the *"Wirtschaftswunder"*). Even so, anticyclical policies continued and were remarkably successful; social insurance was expanded further, and for quite some time reconstruction in Western Europe required some forms of intervention, including macroeconomic planning imposed by the Marshall Plan! After the reconstruction period was over, by 1960 or thereabouts, there was an increased interest for some economicist elements, strengthened also by some difficulties experienced by Eastern Europe, which had been brought under Soviet domination after World War II. With increasingly complicated production processes and higher standards of living, Eastern Europe needed another mix of planning and freedom, or, at least, a larger number of levels of decision making. This was brought by some of the well-known reforms, exploited by free-enterprise propaganda. A number of American authors introduced the concept of "market socialism," without stressing sufficiently that for many activities, even in the West, markets must be regulated and that a considerable sector still requires more rather than less planning. This is true in particular for activities causing ecological disequilibria (air and water pollution, external effects of drugs, etc.). It cannot be denied, however, that around the '60s there was a tendency again to forget about a number of "social" issues. We will discuss their nature and their role in the subsequent section of this paper.

NATURE OF SOCIAL ELEMENTS

Scientific interest in "social elements" in human life has developed, as usual, much later than the intuitive interest shown by society. While economics faculties were given independence from law faculties around the '20s in the more advanced countries, sociology or social science faculties were a much later development. While the League of Nations' Secretariat had a Financial Section and Economic Intelligence Service next to a number of political and technical units, it is only in the United Nations' Secretariat that, even though the International Labour Organization was the oldest "specialized agency," the need was felt to have a Department of Social Affairs separate from the Department of Economic Affairs. As late as 1964, the United Nations Research Institute for Social Development took its modest place in the United Nations family of institutions, reflecting an increased pressure from the Social Commission (now Commission for Social Development) to introduce "social elements" into development planning. This is not to say that many social issues had not already been studied scientifically long before, but it illustrates the lag in comparison with the study of economic problems.

There are *three definitions* of what constitute "social aspects of development." The oldest and, in my opinion, best definition is all measures to correct complete economic freedom. In this sense we have seen the establishment of

Ministries of Social Affairs and similar institutions in order to moderate laissez faire. The second definition has come into use, it seems, in an attempt to define some fields or sectors in which the activities of a social character were concentrated. It is a summing up, as of health, housing, education, nutrition, community development, and perhaps a few other matters. My rejection of such a definition is based on the fact that not all activities directed at better health, better housing, education, etc., are in fact meant to be included. Large portions of these activities are carried out by individuals acting in their own interest, as under laissez faire. It is only to the extent that the results of income distribution under economic freedom have to be corrected that measures on behalf of weak groups are taken in these fields, and this means that the first definition of social aspects is more satisfactory. A third attempt to define social, as distinct from economic, aspects identifies social aspects with the ultimate aims of development and economic aspects with the means applied to attain the aims. While this structure of a development policy, that is, the distinction between aims and means, is very important, we have already got terms to make the distinction ever since the Swedish school started giving attention to it,[a] and I see no need to change the terms chosen. This is why I prefer the oldest definition. I should add that there is perhaps no need at all to use the word social in contradistinction to economic. We may simply speak about socioeconomic development, thus using a term introduced long ago by Cassel [b] simply to indicate the difference between the economics of society and business or enterprise economics.

In any case, it has long been admitted that the aims of development should be the maximum welfare of the population as a whole over a long period and that the elements of welfare are all those things that make people happy. Among them are things of a material nature and of a nonmaterial nature and things pertaining to personal needs as well as to social needs—that is, needs of groups or needs springing from a *comparison* between one individual and others The length of the period is the most difficult aspect of defining welfare; elements involved here are that longer-term interests of one individual should not be forgotten vis-à-vis short-term interests, but also that the distribution between generations should be observed.[c]

Although the economicist view does imply the care for elements of an individual character, the integrated socioeconomic view, while not forgetting the importance of personal incentives and initiatives, brings in the elements of comparison and hence of distribution. For the satisfaction of the poorer groups in any society, the distribution of welfare among individuals and groups is a vital element. It is insufficient to see to it that, as *Rerum novarum* put it, a minimum of decent life is attained, or, rather, this formula is empty to the extent that what is considered decent depends on what others have. Marx

[a] Although the concepts have been used earlier by his compatriots the clearest introduction of the aims and means of economic policy (in his case fiscal policy) can be found with B. Hansen in his Finanspolitikens Ekonomiska Teori (Stockholm 1955); cf. also E. Lindahl, Studies in the Theory of Money and Capital, London 1939, p. 22.

[b] G. Cassel called his well-known textbook on economics *"Theoretische Sozialökonomie"*.

[c] M. Inagaki, The Theory of Optimal Economic Growth: A Contribution, diss. Rotterdam (to be published by North Holland) made a very remarkable contribution to exactly this aspect of the problem.

spoke of a "customary standard" to establish what people think is decent. What really matters is the *distribution* of welfare (or, as the closest proxy, income) and not the absolute level or even the rate of improvement, although both of the latter have some relevance, especially in the short run. The following evidence may be quoted in support of my thesis.

1. Already within the world of labor it is well known that, upon appraising a wage offer, workers pay considerable attention to the *ratio* of that wage to wages of others. As a matter of experience, wages in one industry are kept in line with those in some standard industry, especially in the metal-working industry.

2. More generally, there is the well-known desire not to differ too much from others, as illustrated by the phrases "keeping up with the Joneses" and "conspicuous consumption" as factors of discontent among those not able to consume as much as the standard consumption pattern.

3. As an illustration we see that in recent years all discontent about society shows up in demands for *wage increase far surpassing the average rate of increase in income per capita.* If workers were satisfied that the present distribution of income is "correct," they would not ask for so much more; but they feel they are disadvantaged.

4. Accordingly, public opinion polls reflect *more* dissatisfaction with existing income distribution the *lower* a group's income is.

In conclusion, I want to state that what has been overlooked by the economicist view is primarily the widespread feeling that *income distribution is unsatisfactory*, not only under laissez faire, but even in Western Europe's mixed system [d] and Eastern Europe's centrally planned system. On the other hand, there remains a hard core of truth in the economicist view, namely the need for personal incentives.

RECENT BLOWS TO ECONOMICISM

During the last five years or so, additional blows have been given to purely economicist thinking, on both sides of the Atlantic. In Western Europe, the postwar innovation processes had come to an end around 1960, and there is an increasing feeling that improvements in income distribution have also stopped. Moreover, the desire for more *participation*, which was not widespread in the '20s, has become a new dimension in the thinking not only of students but also of groups of employees. Although in different forms, some of these desires have also come to the fore in Eastern Europe. In some respects the Yugoslav development of workers' councils has attracted much interest. Yet, for a majority of employees, it is probably still income distribution that interests them most. The difficulty surrounding the demand for added income is that it is manifested by a demand for instruments that are not, according to economic analysis, the most appropriate ones. The demands are for higher wages, whereas the instruments that may do the job are to be found in tax structures and education policies.

[d] In Sweden, the Socialist Party and the Trade Union Central Organization, published a report, "Jämlikhet" (Equality), in 1969, which reflects the renewed interest in the subject.

In the United States, similar desires have been expressed, but the problem has been complicated by its race aspect. In fact, what the United States is experiencing is a prelude to what the world at large will be facing soon: how to make up for the poor sections of society whose poverty is partly due to underprivileged positions and partly to the ensuing lag of participation in modern education and knowledge.

Two other features in recent discontent are worth mentioning. One is the tendency of some young people to despise the comfort created by technological progress. This could be a positive force for development if there were not another negative feature, a sort of modern anarchism, implying the untenable view that under the present technological circumstances one can first destroy the existing order and then start building a new one. The lack of understanding of present-day technology (in the widest sense, and including the technology of administration and government) inherent in this view is indicative of ignorance and emotion; our hopes must be for a more mature view on the means to change the society.

THE DEVELOPING WORLD

As stated in the preceding section, the world at large is the scene of similar feelings of discontent characteristic for the industrialized countries. The extreme poverty in which the masses of Asia, Africa, and Latin America are living is no longer accepted as an inevitable destiny. Thanks to the general development of science and technology, ideas are rapidly spreading to all corners of the earth and a confrontation of the poverty of the developing world with the relative prosperity of the developed world becomes more natural every day. The ideas are spreading that the growing gap in welfare is not a necessity and that the economicist view can no longer be the only basis for world economic relations.

Two features enormously complicate the problems of the Third World. One is that the attitudes of their populations are still influenced by cultural and societal elements that stand in the way of an efficient development of their economies. The *cultural* elements are those of a metaphysical origin as described, in their great variety, by cultural anthropologists. The *societal* elements are those of feudal stratification, with all the discrimination and taboos that go with it. The other complicating feature of the underdeveloped countries is the size of the problem, which has been so magnified by the population explosion, caused by the penetration of Western knowledge in the field of health before the other balancing elements of Western knowledge spread.

Even so, the essential problem of the developing countries is a repetition of the social problem in the developed world: the discovery of the neglect of what were called in the preceding sections the social aspects of development. Accordingly, we see a repetition of many discussions, first held about the Western social problem and now held about the world development problem. Most of the remedies found for at least a partial solution of the West's social problem can be recommended for the beginning of the solution of the world's development problem. I am not going to demonstrate this here; I have tried to do so elsewhere. The one aspect I want to highlight here is that again, as in *Rerum novarum,* we find those who maintain that it is sufficient that a minimum of decent living be organized for the masses of the poor.

WHAT LESSONS TO BE LEARNED?

What we have to learn from our own previous experience is that such a minimum of decent living, although desirable and necessary, will not be sufficient. What will really matter, in the long run, is the distribution of income, both within nations and among nations. Temporarily, a quick rate of improvement will satisfy the underprivileged; in the end their satisfaction will be determined by the ratio between their incomes and those of the privileged. While some differences in income are accepted as reasonable by everybody, a considerable portion of present income difference is not considered just. Differences due to explicit privileges—that is, discrimination—will not be accepted. Incomes obtained without effort will not be accepted either. There will be a tendency to ask equal incomes for people who are making equal efforts, even if these efforts do not produce the same results. This is what I think we are heading for. Economic science should try to find out how this can be attained without killing incentive. I think there are more possibilities than are usually considered among economists.

As already observed, the emphasis should be on tax systems and education activities, In regard to taxes, a shift from income taxes to wealth taxes is one of the avenues to follow. A given amount of wealth tax tends to contribute more to equality and to weaken incentives to work less than an equal amount of income tax. Because little is known about the precise reactions of the taxpayers, the shift should be undertaken step by step, so that the community can feel its way. In addition to a tax on financial or physical wealth, the possibility of taxing personal capabilities should be explored. Here we can expect only some modest first steps; thus, we may think of a special tax on persons with high academic scores. There are beginnings of such taxes in that scholarships have to be paid back at a higher annual rate by alumni who obtain good jobs than by those who obtain less good jobs.

In the field of education, the bottleneck has now been found to be the *preschool education* received in the family and the outside environment, and experiments are being made with special supplementary teaching to youngsters in the primary schools coming from underprivileged surroundings. If, along these lines, together with the already existing facilities for scholarships, the supply of trained and qualified manpower can be increased while the supply of less trained goes down, the bargaining position for income of the poorest strata will be improved and less inequality in wage and salary scales will result.[e]

In the international field, we should learn from experience first, something about the eventual aim of development cooperation. This aim should be to reduce the income gap between developed and developing countries. It will not be sufficient to aim at a "decent minimum" of income per capita, although for quite some time to come this aim will be sufficient. Actual living levels are so much below anything that could be called decent that both aims require the same action. Eventually, however, it will be distribution—that is, ratios—that matters.

Much more important at present is, second, what should be undertaken in the short run.

The strategies needed in the international field have recently been described

[e] Some further elaboration of the ideas briefly set out here will be found in J. Tinbergen, "Development Strategy and Welfare Economics," Coexistence 6 (1969) p. 119.

by the *Pearson Report* [f] and the *Report of the United Nations Development Planning Committee*.[g] It would be duplication to make a new attempt to indicate these strategies. Let me therefore briefly list only some of the most important recommendations of the latter committee, of which I happened to serve as chairman.

For developing countries the most important recommendations are:

— to increase the country's savings by 0.5 percent of gross domestic product per annum;

— to carry through land reforms (that is, putting a ceiling on the amount of land owned by one family) or to introduce a progressive tax on land and some other elements of physical wealth which can be easily assessed;

— to eliminate privileges given to some social groups;

— to improve efficiency and cooperation within the government machine;

— to reorient education and methods of education to meet the need for technicians and entrepreneurs;

— to cooperate with neighboring countries to establish larger markets;

— to improve marketing of their export products.

For developed countries the most important recommendations are:

— to make available a financial flow to developing countries of one percent of gross national product;

— to ease conditions for public financial transfers, which should attain 0.75 percent of gross national product and, by 1975, show an "aid content" of about 80 percent;

— to eliminate, in about five years, trade impediments to imports of semimanufactured and manufactured goods for developing countries;

— to conclude commodity agreements for unstable raw material markets;

— to spend, by 1980, 5 percent of their research and development expenditures on problems relevant to developing countries and to transfer 0.05 percent of their gross national product to developing countries for direct support of science and technology.

For the *United Nations system of agencies* the recommendations are that the process of development policies and cooperation be regularly evaluated, at the following levels:

— the level of the single nation;

— the level of consortia or similar groups of countries cooperating with a developing country or a group of such countries;

— the regional level (corresponding with the Regional Commissions); and

— the world level.

[f] Partners in Development, Report of the Commission on International Development, Lester B. Pearson, chairman; New York, 1969.

[g] United Nations Development Planning Committee, Report of the Sixth Session (New York, 1969).

While these evaluation processes are considered part of the administration of implementation, a group of independent experts has been asked to comment on the criteria used and the judgments passed. Their comments should be offered separately to the U.N. Economic and Social Council, where an annual discussion of progress and prospects should be held in preparation for the General Assembly. The importance of evaluation is its feedback on policies, and hence the contribution it can make to increase the efficiency of the operation

POLICY SCIENCES: DEVELOPMENTS AND IMPLICATIONS

Yehezkel Dror

On leave from the Hebrew University of Jerusalem, Israel, to the Rand Corporation, Santa Monica, Calif

The Need for Policy Sciences [a]

From the point of view of human action, scientific knowledge can be divided into three main levels: knowledge relevant to the control [b] of the environment, knowledge relevant to the control of society and individuals, and knowledge concerning the control of the controls themselves, that is, on *metacontrol*.

Knowledge of control of the environment, as supplied by rapid progress in the physical sciences, is the most highly developed of the three. Knowledge about control of society and individuals is much less advanced, but at least the social sciences and psychology constitute recognized components of science, receive significant support, and show some signs of progress. Least developed of all, and scarcely recognized as a distinct focus for research and study, is metacontrol knowledge, that is, knowledge concerning the design and operation of the control system itself.

Scarcity of knowledge relevant to design and operation of the social overall control system—which I call the public policymaking system [c]—has accompanied humanity from its beginnings. While some progress has taken place in the mechanics of control and microcontrol systems of some social components (such as corporations), the essential features of the public policymaking system continue to be beyond penetrating understanding and, even more so, beyond conscious and deliberate design.

This blind area in human knowledge has always caused suffering and tragedy, in terms of human values. But, from a longer-time perspective, the weaknesses of the public policymaking system did not matter very much as long as the operations of that system did not constitute an important variable in shaping human destiny. When most variables shaping human and social fate were beyond influence by the public policymaking system because of the absence of powerful policy instruments, bad decisions on the use of the few available instruments (or, to be more exact, "instrument images") had only

---

[a] Any views expressed in this paper are those of the author. They should not be interpreted as reflecting the views of the Rand Corporation or the official opinion or policy of any of its governmental or private research sponsors.

[b] I am using the term "control" in the sense of regulating, governing, shaping, directing, and influencing. "Monitoring" is one subelement of "control," in the broad sense in which I use the latter.

[c] Comparable terms are "central guidance cluster" as used by Bertram Gross and "societal control centers" as used by Amitai Etzioni. See Bertram M. Gross, *The State of the Nation: Social Systems Accounting*, London: Tavistock Publications, 1966, pp. 72–73, and Amitai Etzioni, *The Active Society*, New York: Basic Books, 1968, p. 112.

very limited impact on basic reality and therefore could not cause long-range harm.

It is this insignificance of public policymaking systems on the long-range fate of humanity that is changing, thanks to rapid progress in knowledge of policy instruments that permits control of environments, society, and individuals. New knowledge supplies increasingly potent instruments for use by mankind. The nuclear bomb and ecology-poisoning techniques and materials are but weak illustrations of the powerful policy instruments supplied by modern science. Altering the gender of children, weather control, genetic engineering, stimulation of altered states of consciousness, and emotion controls—these are only some illustrations of the more powerful capacities for controlling the environment, society, and individuals that the progress of science is sure to supply in the foreseeable future.[d]

It is the growing gulf between capacity to control the environment, society, and individuals, on one hand, and, on the other, knowledge of how to design and operate policymaking systems so they can use these capacities that constitutes the major danger to the survival and development of humanity. The emergence of *controlling man,* who exerts dominance over his environment, over social institutions, and over the very nature of human beings, makes it absolutely essential to improve policymaking systems so as to use wisely the powerful instruments at his disposal.

I purposely use the term "wisely" to emphasize the multidimensionality of required changes in public policymaking systems. Urgently needed, for instance, are new values and belief systems that meet the new global role of *controlling man.* Scientific knowledge cannot supply new values and belief systems,[e] though perhaps some of the conditions of value innovation can be studied and consciously encouraged. But science can and should supply knowledge on preferable designs and patterns for the rationality components of public policymaking systems, including rational means for improving the designs and patterns of the essential extrarationality components.[f]

In short, a main problem faced by humanity can, I think, be summed up in what I aphoristically call the Second Dror Law: [g] *While human capacities to*

---

[d] For a careful discussion, see John McHale, *The Future of the Future,* New York: George Braziller, 1969. For longer-range and more speculative explorations, see Gordon Rattray Taylor, *The Biological Time Bomb,* New York: Signet Books, 1968, and Burnham Putnam Beckwith, *The Next 500 Years,* New York: Exposition Press, 1967.

[e] For somewhat different and stimulating views, see Hazan Ozbekhan, "Toward a General Theory of Planning," in Erich Jantsch, Ed., *Perspectives of Planning,* Paris: OECD, 1969, and Erich Jantsch, "From Forecasting and Planning to Policy Sciences," *Policy Sciences,* Vol. 1, No. 1, Spring 1970, pp. 31–42. Completely unacceptable, in my opinion, are the naïve proposals made from time to time by physical scientists to achieve deliberate and systematic value innovations aimed at the long-range future through quasi-rational mass movements. See, for instance, Gerald Feinberg, *The Prometheus Project,* Garden City, New York: Doubleday, 1969.

[f] For an extensive discussion of the roles of rationality and extrarationality components in preferable policymaking, see Yehezkel Dror, *Public Policymaking Reexamined,* San Francisco: Chandler Publishing Company, 1968, pp. 154–196.

[g] The First Dror Law states: *While the difficulties and dangers of problems tend to increase at a geometric rate, the knowledge and manpower qualified to deal with these problems tend to increase at an arithmetic rate.*

shape the environment, society, and human beings are rapidly increasing, policymaking capabilities to use those capacities remain the same.

A large number of dispersed efforts to develop knowledge relevant to policymaking improvement do take place. These include work under the auspices of a number of new disciplines, approaches, and interdisciplines, such as operations research, praxeology, systems analysis, organization theory cybernetics, information theory, theory of games, organizational development approaches, strategic analysis, future studies, system engineering, decision theory, and general systems theory. Also important is some work in new directions within more traditional disciplines, especially economics, some branches of psychology, and some parts of political science.[h] This work supplies important insights, promising concepts, and stimulating ideas. But, in general, present endeavors to develop scientific knowledge relevant to the improvement of policymaking tend to suffer from the following weaknesses: [i]

1—Microapproach, with applications to some types of decisions but very limited relevance to the policymaking system as a whole.

2—Disjointedness, resulting in fragmented views limited to single dimensions of policymaking. Thus, systems analysis is quite isolated from organization theory, operation research from psychology of judgment, and decision theory from general systems theory.

3—Preoccupation with the rationality components of policymaking, with little attention to the fusion of rationality with extrarationality and the improvement of the latter.

4—Incrementalism, with nearly complete neglect of the problems of policymaking systems novadesign (i.e., design anew), as distinguished from slight redesign.

5—Narrow domain of concern, which neglects consideration of possible improvement needs and improvement possibilities of some critical elements of the policymaking system, such as politicians.

6—Sharp dichotomy between the behavioral approaches, which study some segments of policymaking reality, and the normative approaches, which design abstract rationality-based microdecision models. Therefore, no comprehensive approach to understanding and improvement of the policymaking system as a whole.

7—In the normative approaches: strong dependence on metric quantification and therefore inability to handle "qualitative" variables.

8—In the behavioral approaches: lack of interest in prescriptive methodology and gaps between lack of interest in application and partisan advocacy.

9—Fixation on conventional research methods and consequent inability to utilize important sources of knowledge (such as tacit knowledge of policy

---

[h] For selected bibliographic references to relevant work until 1967, see "Bibliographic Essay" in Yehezkel Dror, *Public Policymaking Reexamined, ibid.,* pp. 327–356. For a survey of more recent relevant literature, see Yehezkel Dror, The Bibliographic Appendix in *Design for Policy Sciences* (N. Y.: American Elsevier 1971).

[i] For an extensive discussion of such weaknesses of applied social sciences and of analytical decision approaches, see Yehezkel Dror, "Systems Analysis and Applied Social Sciences," to be published in the proceedings of the Rutgers University and *Trans-Action* Magazine Conference on Public Policy and Social Science (Carpender Conference Center, Rutgers University, New Brunswick, N. J., Nov. 23–26, 1969), edited by Irving L. Horowitz.

practitioners) and difficulties in designing new research methods to meet the special problems of policymaking study and improvement (e.g., social experimentation).

I could go on adding additional items to the list of inadequacies of most contemporary efforts to build up policymaking knowledge. But I think the problem goes beyond a shorter or longer list of discrete weaknesses. The problem is not one of accidental omissions that can be easily corrected. Rather, I think that the overall lack of saliency of contemporary scientific endeavors to the improvement of policymaking reflects a basic discongruency between the paradigms of contemporary sciences in all their heterogeneity and the paradigms necessary for building up policymaking-relevant scientific knowledge.

To put this opinion into a positive form, it seems to me that in order to build up a science of policymaking, we need a new type of science based on a new set of paradigms.[j] Following the pioneering suggestion of Harold D. Lasswell,[k] I propose to call this new area of study, research, teaching, professional activity, and application "policy sciences"; but the name does not really matter.

As a matter of fact, policy sciences are at present in *status nascendi* and hopefully approach a taking-off stage. Among the signs of their emergence, let me mention the following:

1—The already mentioned proliferation of research and study of various policymaking issues within new and traditional disciplines. This testifies to widespread interest and serves to build up important, though disjointed, subcomponents of policy sciences.

2—The invention and development of new types of policy research organizations that, in effect, engage in the development and application of policy sciences. The Hudson Institute, the Urban Institute, parts of the Brookings Institution, the new Woodrow Wilson Foundation, the Institute for the Future, the Rand Corporation, and the New York-Rand Institute illustrate this trend in the United States.

3—The self-education of outstanding individual policy scientists who, thanks to personal multidisciplinary background, accidents of opportunity, and interest in application of scientific methods to acute problems, got into the pioneering of policy sciences and thus demonstrated the feasibility of policy sciences and its promises.

4—The recent establishment of new university programs devoted to policy

[j] My terminology follows Thomas S. Kuhn, *The Structure of Scientific Revolutions,* Chicago: University of Chicago Press, 1962.

[k] The concept of "policy sciences" was first proposed in 1951 by Harold D. Lasswell, in Daniel Lerner and Harold D. Lasswell, eds., *The Policy Sciences: Recent Developments in Scope and Methods,* Stanford: Stanford University Press, 1951. For recent versions of Lasswell's views, see Harold D. Lasswell, "Policy Sciences" in *International Encyclopedia of Social Sciences,* Vol. 12, pp. 181–189, and Harold D. Lasswell, "The Emerging Conceptions of the Policy Sciences," *Policy Sciences,* Vol. 1, No. 1, 1970, pp. 3–14. The subject is extensively treated in a forthcoming book by Harold D. Lasswell, *A Preview of Policy Sciences* in Yehezkel Dror, *Ventures in Policy Sciences: Concepts and Applications* (N.Y.: American Elsevier, 1971) and in Yehezkel Dror, *Designs for Policy Sciences,* op. cit.

sciences, with or without use of that term. In the United States alone, more than ten such programs have been initiated during the last two or three years.[1]

5—The rapidly increasing number of conferences, books periodicals, "invisible colleges," and similar expressions of professional activity and interest devoted in effect to the advancement of policy sciences as a whole or of some of its major aspects.[m]

These are some of the signs of search, concern, experimentation, and interest that, I think, indicate the emergence of policy sciences. Nevertheless, at best, we are only in the first stages of the required scientific revolution, and there is no assurance that it will be successful in bringing forth a viable and significant new kind of science. The challenge may be beyond our intellectual abilities, charlatans may discredit the idea of policy sciences before it really gets started, political culture may inhibit the efforts, or the conservatism of "normal" scientists may choke it. Even if policy sciences do emerge as a new type of scientific endeavor, it is doubtful that one can predict now their future characteristics and implications. Therefore, the following exploration of the new paradigms of policy sciences and of their applied implications should be regarded as a normative forecast, directed at least as much at shaping the future as at foreseeing it.

Subject to this qualification, I think that preliminary examination of some of the unique paradigms of policy sciences, as I see them, will serve to illuminate both the current effort and the urgent need. It will also serve as a basis for examining some applied implications. As our analysis is a rough one, mistakes in some specifications do not matter. It is in the overall *Gestalt* of policy sciences that we are interested.

Some New Paradigms of Policy Sciences [n]

It seems to me that the main paradigmatic innovations to be required of and expected from policy sciences can be summed up as follows:

1—The main concern of policy sciences is with understanding and improvement of macrocontrol systems: that is, public policymaking systems. In addi-

[1] The graduate university programs about which I happen to know include, in no particular order: The program in public policy at the John F. Kennedy School at Harvard University; the Doctoral Program in Policy Sciences at the State University of New York at Buffalo; the Graduate School of Public Affairs at the University of California, Berkeley; the Doctorate Program in Social Policy Planning, also at the University of California, Berkeley; the Graduate Program in Planning at the University of Puerto Rico; the Institute for Public Policy Studies at the University of Michigan; the School of Urban and Public Affairs at Carnegie-Mellon University; the Doctorate Program in Public Policy Analysis at the Fels Institute of Local and State Government at the University of Pennsylvania; the Program in Planning and Policy Sciences, also at the University of Pennsylvania. Also moving in the same direction seem to be the Lyndon B. Johnson School of Public Affairs at the University of Texas, a proposed Center for the Policy Sciences at Brown University, and a proposed new school at the University of Hawaii.

[m] To illustrate, let me mention some relevant recently founded periodicals: *Futures, Long-Range Planning, Policy Sciences, The Public Interest, Public Policy, Socio-Economic Planning Sciences,* and *Technological Forecasting.*

[n] This and the following section lean in part on Yehezkel Dror, "Prolegomenon to Policy Sciences," *Policy Sciences,* Vol. 1, No. 1, Spring 1970, pp. 135–140.

tion to overall improvement-oriented study of such systems, main foci of policy sciences include, for example (a) policy analysis, which provides heuristic methods for identification of preferable policy alternatives, (b) alternative innovation, which deals with the invention of new designs and possibilities to be considered in policymaking, (c) master policies, which provide guidelines, postures, assumptions, strategies, and main guidelines to be followed by specific policies, (d) evaluation and feedback, including, for instance, social indicators, social experimentation, and organizational learning, and (e) improvement of the policymaking structure through redesign and novadesign of its organizational components, selection and training of its personnel, and reconstruction of its communication and information network. While the main test of policy sciences is better achievement of considered goals through more effective and efficient policies, policy sciences as such are, in the main, not directly concerned with the substantive contents of discreet policy problems (which should be dealt with by the relevant normal sciences), but rather with improved methods, knowledge, and systems for better policymaking.

2—Breakdown of traditional boundaries between disciplines, and especially between the various social sciences and the decision disciplines. Policy sciences must integrate knowledge from a variety of branches of knowledge and build it up into a supradiscipline focusing on public policymaking. In particular, policy science is based on a fusion between social sciences and analytical decision approaches. But it also absorbs many elements from decision theory, general systems theory, organization theory, operations research, strategic analysis, systems engineering, and similar modern areas of study. Physical and life sciences are also relied on, insofar as they are relevant.

3—Bridging of the usual dichotomy between "pure" and "applied" research. In policy sciences, integration between pure and applied research is achieved by acceptance of the improvement of public policymaking as its ultimate goal. As a result, the real world becomes a main laboratory of policy sciences, and the test of the most abstract theory is in its application (directly or indirectly) to problems of policymaking.

4—Acceptance of tacit knowledge and personal experience as important sources of knowledge, in addition to more conventional methods of research and study. Efforts to distill the tacit knowledge of policy practitioners and to involve high-quality policymakers as partners in building up policy sciences are among the important characteristics distinguishing between policy sciences and contemporary "normal" sciences.

5—Policy sciences share with normal sciences main involvement with instrumental-normative knowledge, in the sense of being directed at means and intermediate goals rather than absolute values. But policy sciences are sensitive to the difficulties of achieving "value-free sciences" and try to contribute to value choice by exploring value implications, value consistencies, value costs, and the behavioral foundations of value commitments. Also, parts of policy sciences are operative in invention of different "alternative futures," including their value contents. Furthermore, "organized creativity"—including value invention—constitutes important inputs into parts of policy sciences (such as policymaking-system novadesign and redesign, policy design, and policy analysis), and encouragement and stimulation of organized creativity is therefore a subject for policy sciences. As a result, policy sciences should make a breach in the tight wall separating contemporary sciences from ethics and philosophy of values and build up an operational theory of values (including value morphology, taxonomy, measurement, etc., but not the substantive absolute norms themselves) as a part of policy sciences.

6—Policy sciences are very time-sensitive, regarding the present as a "bridge between the past and the future." Consequently, they reject the ahistoric approach of much of contemporary social sciences and analytical approaches. Instead, they emphasize historic developments on one hand and future dimensions on the other hand as central contexts for improved policy-making.[o]

7—Policy science does not accept the "take it or leave it" attitude of much of contemporary social sciences, neither does it regard petition signing and similar "direct action" involvements as a main form of policy science's contributions (in distinction to scientists acting as citizens) to better policy-making. Instead, it is committed to striving for increased utilization of policy sciences in actual policymaking and to preparation of professionals to serve in policy-sciences positions throughout the macrocontrol system (without letting this sense of mission interfere with a clinical and rational-analytical orientation to policy issues).

8—Policy science deals with the contribution of systematic knowledge and structured rationality to the design and operation of macrocontrol systems. But policy science clearly recognizes the important roles both of extrarational processes (such as creativity, "intuition," charisma, and value judgment) and of irrational processes (such as depth motivation). The search for ways to improve these processes for better policymaking is an integral part of policy sciences, including, for instance, possible policymaking implications of altered states of consciousness. (In other words, policy sciences faces the already mentioned paradoxical problem of how to improve extrarational and even irrational processes through rational means.)

Some Implications of Policy Sciences

Any policy science the *Gestalt* of which resembles the image conveyed by the proffered policy sciences paradigms will have far-reaching implications. Of relatively minor importance are various implications for the organization of science, its research, and its teaching. These include, for instance, transfer of some major research and teaching functions from universities to policy-research organizations; participation of experienced politicians, executives, and similar policy practitioners in scientific activities; novel teaching designs;[p] and new career patterns entailing transitions between abstract policy sciences research, long-range policy research, and policy analysis of pressing issues—accompanied by movement between universities, policy research organizations, and a variety of new roles in various branches of government and in public, quasi-public, and private organizations.

Those are implications of much importance for academia. But from an

---

[o] On the relations between future studies and policy sciences, see Yehezkel Dror, "A Policy Sciences View of Future Studies: Alternative Futures and Present Action," *Technological Forecasting and Social Change*, Vol. 2, No. 1 (Winter 1970), pp. 3–16.

[p] For an illustration, see Yehezkel Dror "Teaching of Policy Sciences: Design for a University Doctoral Program," *Social Science Information* Vol. IX, No. 2 (April 1970), pp. 101–127.

overall social point of view the critical significance of policy science is in basic changes that it brings about in the age-old dilemma of *scienta et potentia,* knowledge and power. These, in turn, have far-going implications for the exercise and structure of social power, that is, for politics.

The relevant unique feature of policy science is that it presumes to deal with the internal processes of policymaking and presumes to tell the policy-makers how to arrive at decisions. This is a degree of penetration into the innermost processes of politics removed by a step-level function from the contributions of contemporary "normal" sciences to policymaking. Contemporary "normal" sciences supply inputs to be taken into account in policy-making and sometimes propose solutions as stipulated outputs of policymaking; but contemporary "normal" sciences do not open up the black box of how policy decisions are made and do not claim to develop scientific models for rewiring the box.[q]

In blunt language, the more policy science indeed does develop, the more the policymaking system should be redesigned to avail itself of policy sciences knowledge and the more politics should be reformed to permit full symbiosis between political power and policy sciences knowledge. The basic roles of elected politicians in a democratic society will not be impaired. Indeed, the critical functions of value judgment, interest presentation, consensus mainte-nance, and trans-scientific judgment not only will not be weakened but will be strengthened, thanks to clearer presentation of alternatives, better control of implementation, more reliable feedback, fuller explication of tacit theories, and similar contributions of policy analysis. But essential are policymaking arrangements that will assure that policy-science knowledge will be correctly appreciated and taken into account and that both its underutilization and its overutilization will be avoided.

Somewhat to concretize this general idea, let me present some implications for changes in the policymaking system that seem to result from initial work in policy sciences.[r] To provide variety in my illustrations, some are presented as a short enumeration while others, which are less technical, are discussed at some length:

1—Pervasive utilization of policy analysis for consideration of issues, ex-ploration of alternatives, and clarification of goals.[s]

[q] Some exceptions are provided by political science and public administration, both classic and modern. But the relevant work in political sciences tends to suffer from one or more of the following characteristics, which make them inadequate surrogates for policy sciences: (1) mainly ideological orientation; or (2) mainly technical orien-tation, dealing with "administrative efficiency"; (3) focus on specific components of the policymaking system, without an overall systems view; (4) absence of empiric basis; or (5) absence of decision theory basis.

For discussions of administrative reforms that clearly bring out these and addi-tional weaknesses of administrative reforms theory and reform practice alike, see: Ralph Braibanti, ed., *Politics and Administrative Development,* Durham, N. C.: Duke University Press, 1969; Gerald E. Caiden, *Administrative Reform,* Chicago: Aldine Publishing Company, 1969; and the still unique Dwight Waldo, *The Administrative State,* New York: The Ronald Press, 1948.

[r] For a detailed discussion of some of these recommendations and their policy sciences theoretic bases, see Yehezkel Dror, *Public Policymaking Reexamined,* op. cit., esp. Part·V, pp. 217 ff.

[s] See Yehezkel Dror, *Design for Policy Sciences,* op. cit., chapter 9, 1969.

2—Encouragement of explicit policy strategy decisions, in distinction from discrete policy determinations. Explicit strategy decisions (including mixed strategies) are needed on the following issues, among others: degrees and locations of acceptable innovations in policies; extent of risk to be accepted in policies and choice between a maximax strategy or/and maximin strategy and/or min-avoidance strategy;[t] preferable mix between comprehensive policies, narrow-issue-oriented policies, and shock policies (which aim at breakthroughs accompanied by temporary disequilibration); and preferable mix between policies oriented toward concrete goals, toward a number of defined future options, and/or toward bulding up resources better to achieve as yet undefined goals in the future.

3—Encouragement of comprehensive master policies, in which discrete policy issues are considered within a broader context of basic goals, postures, and directives.[u]

4—Systematic evaluation of past policies to learn from them for the future. For instance, methods and institutions should be established every fixed period to provide an independent audit of the results of legislation.

5—Better consideration of the future. Special structures and processes should be designed to encourage better consideration of the future in contemporary policymaking.[v] This includes, for instance, dispersal of various kinds of "lookout" organizations, units, and staff throughout the policymaking system and utilization of alternative images of the future and scenarios in all policy considerations.

6—Search for methods and means to encourage creativity and invention in respect to policy issues. This may entail, for instance, no-strings-attached support to individuals and organizations engaging in adventurous thinking and "organized dreaming"; avoidance of their becoming committed to present policies and establishments; and opening channels of access for unconventional ideas to high-level policymakers and to the public at large. Creativity and invention may also be influenced within policymaking organizations by in-

---

[t] I use the term "min-avoidance" to refer to policies directed at avoiding the worst of all possible situations. One important advantage of such a strategy concerns support recruitment: it is often much easier to achieve agreement on ills to be avoided than on operational positive formulations of "good life" to be realized.

Some success in min-avoidance would constitute a significant improvement over reality. However simple this may sound, human capacities to approximate minimin are amazing. Still well worth reading in this connection is Walter B. Pitkin, *A Short Introduction to the History of Human Stupidity,* New York: Simon and Schuster, 1932. Recent policies around the world could provide a long second volume for such a history.

[u] President Nixon's First Annual Foreign Affairs Message, *United States Foreign Policy for the 1970s: A New Strategy for Peace,* well illustrates such an effort. It is relevant to observe that this innovation in comprehensive master policies is closely related to the existence of a new type of policymaking improvement-oriented policy analysis unit in the White House, namely, Dr. Kissinger's staff.

Preparation of similar master policies for, say, urban problems would require more than establishment of a parallel urban policy analysis unit in the White House. The basic-concept package and integrative framework have first to be developed. Among the urgent tasks awaiting policy sciences is work on overall policy-concept packages, on integrative problem mappings, and on issue taxonomies.

[v] The recently established National Goals Research Staff in the White House is an interesting step in this direction.

stitutionally protecting innovative thinkers from organizational conformity pressures. Requiring careful study also are creativity-amplifying devices and chemicals and arrangements for their possible use in policymaking.

7—Establishment of a multiplicity of policy-research organizations to work on main policy issues. Some of these policy research organizations would work for the central Government, some for the legislature, and some for the public at large, diffusing their findings through the mass media of communications.

8—Development of extensive social experimentation designs and of institutions able to engage in social experimentation (including reconsideration of related ethical problems). It seems quite clear that social experimentation is essential for finding solutions to present and emerging social issues. For instance, new experimental cities may be needed to develop suitable habitations for the 100 million additional Americans expected by the year 2000. Careful social experimentation requires invention of new research designs and of new legal-political arrangements. Also important and very difficult is the requirement for a political and social climate in which careful research and experimentation on social institutions is encouraged. (To take a United States illustration: A change is needed in attitudes that expressed themselves, for instance, in the legislative prohibition of studies on the operation of juries.)

9—Institutional arrangements to encourage "heresy" and consideration of taboo policy issues, such as the possibilities of long-range advancement of humanity through genetic policies and of changes in such basic social institutions as the family.

10—Improvement of one-person-centered high-level decision making. Even though of very high and sometimes critical importance, one-person-centered high-level decision making is very neglected by both contemporary research and improvement attempts. This, in part, is due to difficulties of access, on one hand, and dependence of such decision making on the personal characteristics and tastes of the individual occupying the central position and on the other hand, the consequent difficulties in improving such situations. Thus, neglect of the study and improvement of one-person-centered high-level decision making is illustrated in the lack of suitable research methods, conceptual frameworks, and instrumental-normative models in contemporary normal sciences. With the help of the novel approaches of policy sciences, one-person-centered high-level decision making can be improved. Many conditions of better decision making can be satisfied by a variety of means, some of which may often fit the desires of any particular decision maker, e.g. information inputs, access of unconventional opinions, feedback from past decisions, and alternative predictions can be provided by different channels, staff structures, mechanical devices, and communication media. This multiplicity of useful arrangements provides sufficient elasticity to fit the needs, tastes, preferences, and idiosyncrasies of most, if not all, top decision makers.

11—Development of politicians. The idea of developing the qualifications of politicians is regarded as "taboo" in Western democratic societies. But this is not justified. The qualifications of politicians can be improved within the basic democratic tenets of free elections and must be improved so as to permit the required new symbiosis between power and knowledge. Thus, for instance, politicians need an appreciation of longer-range political, social, and technological trends, need capacities to determine policy strategies, and should be able to critically handle complex policy-analysis studies. One possible approach to the problem is to encourage entrance into politics of suitably qualified persons and to vary the rules of presentation of candidates to permit better judgment by the voter. Other, less radical proposals are to establish policy-sciences programs in schools where many future politicians study (such as law schools), and to grant to elected politicians (e.g. members of a state legislature) a sabbatical to be spent in self-developing activities, such as studying and writing. Suitable policy-sciences programs can be established at universities and at special centers for active politicians to spend their sabbaticals in a productive and attractive way.

12—Advancement of citizen participation in public policymaking.[w] Here, modern technology may be very helpful by providing tools for much better presentation of policy issues before the public (e.g. policy analyses of controversial issues on TV and citizen involvement through active participation in policy games through cable TV) and for more intense participation of the public in decision making (e.g. systematic opinion polling with the help of computer home consoles).

13—Education of adults for more active roles in public policymaking. I just mentioned the intensification of citizen participation in public policymaking as one of the possible policy sciences-recommended improvement. But in order for increasing citizen participation to constitute in fact an improvement, changes in the quality of that participation are needed. At the very least, more knowledge on policy problems, better understanding of interrelations between different issues and various resolutions, and fuller realization of longer-range consequences of different alternatives are needed. Also highly desirable are better value explication and sensitivity to value trade-offs, increased propensities to innovate, and capacities to face uncertainty.

The slogan of "enlightened citizen" as a requisite of democracy has been with us for too long to be taken seriously. Nevertheless, increasing demands for citizen participation based on both ideological reasons and functional needs do combine and make "citizen enlightenment" a hard necessity. Indeed, because of the growing complexity of policy issues, increased quality of citizen contributions to public policymaking is essential to preserve the present level of citizen participation in public policymaking. In other words, if the quality of citizen inputs into public policymaking remains as it is now, meritocracy may well become the only chance for survival. Therefore, building up the policy-contribution capacity of citizens is essential for the continuous viability of democracy.

This is the challenge facing adult education from the point of view of public policymaking improvement. To meet this challenge, radical novadesign of adult education is required.

To concretize, let me mention these main policy sciences-related directions of novadesign of adult education:

(a) Policy sciences must develop new formats for presenting and analyzing public issues in the mass media of communication in ways conducive to formation of informed individual opnions. For instance, policy issues should be presented in the form of policy-analysis networks, with clear alternatives, explicit sensitivity analysis, uncertainty explication, and assumption visibility.

---

[w] For elaboration of this and the next two points with specific reference to urban problems, see Yehezkel Dror, "Urban Metapolicy and Urban Education," *Educational Technology* (September 1970), pp. 18–21. Vol. X, N. Y.

Techniques are required for presentation of such programs on TV in ways to combine audiences appeal with improvement of citizen comprehensions of complex issues.

(b) Training tools that are simultaneously interesting and beneficial must be developed. Such tools include, for instance, cases, projects, policy games, and individual policy-exploration programs. In particular, policy games and individual policy-exploration programs are very promising. Based on computers and brought to each house through cable TV and home computer consoles, suitable games and policy exploration programs should be able to combine education for better policymaking with inputs into ongoing policymaking.[x]

(c) Incentives for participation in policy-oriented educational activities must be provided. It is to be hoped that increased opportunities to participate in public policymaking, together with availability of clearly relevant learning opportunities, will provide basic motivation. This may be the case all the more because of the possibility—illustrated by the proposed techniques—of combining the useful with the attractive. But additional incentives may be necessary. Competitive games and exercises may provide one set of incentives; public attention and dramatizaton may provide a second. If this is not successful, reservation of some special opportunities to participate in public policymaking (other than the basic rights of voting, expression of opinion, etc., reserved, of course, for all) for those who do undergo a set of learning activities might prove necessary in some circumstances in the longer run. But adoption of suitable programs in schools—to be discussed later—should make such distasteful distinctions unnecessary.

These are only some illustrations that do point out the possibility for redesign of education to serve, *inter alia,* the needs of increasing citizen participation in public policymaking. This is a problem in need of much research and creativity.

14—Preparation of children for future roles in public policymaking. On a more fundamental level, preparation for increased participation in public policymaking must take place before maturation. The best location to prepare the citizen for increased policymaking roles is in school, when the necessary knowledge and capacities should be developed as a basic part of the equipment needed by every citizen in a modern, urban, democratic society.

The necessary knowledge and capacities to be conveyed and developed at school do include, among others, some knowledge and understanding of the social system and of social dynamics; a feel for alternative social futures; abilities to handle uncertainty and probabilities; basic skills in logic and semantics; understanding of the elements of policy analysis and capacity to handle problems with the help of policy-analysis networks; tolerance of ambiguity; appreciation of main concepts of social sciences, economics, and decision theory and their application to policy issues; and ability to search for information on new problems and issues and absorb that information within one's frame of appreciation.

This formidable list may look prohibitive, unless we bear in mind that no technical skills and professional knowledge are aimed at. Some familiarity with fundamental concepts, some appreciation of their use and—most im-

---

[x] E.g. see Stuart Umpleby, "Citizen Sampling Simulation: A Method for Involving the Public in Social Planning," Paper presented at the International Future Research Conference, Kyoto, Japan, April 10–16, 1970.

portant of all—some skill in application of the knowledge and concepts to concrete issues as a main mode for making up one's mind are the only aims.

Even so, this is an ambitious program, which can be approximated only through far-reaching changes in school teaching. Much of the required knowledge and capacity should be developed through new approaches and novel teaching methods in traditional subjects. Thus, the study of history should include the history of policy issues, should be problem-oriented, and should be supplemented by treatment of alternative futures. To add another illustration: mathematics should be taught as a problem-solving approach, with emphasis on probability theory, Boolean algebra, and theory of games. Some new subjects also have to be added, devoted explicitly to policy problems and policy analyses. In the new subjects and in the new contents of the traditional subjects, new teaching methods will play a major role. Such new teaching methods will include, for instance, gaming, computer interaction, and internships. Existing methods, such as projects and essays, can also be very useful, if suitably adjusted.

All this depends on the development of policy-sciences knowledge, which can serve as a basis for suitable teaching material and teaching methods. Here we meet another innovative facet of policy sciences: it should not constitute esoteric knowledge monopolized by an initiated few; instead, conscious and intense efforts must be made to transform at least the basics of policy-sciences knowledge into forms that can be widely communicated to different policy-making actors, to the interested broad public, and even to schoolchildren.

Lest the impression created by these illustrative policy-sciences implications for redesign of policymaking is that most of the burden of change lies on politics, the public, and education, let me add a word on implications for the scientific community, which go beyond the earlier-mentioned reorganization of research, teaching, and career patterns.

The emergence of policy sciences leads not only to many requirements for repatterning politics, education, etc., but also for repatterning the contributions of scientists to policymaking. At present, many of the pronouncements of scientists on policy issues suffer from serious defects, as can easily be illustrated from the debates on such issues as pollution, the nuclear test ban,[y] and the ABM.[z] These defects are related to failure to distinguish—first of all, for oneself, and then in one's pronouncements—between highly reliable scientific facts within the professional competence of the actor, doubtful scientific facts within the area of competence of the actor, issues that belong to science but are not within the competence of the actor, and issues that are outside the domain of science, such as judgment of values to be pursued and of value priorities, judgment in risks to be taken, and judgment on time preferences and metaphysical assumptions.

As a result of the failure to make these distinctions, recommendations are often presented "in the name of science" that, in fact, are based on assumptions and preferences in large part outside the domain of competence of the actor.

Even with the present very limited policy-sciences knowledge, this state of

[y] Cf. Robert Gilpin, *American Scientists and Nuclear Weapons Policy* (Princeton, N.J.: Princeton University Press, 1962) and Robert A. Levine, *The Arms Debate* (Cambridge, Mass.: Harvard University Press, 1963).

[z] See Yehezkel Dror, *Energy States: A Counterconventional Strategic Problem* (Boston: Heath & Co.), 1971 (in press), chapter 1.

affairs is not only regrettable but inexcusable: knowledge in policy analysis already available permits presentation of recommendations by scientists in formats that clearly distinguish between the different bases of their recommendations. Such formats would enable those entitled to it—whether the elected politicians or the public at large—to exercise their judgment in respect to those issues not included within the area of competence of the recommending scientist; therefore they should be widely used even now. When policy sciences are more developed, the demand on scientists to be self-sophisticated and self-restrained in their contributions to policymaking becomes more than a recommendation; it becomes, I think, a moral absolute imperative, deviation from which may well destroy democracy, science, or both. Thus, the emergence of policy sciences will be accompanied by very strict and, in some respects, restrictive, demands on scientists, not less so and perhaps even more so than on politicians and other actors in the public policymaking system.[aa]

Conclusion

Policy sciences holds forth the hope of improving the most backward of all human institutions and habits—policymaking and decision making. It constitutes a major attempt to assert and achieve a central role for rationality and intellectualism in human affairs and to increase by jumps the capacity of humanity to direct its futures. Important first steps to build up policy sciences are being attempted now. There is no assurance that these steps will lead anywhere and that the endeavor to build up policy sciences will succeed. But the expected benefits of policy sciences, and—even more—the gloomy results of failure to advance policy sciences, make this endeavor one of the more critical challenges ever faced by science. It is also one of the most difficult challenges because of the intrinsic difficulties of the subject, because of the needed revolution in scientific paradigms, and because of the far-going and in many respects radical implications. Therefore, policy sciences needs and deserves all the help it can get, including first of all strong support and intense personal commitment from the scientific community.

[aa] Especially vexing are the moral issues facing policy scientists. While all knowledge can be used for "good" and for "bad," the high potentials of policy sciences require special safeguards to reduce the probabilities of misuse. This problem is beyond the confines of this paper, but I want explicitly to point it out.

THE TIMETABLE PROJECT

Center for Integrative Studies
School of Advanced Technology
State University of New York
Binghamton, N.Y

The major problems and crises that human society now faces are not intrinsically new in our historical experience. Human development is a continuous record of the struggle against hunger, disease, war, and ignorance.

The present gravity and critical nature of these familiar aspects of the human condition is their expanded dimension. Their vastly increased scale and magnitude has been compounded, paradoxically, by those successful advances in the sciences and technologies that are the major means of combating them. By making man more secure against hunger and disease, we have added astronomically to the numbers of men; by shrinking the physical distance between peoples, we have increased the critical interdependence of all human societies; by communicating the material possibilities of a better life, we have enormously increased the expectations and demands of all peoples to share in our accumulated wealth and knowledge; and by the prodigal exploitation of our physical resources we have produced grave imbalances in our life-sustaining natural environment.

At the time, then, when we suddenly possess the developed skills and resources with which to solve many of our major human problems, they have, in turn, become seemingly intractable; and the recurring crises that accompany delays in providing solutions lead to frustration and conflict in major sectors of society. Disillusionment with rational means is encouraged, and an antiscientific and antitechnological bias begins to affect even the allocation of resources to those directions in which problem solutions might best be sought. This situation has been further obscured by the many conflicting ideas, plans, and programs for solution of major problems at the local and world level; such "solutions" often pay little regard to the actual, or potential, realities and limitations of the physical and human resources currently available to us.

Both the seeming "intractability' of the problems themselves and mounting public pressures for more immediate solutions arise, in part, from *the lack of a coherent and integrated assessment* of the resources, facilities, and time that may be required for even minimal solutions to many of our problems. Some problems could be solved tomorrow by legislative fiat. Others may take one year, five, ten, or twenty years, to reduce to less critical dimensions; and there are, obviously, several problems in human society that it may not be possible to solve, or even mitigate, in less than two or three generations. It is also increasingly evident that many problems approaching critical magnitude at various local national levels, e.g. population, health, education, environmental deterioration, etc., can no longer be solved even on national terms.

In similar fashion, the scale of many of our global technological systems, e.g. production, distribution, transportation, and communications, has gone beyond the capacities of any single nation or group of nations to sustain and wholly operate. They require, and are dependent on, the resource range of the entire earth for the metals and materials of which they are built and the energies to run them—in which no single nation is now self-sufficient. The whole planetary "life-support system" is also increasingly dependent on the global interchange not only of physical resources and finished products but of the "knowledge pool"—research, development, technical and managerial expertise—and the highly trained personnel who sustain and expand this.

The scale and range of our technological intrusions into the planetary biosphere is now such that all large scale technoindustrial undertakings need, increasingly, to be gauged in terms of their long-range consequences and implications for the global community. In these combined senses, and at this scale, there are few wholly "local" problems that may be left entirely to the short-range "economic" expediency or temporal ideological preference of some exclusively national concerns or narrow self-interest groups.

The "problem of problems," therefore, lies within this realm: initiation of a series of authoritative and comprehensive reviews of major problem areas and their *interdependent* aspects, together with a sober evaluation of the resources, facilities, and priorities that set the *time* and *scale* of their possible solutions within the existing and projected state of our knowledge. From these reviews and evaluations would also emerge a long-range predictive model of the *"crisis-points range"* within which various interrelated world problems may reach dangerous proportions in the near future.

Such a task, though seemingly overambitious and beyond the scope of our currently organized national and international scientific programs, is now most urgent. Whether it *may* be feasible or not within the present state of our knowledge and our capacity to organize and apply that knowledge on such a massive scale, it is essential that we begin now. U.N. Secretary General U Thant has, indeed, suggested that our effective margin for engaging in such cooperative global actions is limited to the next 10 years—if we are to survive our present dilemmas:

> I do not wish to seem overdramatic, but I can conclude from the information that is available to me as Secretary General that the Membership of the United Nations has perhaps ten years left in which to subordinate their ancient quarrels and launch a global partnership to curb the arms race, to improve the human condition, to defuse the population explosion and to supply the required momentum to development efforts. If such a global partnership is not forged within the next decade, then I very much fear that the problems I have mentioned will have reached such staggering proportions that they will be beyond our capacity to control.[1]

The highly critical nature of our current transitional period is such that supportive research programs towards such global action should be given the highest priority by the international scientific community. This is strongly implied by the theme of this Conference and volume, *"Environment and Society in Transition,"* and might be considered as one of the central study and research foci for the development of the World University, which is also being discussed here.

Research and documentation for the organization of such a collaborative global network for the study and solution of world problems was carried out during the "World Design Science Decade" activity,[2] by R. Buckminster Fuller and myself, during the period from 1961 to 1968. This work focused on the assessment of the total energy, material, and human resources now available to man on the global scale and the ways in which these might be redeployed and redesigned to the maximal advantage of the global community.

The role of scientific task forces to spearhead such a global undertaking has been particularly advocated by the biophysicist John R. Platt:

> There is only one crisis in the world. It is the crisis of transformation. The trouble is that it is now coming upon us as a storm of crisis problems from every direction. If we look at the historical course of our changes in this century, we can see more clearly why this is happening so suddenly, and why it has now become urgent for us to mobilize all our intelligence to solve these problems if we are to keep from killing ourselves in the next few years.[3]

My specific proposal concentrates attention on the *Timetable* aspect of problem assessment. Though framed at the global level, it should be noted that such study(s) could, and should, encompass within its scope specific timetable assessments at various national levels. It is essential, however, that such "local" studies proceed from their position in a world context. Unfortunately, although much excellent work has been done in national planning both in the short and longer ranges, particularly during the past decade, it has inevitably been vitiated by unanticipated changes in the global context.

> Because of the high degree of autonomy, more than 140 governments follow their own economic policy and do their planning with inadequate means at the level of the nation-state or even some lower level. This leads to unadjusted imbalances which grow much faster than the means and capacity to stabilize them . . . we need more than economic planning. Planning future technologies to tackle imbalances and integrate their results into the political machinery, governments and people, this should give the aggressiveness of mankind new goals.[4]

No mandate or institutional setting at present exists for the level of supranational assessment, planning, and goal setting that is envisaged here. The pioneer work of the United Nations and its ancillary agencies is still constrained by specifically political and national representation and the procedural directions imposed by these constraints. Other emergent transnational agencies and associations, multinational corporate organizations, etc., do not yet furnish a common constituency or advocacy group for such undertakings. The initiative must then be assumed *de novo* by those individuals, groups, or institutionalized interests perceiving the urgent necessity of doing so.

In summary form, the Timetable project would, therefore, combine:

(1) Realistic appraisal of the current global status of given problems in themselves.

(2) Systematic analysis of their causally connected interactions and relationships.

(3) Projection, in various time ranges, of the estimated crisis points of key world problems.

(4) Assessment of the human and physical resources and the development and potential capabilities available for the solution of the problems in the various time ranges and scales of magnitude.

(5) Identification of value and goal priorities, options, and alternatives within the various areas considered above, and of their local and transnational policy implications.

Such an overall study would be grounded within a framework of the actual distributions, allocations, and flows of the physical world resource base and of the developed knowledge and technological capabilities available (or in projected development) for any preferred implementation of solution directions. This would function as a material balance sheet within which to set up the array of priorities, goals, and options available in any time period and would be so projected as to allow for anticipation of the consequences and implications of problem solution actions in relatively long-range terms.

In effect, what we are discussing is an emergency agenda for the world community! Although it may be suggested that such an undertaking is too idealistic and too far removed from present socioeconomic and political realities, it might also be underlined that we have no choice but to proceed on the most idealistic premises if we are to envision at all the basic survival of our world.

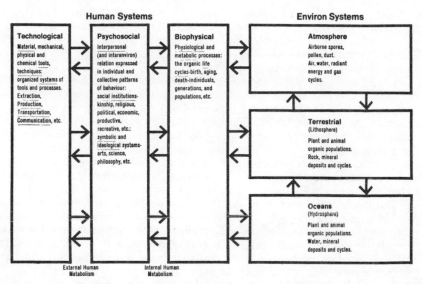

FIGURE 1. Relations among human and environmental systems.

The overall "whole systems" for such study(s) should be one that encompasses the interactions of the various subsystems in a manner that allows and accounts for their most indirect relationships and feedback operations.

In using such models we recognize that all of our large-scale problems must be viewed as occurring within a relatively small global ecosystem in which human activities have explosively increased as a major organic sector of the biosphere, particularly within the last century, and now assume a crucial role in the maintenance of life on the planet. Nature is modified not only by human action, as manifested in science and technology—through physical transformations of the earth for economic purposes—but also by such factors, less amenable to direct perception and measurement, as political-ethical systems, education, needs for social contiguity and communication, art, religion, etc.

We should, of course, emphasize the wholly denotative convenience of labelling different parts of what is essentially a dynamic whole. When we deal with human activities, their complexity tends to force the use of simplistic models that can be fitted into some neat disciplinary forms. This can become dangerously misleading if we then assume that such problems can be solved within the artificial divisions set up for intellectual convenience. No social problem of small or large scale—and all human problems are axiomatically social—can be solved within the terms of any one discipline.

A wholly technological solution, however logical and seemingly efficient, may fail by overlooking some elementary sociocultural requirements. Conversely, solutions conceived solely in socioeconomic terms may fail through lack of adequate technological considerations. Also, the lack of a systemic whole-system approach to such problems may solve one problem—only to increase the critical range of another. Many of our crucial national and international problems have been caused by short-range and expedient economic and technological decisions taken with little or no regard for their longer-range deleterious effects on man and his environment. These points may seem obvious, but many of our current dilemmas arise from piecemeal and unsystematic approaches to the satisfaction of human needs and emergency solutions of human problems in our recent past.

Major world problems themselves [5] derive specifically from accelerated changes in the past century, when man approached an historical watershed comparable only to the earlier agricultural revolution. In terms of problem definition and priorities, the continued disparities between the so-called advanced nations and the lesser-developed ones may be viewed as the gravest threat to the global community. The explosive rises in population, the pressures on food, lands, and other resources, the scale of wastage, disorganization, and pestilence now accompanying our "local" wars are also linked in due measure to the revolutions in human expectations that now circle the globe.

This disparity between *have* and *have not* peoples may also be defined as part of a growing ecological imbalance in which the hyperactive advanced economies extract more, produce and consume more, and produce more waste, than the lesser-developed; and because of the increased dependence of the more advanced countries on raw materials from the lesser-developed countries, the former now exist in a directly parasitic relationship.

In terms of energy use, the more fortunate individual in the industrial countries consumes more than 50 times that of his counterpart in the poorer regions—and contributes in due measure much more than 50 times the by-product pollutants now critically affecting the global environ systems.

Overall, the advanced countries in the past decade consumed 77 percent of all the coal, 81 percent of the petroleum, and 95 percent of the natural gas for less than one quarter of the world's population; one nation alone, the United States, used one-third of the world's total industrial energy and consumed approximately 40 per cent of the world's output of raw materials.

In round terms, approximately 20 percent of the world's population currently enjoys 80 percent of the world's income and uses more than half of all the earth's resources, producing a concomitant balance of the biospheric pollutants that have become critically apparent in the past period.

This gap between rich and poor has other salient features that make it one of the most critical problems facing man in the next decade. When treated in terms of population and food, the widening gap has even more threatening connotations. Present world population totals roughly 3.5 billion, and the current rate of increase about 1.8 percent per year will double world population in just 30 years, i.e., an agreed median estimate for 6 billion people by the year 2000. More than 80 percent of the current increase of population will be in those world regions that now have inadequate food supplies: Asia, Africa, and Latin America.

So the range of crisis points in terms of the population explosion alone is that not only does this doubling within one generation mean not only a necessary doubling of food supply but, if other material standards are to be achieved as well, it will require a doubling, and tripling, of housing, of city sizes, highways, consumer goods, etc., with their equivalent expansion of industrial materials extraction, production, and distribution, which, in turn, will necessitate more than tripling the energy consumption of the various fuels—and so on! In terms of *current* socioeconomic organization, the byproduct pollution and deterioration of the world environment might be near catastrophe.

Though much attention is given to closing the gap between rich and poor nations in traditional terms, through foreign aid, favorable trade balances, etc., it is rather sobering to consider more realistically what this means in material-resource-usage terms. To bring the total world population up to an advanced standard of Western material consumption would require, for example, more than five times the present world production of metals and minerals, which is far more than we can attain with current levels of materials and energy-conversion capacities.

Using one example—extending full-scale electrification to the lesser developed nations—we may note that the average use of copper in the industrialized nations is approximately 120 pounds per capita. The increase of even one pound per capita consumption for full-scale world electrification would require a 36 percent increase in world copper production.

In terms of the energy required for such living standard increase, we may underline that in dealing with higher metals extraction and use from increasingly lower-grade ores, the energy required for this alone would be overwhelming, i.e.,

> The conclusion to be drawn is that 5 billion persons living at present U.S. standards would require 25 times as much energy as the U.S. does today . . . (and) at 25 times the present U.S. consumption of coal, oil, and gas the human race would burn up the earth's estimated reserves of fossil fuels not in a matter of centuries, but in a few decades.[6]

Clearly we face many rigorous limitations on closing the gap between the rich and poor nations. Given our developing scientific and technological capabilities, it is possible, but the approaches must be examined more realistically than in terms of pious hope and traditional practices. Such gap closing is only one facet of the world problem. Even when we refer to the so-called *advanced* nations as enjoying higher standards of living, the quality of life itself is in question where such societies are already faced with severe dislocation, deterioration, and obsolescence in critical areas of their socioeconomic and political structures. Many of their internal institutions are archaic, strained toward breakdown, and their physical environments are still suffering from the backlash of the intial developmental phases of unrestrained industrial exploitation.

And although we refer glibly to advanced scientific and technological societies, not one of these yet has approached the beginnings of what might be termed a "scientific" society, i.e., one whose motivations, goals, and orientations are congruent with and permeated with the scientific outlook in the larger sense.

Returning more explicitly to the Timetable project, I would emphasize that such studies should not be confined within traditional disciplinary perspectives or within conventional "single-track" planning programs. The scale and range of problems the project is intended to confront, problems that are highly characteristic of the period itself, have a hitherto unknown degree of complexity. It will require the development of a new perspective and approach to the planning function in itself, i.e., "planning which cuts across a multitude of dimensions inherent in such a system—in particular, social, economic, political, technological, psychological and anthropological dimensions—has become known as *integrative planning*. Integrative planning is, above all, planning in terms of the quality of life." [7]

In light of the demand for relevance in the university structure and for a deeper commitment to problem-focused educational programs, this type of project may serve ideally as a nucleus for reorienting various aspects of university training. The aims of the project itself and the cross-disciplinary work that it will entail by both faculty and students should give a renewed sense of social purpose and of a deeper long-range involvement and commitment to the larger society.

Obviously, as framed in this brief paper, such a project could occupy a great number of professional experts for a long time. Its more pragmatic implementation in combination with the educational process itself may be the most efficacious way to begin and to take direct, gainful advantage of:

(1) A. The direct experience in problem-oriented education that would be derived from involvement in such a study by both graduate and undergraduate students in the university.

   B. The postgraduate interdisciplinary work it may afford to a smaller number of those professionals in various fields—who would constitute the core staff.

   C. The individual and collective expertise and experience available through a distinguished advisory board that would also assist in, and guide, the project, and through the collaborative involvement of a number of key institutes and organizations, both in the U.S.A. and abroad.

This initial pilot phase will occupy the first two years of the study. At the conclusion of this period its findings may then be assessed for the feasibility of continuing the study in greater detail and range, via a collaborative network of other centers and organizations on a larger national and international basis.

(2) A. *Relevance to "Social Navigation"*

   Aspects of specific social and educational relevance.

   (a) General social awareness and understanding of the problems accompanying the accelerated changes in the past century has not kept pace with the changes themselves. Attention has been most easily focused on the more newsworthy aspects of local tensions and societal crises that are, usually, the surface manifestations of larger problems rather than prime causal agencies. "Literacy" concerning the purview and perspective of larger issues has been inadequate and productive of insecurity and social. tension

(b) The conduct of social policy under these conditions has, in many cases, been forced into continuing postures of crisis management. Decision making at the regional, national, and international levels has specifically lacked the longer-range *social navigation* aids and "early warning" systems that might have made for smoother transitions in critical change areas and allowed for the degree of anticipatory long-range planning necessary to our forms of complex sociotechnological society. The scale of human "intrusion" into the environment has reached critical magnitude only in the past century—and the majority of institutional arrangements have evolved in earlier periods, when the time, scale, and margins for error were much greater.

(c) In recent years, this awareness of inadequacy in our "societal feedback" and anticipatory control systems has led to the proposal and development of various types of longer range "social accounting" and forecasting, etc. An example of the most recent of these is the "Social Indicators" movement.[8] The inauguration of the National Goals Research Staff at the highest legislative level was also a further manifestation of the urgent need to increase *integral* social awareness of the complexity, long range consequences, and implications of all aspects of the local society's problems and to focus attention on the need for more adequate measures of social development within a larger framework of specific goals and values and their policy options.

(d) The Timetable study should be viewed within this outlined context of societal relevance as attempting to develop a more precise overview and awareness of the interrelated complexity of our major problems and to communicate the *time, range,* and *scale* of their critical proximities. Its findings would, therefore, be addressed in various forms to the largest audience, and its ongoing synthesis of social navigational trend information would be of continuing value to other agencies and institutions engaged in other problem-solving and goal-definition areas.

B. *The educational relevance* of such a study, as conducted within the university, with a relatively large group of students, lies in its provision of an opportunity to confront "problem-oriented education"[9] in an innovative manner that would allow students to work at such a study in an active research-team project design rather than in their customary passive course-taking role and that would encourage the development of a genuine *transdisciplinary* educational base that would draw on, for example, a generalized knowledge of the basic sciences—social, as well as physical, chemical, and biological—and of the ways in which these contribute to the overall conceptual whole of the study. Assessing the scale and magnitude of human trends in the past period would lead to involvement with socioeconomic, scientific, and technological histories and case studies in a somewhat different manner than through more specialized course work in these areas

Considerable attention would also be given to the current range of forecasting methods and techniques and to their validation and practical use, for example, in ongoing policy decisions in various problem areas. Students would also be introduced in a real-work situation to the various sophisticated systems-analysis and forecasting methods available. The compilation, indexing, and coding of human and physical-resource base information will require student work teams to review and develop efficient information storage, access and retrieval —including E.D.P. systems and computerized techniques.

Because the presentation of the results of the study would be designed for a variety of audiences, there would also be the concomitant exposure to, and use of, the range of communications media that may be employed for different audience ranges and purposes.

In these various ways, the educational framework of this study would also be unique in that the students would be initiating and organizing their own work-group situation and allocating various organizational member functions; directly consulting with, and being exposed to, various individuals and authorities in the university, industry, and government; and coordinating their own work internally in close collaboration with other groups working in contingent areas.

Assuming that the above first-year phase was confined to graduate students, one could in the second year of the program use a *multiplier* effect to seek the largest feasible real involvement of undergraduates by attaching to each graduate researcher a number of undergraduate students who would then form an extended group engaged through seminars and direct work in the ongoing program. In this fashion the project could be gradually expanded to become an important self-generating focus of creative involvement for the university at large. A not inconsiderable aspect of such direct involvement and confrontation with world problems in a systematic way is that it would provide meaningful answers to the searching and often turbulent questions that are today asked both of the university and of the society.

References

1. Thant, U. 1970. Introductory remarks to "The Challenge of A Decade, Global Development or Global Breakdown," Prepared for the U.N. Center for Economic and Social Information by Robert Theobald.
2. Fuller, R. B. & J. McHale. 1963. Doc. 1. Inventory of World Resources Human Trends and Needs.
   Fuller, R. B. 1964. Doc. 2. The Design Initiative.
   Fuller, R. B. 1965. Doc. 3. Comprehensive Thinking.
   McHale, J. 1965. Doc. 4. The Ten Year Program.
   Fuller, R. B. 1967. Doc. 5. Comprehensive Design Strategy.
   McHale, J. 1967. Doc. 6. The Ecological Context: Energy and Materials.
3. Platt, J. R. 1969. What we must do. Science 166: 1115–22.
4. Menke-Gluckert, P. 1969. Proposals for an international program for joint tech-

nological endeavors for peaceful purposes. *In* Mankind 2000. R. Jungk ano J. Galtung, Eds. Oslo Universitsforlag. Oslo, Norway.

5. A more detailed review and analysis of these problems may be found in the World Facts and Trends booklet prepared for this Conference by the author, and staff of the Center for Integrative Studies, State University of New York at Binghamton. New York.

6. Population Bulletin No. 2. Vol. XXVI. 1970. Population Reference Bureau, Inc. Washington, D.C.

7 Jantsch, E. 1969. Integrative Planning for the "Joint Systems" of Society and Technology—The Emerging Role of the University. Paper for Alfred P. Sloan School of Management, Massachusetts Institute of Technology.

8. McHale, J. 1969. Science, technology and change. *In* Social Intelligence for America's Future—Explorations in Societal Problems. Bertram M. Gross, Ed. Allyn and Bacon, Inc. Boston, Mass.

9. Steinhart, J. S. and S. Cherniak. 1969. The Universities and Environmental Quality—Commitment to Problem Focused Education, A report to the President's Environmental Quality Council. Office of Science and Technology, Executive Office of the President.

Rapporteur: Douglas M. Johnston

The Problem of Environmental Resources

The electric light bulb was invented in 1885, and in 1969 man set foot on the moon. In historical perspective, this cannot be described as merely technological progress but should be viewed as a quantum jump or almost instantaneous change. What is happening now is new in the annals of human history. We are in the midst of an experience qualitatively different from anything that has happened before to the human race. Even more disturbing, this particular experience will probably not happen again. Thus, we as a race have one and only one opportunity to make the choices that lead to survival. In our age, a production economy is needed to sustain the standard of living we have all come to expect, but production causes waste and pollution. We are rapidly depleting the earth's resources and polluting its environment.

At the same time, we cannot turn back from the technological revolution. Knowledge exists in the mind of man, and if a catastrophe were to destroy all our physical constructions and tangible evidence of technology, we could rebuild within a few generations. Further, we have come to expect a high degree of individual freedom as well as the material benefits that technology permits. The basic problem thus becomes: How can we preserve the resources of the physical environment and at the same time preserve individual man and his aspirations? Although we are still in the midst of a period of high technological change, trend-curve analysis suggests that the revolution of physical science and technology may approach an asymptote by the turn of the century. By that time, our goal should be the achievement of some kind of steady-state solution in which population, technology, and resources come into balance. Between now and the end of this century the current revolution of physical science and technology will most probably have run its course. Today, new revolutions based on the biological and social sciences are under way and must be factors in our planning for the future.

Consideration of Alternatives

The group considered two mutually inconsistent approaches to the problem of environmental quality. First, an alternative sometimes advocated today, to check further technological evolution and population growth, reduce production, and lower standards of living in the developed nations until the rate of consumption of resources and the rate of environmental pollution come within tolerable limits. Second, we can seek a nearly closed ecological cycle in which resources consumed are later recycled and even pollutants find their way back into the cycle as useful raw materials for production. Such a cycle, of course, would also depend on arriving at a stable population level.

It was the consensus of Working Group A that the second alternative not only is the most feasible from a human and technical point of view but is also the most conducive to the preservation of a maximum level of individual freedom

Requirements to Achieve a Closed Ecological Cycle

In order to achieve a closed, or nearly closed, ecological cycle of resources by the end of the century, we must identify what is needed in the way of information, research, technology, and social and political action during the intervening years.

The Group identified the following requirements:

1. The quick availability of data in readily usable form about all aspects of the physical and biological environment.

2. Indices of environmental quality and of resource loss or gain in a form suitable for management decisions.

3. The technological means of controlling different elements of the environment as required to maintain the cycle.

4. Institutional mechanisms for conducting analyses of the environment and reaching decisions on necessary action.

5. Social, political, and other means of bringing about such necessary action.

Specific Recommendations

Environmental Quality Indices

It is clear that the problems of pollution of the total living environment cannot be divorced from the larger context of environmental quality. Even the major parameters of acceptable quality are not yet sufficiently known or clearly defined. Only after analysis of all the important factors affecting environmental quality will it be possible to establish criteria and indices of environmental quality for different regions and for different environments (outer space, atmosphere, land masses, and sea). In order to accelerate this process, the group believes that an appropriate working committee should be formed by the World Academy to undertake the following tasks:

1. To list the critical environmental factors according to their relative importance in developed and developing regions;

2. To identify areas of required research to permit the development of appropriate indices of environmental quality on a worldwide basis; and

3. To prepare a report before the Stockholm Conference in 1972.

Resource Development

It is imperative that those responsible for resource development employ greater care in reducing the dangers of waste through underuse and misuse and, in the case of renewable resources, through overuse. A long-range goal should be the establishment of a steady-state ecology on a global basis that approximates a closed cycle.

Disposal of Waste

Among the many aspects of the waste disposal problem, the Group emphasized the immediate need for more imaginative economic policies to provide a positive incentive to industry for more easily disposable products through bonus or compensation arrangements and by penalties for those industries that produce less costly but less easily disposable products. A long-range goal must be complete recycling or reuse of waste products, with the intermediate goal of storing waste in a manner that permits its future availability, except in cases where such storage would constitute a hazard.

Antipollution Measures

In addition to the conventional research to reduce pollution hazards, more attention should be given to biological approaches: for example, the anti-water-pollution experiments of Dr. Boyko, using higher plants to destroy pathological microorganisms and toxic substances, that may result in low-cost antipollution measures, at least in smaller communities in both the developed and the developing areas of the world. A long-range goal should be a reusable cycle of all pollutants.

Satellite Surveys of Earth's Resources

In view of the many potential benefits available to the international community from satellite observations of the earth's terrestrial environment, the Group strongly recommends that in all future planning of human environmental activities, the actual and potential implications of these surveys be considered. Organized efforts should be made to ensure that the actual results of experiments will be made fully available.

Problems of Communication

Interdisciplinary Cooperation. Perhaps the most serious dangers to the future state of environmental studies will be those that arise from inadequate or ineffective communication. The interdependent character of environmental studies requires close and continuing collaboration among the scientific disciplines, with full participation on a personal and institutional basis at all levels of professional and research activity. For example, the gap between the medical profession and other scientific disciplines is an especially serious obstacle to progress in the medical sciences in most countries.

Professional Attitudes. Many scientists have a strong conviction that basic research can be considered of professional significance only if it lies within the confines of their given specialty and has as its purposes the acquisition of knowledge, without regard for ultimate application. In meeting world environmental objectives, leaders of disciplinary research must consider environmental problem solving, which will cross the boundaries of scientific specialities, as of comparable professional status to that of purely disciplinary research. Such a change in attitudes would not only advance progress towards essential environmental objectives but also help improve the public image of science and bring awareness of new tools and techniques to the individual specialties.

Education. Efforts should be made to publicize the dangers of premature specialization in the basic and applied sciences and to experiment boldly with flexible courses of instruction designed to broaden the general understanding of present and anticipated problems of environment and society. It is recognized that this may necessitate a slight reduction in the length of training in highly specialized areas, but it is believed that this sacrifice would be more than offset by the advantage of having a larger number of broadly based specialists.

Training. In designing training programs for scientists in the developing areas of the world, it is often overlooked that in addition to obvious differences in race, culture, religion, and economic level, their extreme climatic conditions affect disease patterns, working efficiency, and other aspects of behavior. Simple improvements can be made within the microclimatic environment and in the use of available resources to reduce the adverse effects of such conditions. The group recommends that the World Academy participate in the promotion of environmental studies training centers in developing areas of the world.

General Recommendations

The group requests the World Academy:

1. To bring these various considerations to the attention of the appropriate specialized agencies, such as FAO, ILO, UNESCO, WHO, and WMO.

2. To accept responsibility for maintaining liaison with these agencies and assisting in the realization of these recommendations by establishing a permanent committee as the continuation of this working group;

3. To encourage members of the World Academy to recommend implementation of the above proposals to their respective national governmental bodies; and

4. To increase its effort to bring to international conferences scientists from countries often unrepresented because of financial and political difficulties.

* * *

Rapporteur: John Norton Moore

In responding to threats to man and his environment, a central concern is to make optimum provision for individual man while moving toward more effective and comprehensive planning for protection of the biosphere. In keeping with this emphasis on individual man, the group on Population, Health and Family recommends that the Conference consider the following goals and strategies for their implementation.

Goals for the Well-Being of Individual Man

The International Covenant on Human Rights adopted by the General Assembly of the United Nations codifies the common aspirations of mankind. Such development includes rights to adequate food, education, and employment, improvement of living conditions, receipt of benefits of scientific progress, and the highest attainable standards of physical and mental health. It seems useful to supplement these statements on human rights with a more explicit biological bill of rights for the individual man. Such a bill of rights should include:

1. The right to an adequate food supply, free from toxic additives or pollutants.

2. The right to have clothing and shelter consonant with physiological and aesthetic requirements.

3. The right to live in an equitable physical environment, aesthetically attractive and physiologically healthful.

4. The right to an upbringing that does not emotionally malform and does lay the basis for healthy psychological development.

5. The right to education continuing as desired throughout life, the aim of education to be the growth and development of the individual to the fullest extent of capacity and interest.

6. The right to live in an equitable psychological environment, characterized by respect for human dignity and diversity.

7. The opportunity for creative work and self-development at all stages of life.

8. The right to adequate medical care.

Strategies for Implementation

The range of problems in realizing the full capacity of individual man suggests a need for continuing inquiry into strategies for implementation of all the desired goals. As preliminary recommendations, however, the Working Group on Population, Health, and Family would single out the need for comprehensive planning and research for dealing with population problems, the need for internationally based research for investigation of physiological and behavioral effects of environmental insults, the need for continued development of adequate health-care delivery systems, and the need for continued inquiry into the role of the family in individual and social development. The Working Group also recommends the establishment of a Commission within the structure of the World University for continuing inquiry into problems of man and his environment, in collaboration with the International Council of Scientific Unions, if possible.

A. Population Policy

The task of meaningful implementation on a global basis of minimum human rights to psychological and physiological development is staggering. Even at present world population levels, such implementation will require unprecedented scientific, technological, and political efforts. Despite expected scientific and technological advances, however, the present exponential increase in world population, if unchecked, threatens within a generation to make the realization of these rights unattainable and to condemn a large segment of mankind to a stunted existence.

Accordingly, it is urgent that decision makers at every level—national, regional and global—give immediate attention to the development of institutions and strategies for ensuring a stable population or a rate of increase that will permit a sustainable realization of the fundamental biological rights of man. It is particularly recommended that every nation establish institutional mechanisms charged with the responsibility of developing and implementing an appropriate national response to population problems.

In implementing strategies for population control, it should be recognized that the realization of fundamental human rights may be served by differing policies of population numbers and rates of change in different nations and regions. All such policies, however, should take into account the ultimate interdependence of all mankind in sharing the limited resources of our earth-space environment and should be based realistically on regional differences responsive to the basic biological rights of man. In implementing strategies, it is also preferable to utilize noncoercive techniques of control that promote informed individual choice and to recognize that some strategies, such as family planning, may promote important human rights quite apart from their utilization as techniques of population control. We recommend that necessary research to develop and implement more effective and acceptable techniques of conception control consistent with safety be encouraged.

Consideration of the complexities of the cultural, psychological, and political factors in formulating effective population policies must not be ignored in a rush to realize instant panacea solutions. Indeed, cultural and psychological factors may be the most critical variables in effective implementation of population policies. There is also a responsibility to consider the social implications of societal intervention to control population and its linkage with other problems, and to design sufficiently broad-based policies and institutions to make these policies effective in achieving the basic goals. In spite of the difficulties in formulating optimum policies, however, it is critical that national and global decision makers accurately perceive the magnitude of the present crisis and the limited time available for solution.

B. Research in Effects of Environmental Insults

Though there is substantial research directed at gross pathological changes, there is a need for greater emphasis on research into the genetic and behavioral effects of low levels of environmental agents. Such research could study the more subtle effects of chemical and other factors introduced into the environment, including the synergistic effects resulting from interactions among potentially deleterious substances. To generalize, the environment has in large measure been medically cleansed of the grosser natural environmental hazards, or the information is available to do so. The task for the future, then, is both to cleanse the environment of the most subtle natural contaminants and to protect it from the introduction of new artificial agents.

The scale of investigations necessary to assess adequately the effect of low dosages and complex interactions among factors suggests that internationally based research may be more effective. Possibilities for consideration are the establishment of global data banks focused on the pathological effects of low-level environmental pollutants, establishment of new international laboratories, and conversion of certain existing national laboratories to international use.

C. Health and Well-Being

The Working Group discussed the future development of effective services concerned with human health and well-being. It concluded that the topics most in need of further study were:

1. The need to restructure health care in both developed and developing regions so that satisfactory services could be provided. Such services should focus both on healthy development and on care of the sick. In doing so, they should not make too great demands on the capacity of the educational system to produce high-level specialists.

2. The great importance of more adequate education of the public in basic biology, including physiology and genetics.

3. A more satisfactory balance of the needs for specialists and generalists in the health-care professions.

4. The desirability of a greater emphasis in research policy on problems of normal life (maturation, aging) as compared with disease-oriented processes.

D. The Role of the Family

The Working Group discussed the role of the family in individual and social development. It concluded that family structure is important for a range of functions, including childbearing, child rearing, and psychological and economic needs of family members. It also concluded there is a need for continued research into the range of functions performed by the family in particular cultures and exploration of alternative modes of family organization and other strategies for effective realization of those functions. The Working Group was particularly concerned that attention be focused on the ethical implications of existing and alternative types of family organization. Similarly, attention should also be given to the adequacy of the present legal structure for the regulation of the family to ensure that changes needed for full realization of individual and social potential are realized.

Establishment of a Continuing Mechanism for Inquiry into Environment and Well-Being

The urgency and complexity of the problems of environment and health care suggest a need for a continuing mechanism of inquiry into their social and policy implications. One appropriate action mechanism would be an interdisciplinary commission established within the framework of the World University. Accordingly, the Working Group recommends that an interdisciplinary commission be named as soon as possible to consider both environmental and health care problems and to make appropriate policy recommendations. The Commission could function on a continuing basis within the framework of the World University and could carry forward its task by commissioning appropriate studies, establishing relevant committees, holding periodic meetings, or engaging in other activities deemed useful to its principal functions. Among other issues, the Commission might address itself to implementation of the recommendations made by the Working Groups and to clarification of the range of issues adverted to but not fully explored at the Joint Conference.

If possible, the Commission might also collaborate with the International Council of Scientific Unions and should explore the possibility of establishing other linkages with concerned institutions and professional associations.

Rapporteur: W. Michael Reisman

1. We are committed to the creation of a world order of human dignity in which social and political institutions exist and operate to ensure conditions for the optimal self-realization of every member of the world community.

2. Our demand for human dignity is posited on a conception of the human being as an integrated biological, environmental, social, and psychic creature, appropriately equipped to make and influence the entire spectrum of personal and group decisions of the various communities in which he lives.

3. The systematic search for and organization and utilization of knowledge will play an important role in the realization of human dignity. It is indicative of the deep fractionation of thinking about man in the environment that no single word is readily available to designate inclusively the foci of the natural sciences, the social sciences, the humanities, and the arts. We will use the word *knowledge* to refer to all these cognitive activities that bear on making decisions about man in nature.

Knowledge, so understood, does not make decisons but is a requisite component of efficient, economic, and rational decision making. Where practices abusive to man and his environment have been perpetrated under the guise of science, it is retrospectively clear that the knowledge purportedly applied was quite imperfect, often because of the restrictive focus of a single "discipline." A number of points are patent. Along with errors and damage to the physical and political environment, many radical improvements to these same environments have also been made. As for the serious damage that has been done, a markedly improved flow of knowledge will be necessary for rectification. Hence, the movement toward and ultimately the maintenance of a world order of human dignity is not conceivable without a plenary flow of accurate knowledge to all those concerned with making and appraising decisions—at some point in the future, all mankind.

4. In view of the integrality of natural phenomena and natural and social phenomena and the necessarily global scope of human efforts to harness or accommodate to them, we urge the establishment of a global network of institutional arrangements for the purpose of making knowledge, understood most broadly, about all aspects of social process, available in a comprehensive, timely, and efficient manner to those who make, and to those influenced by, the decisions of mankind. All existing institutions should be considered as resources for this task; new institutions will be created and maintained only as they serve the needs of the world community. Institutions, it must be remembered, are not ends in themselves, but only instruments for the realization of shared goals.

Knowledge goals for global planning and decision may be most economically explored in terms of a world knowledge process, composed of focusing on problems, gathering and processing information, and disseminating knowledge to all concerned. There must also be feedback to the knowledge process itself, tor an ongoing appraisal of the correspondence of practice to goal. The working group has considered and organized its recommendations for each of these successive sequences.

Focus

The sustaining focus for the global process of knowledge is the human being, growing in those conditions that stimulate and facilitate the realization of his manifold potentialities and simultaneously take full account of his biological, environmental, psychic, and social character. An informed focus on mankind includes, at a minimum, comprehensiveness, contextuality, and future orientation. Because physical and cultural realities are not a series to discrete events but are integral, interstimulating components of a whole, meaningful observations and beneficial manipulations can be made only with a disciplined and sustained comprehensive focus. What we need, as one scholar has felicitously put it, is a "macroscope," a focusing instrument that will discard trivia and create purposive attention commensurate with the range of interacting data. Contextuality places a particular event in its full context in order to derive its significance. Single technical solutions, whether they are initially physical or social, must be placed in context so that they do not cause new and greater problems. Future-orientation, an aspect of contextuality, extends the interrelation of events through time, obligating decision makers of today to consider the impacts of alternative choices on future generations in future environments. An appropriate focus implies a responsibility to inspect new social institutions in all their probable effects and to consider the diffusion and social effects of new inventions.

Problem-orientation is also an aspect of focus. We refer to the assembly of knowledge tools to deal with the problems generated by demands for the conditions of human dignity. Some scholarship in the past has included a reflexive, artificial reality in which problems were assembled to fit the limited tools that an investigator happened to have. Real problems are shared; their solutions require pooling resources and sharing data.

Gathering and Processing

The acquisition of data about man and his environment and its processing, storage, retrieval, interpretation, and adaptation will necessitate a worldwide network, integrating existing and newly created institutions, adapting available technology and introducing new techniques for more ambitious and increasingly novel knowledge goals. Although improved communications may permit centralized data banks, processing phases should be widely dispersed to optimize data collection and to share the educational potential of these functions in all sectors of the globe. Within an integrated global knowledge system, we recommend that local and regional units gather and process data of inclusive interest and, at the same time, serve as feedback points for the utilization of information at the local level. Given the fluidity of basic research and applied technology in this area, we refrain from more specific recommendations, which often tend to "freeze" the art and offer, instead, this overall set of policy preferences.

The emphasis on policy and function rather than specific techniques gives perspectives in several features of the gathering and processing sequences.

There has been a tendency to confuse artifact and function. The important function, of course, is the efficient processing, storage, and delivery of information to human beings. For several thousand years, civilizations—as opposed to ancient and contemporary folk cultures—have used script as the primary modality for these sequences, with the primary sensory channel linear and visual. These processing techniques have had a diffused and remarkably deep effect on perception in general. Electronic technolgly has opened up a variety of other modalities, with an entirely new range of influences on perception. In the future, the full range of modalities and channels should be inventoried, the full range of their contextual effects systematically investigated, and choices of method at particular times should turn on the needs of knowledge consumers, their abilities, the exigencies of the context, and the probable impacts on future behavior.

Many of the problems and possibilities of languages entail both processing and disseminating. Problems of changes and trends in language dominance are particularly topical. The development of new communicational systems tooled to the processing and dissemination of specific types of knowledge should be explored and set in the context of available and protential information systems. There is a full range of communicating techniques using other signs and symbols than words. The exploration of this area may provide a set of rapid and economic ways to increase the processing and dissemination of knowledge.

Dissemination

Knowledge is of scant utility to mankind if it is not delivered to those charged with making and appraising decisions in a timely, economic, and accurate manner. We identify four major dimensions to the global dissemination of knowledge: communication among scientists, communication with political leaders, education, and media communication. These important areas must be considered separately.

1. The integral aspect of natural and social phenomena and the unfortunate fractionation of individual disciplines require constant and intense interchange between knowledge specialists, within and across disciplines. Verbal declarations of the interdisciplinary faith and sporadic and comparatively primitive multidisciplinary conferences are a far cry from the needs of a global community. Entirely new techniques must be explored. Electronic communication portends a new range of channels, and they must be explored and expanded. The potential of the telephone, and conference call, the telelecture, closed TV, teletyping, flying seminars, as well as other adaptations of technology, must be explored.

The techniques of interpersonal communication cannot, of themselves, surmount internal psychological and emotional obstacles, nor can they be absorbed and usefully integrated without a frame of reference sufficiently comprehensive to indicate the relevance of the research of others. We note with sadness the ludicrous "sovereignization" of fragmented and refracted disciplines and the petty arrogance of individual scholars regarding the utility of the intelligence products of other disciplines. The word *discipline* in this connotation is unfortunate, for it affirms by denial the separateness of disciplines, including the capacity to generalize. Let the point be utterly clear. Anyone with any understanding of the process of enlightenment and any commitment to the life of the mind automatically confesses his permanent inability to know everything about everything, and, indeed, even to know everything about a small subject. The intellectual life is, in the deepest sense, a community experience; a constant sense of one's own incompleteness and dependence on others with special knowledge of other areas of a varied, manifold, but integral reality.

2. Because a complex society charges certain individuals with major public responsibilities and functions, these decision makers will have exceptional requirements for knowledge, and an understanding, in the most comprehensive sense, of the process of change. In the past, without being so equipped, political leaders have even defined the goals of knowledge specialists and used the knowledge acquired for their own special interests. We believe that knowledge must be used for all mankind, but we recognize the unique demands of specialized political communication. A number of points are clear. First, that political systems are inextricably intertwined with the knowledge process and that a commitment to a global system of human dignity carries with it a commitment to political structures posited on, working toward, and incorporating principles of human dignity.

Whatever the specific structures such systems may take, the special educational needs of decision makers should be emphasized. Decision makers must be equipped with a broad frame of reference that permits them to identify problems and to trace through time the aggregate consequences on man and his environment of the alternate courses of action open to him. We would also recommend the development of education for policy making and for decision making and, in particular, techniques for processing, storing, retrieving, interpreting, and applying the decision experience of individuals; this is a resource that, it seems, is not systematically exploited. In the future, planning goals for increasingly available conditions of human dignity may require more knowledge, and this may become an institutionalized component of global decision making. We note, in this context, Warren S. McCulloch's principle of the redundancy of potential command when information constitutes authority.

3. Every phase of the global knowledge process comprehends learning and relearning. But the sequences of dissemination turn on education in a special sense. First, it is clear that the availability of knowledge assures neither that it can and will be used nor that, if used, it will be used well. Appreciation of the relevance of science to decisions about man and society and the environment and the ability to use that knowledge implies a breadth and depth of education that go far beyond our current goals. Hence, unless the general member of the world community of the future is equipped with a sufficient education, the availability of kowledge will be of no personal use to him, nor, in particular, will he be able to share in the power processes of the various communities in which he lives. Second, a focus on mankind implies a concern with the optimal development of the manifold talents of man, so that he can serve himself, so that he can change or accommodate himself to his plenary environment.

One of the first priorities is more learning about learning. We recommend that man's learning be consciously extended beyond the limits of institutions, through all his life. We emphasize as well that it must be extended forward to birth and even to the fetal state. Because of institutional myths and administrative conveniences, the fantastic learning capacities of the preprimary years have been squandered. We urge that they be utilized.

Education does not mean the colonization of data bits in the neurons of unwilling subjects. It aims at the development of a creative attitude toward knowledge and experience—not a mere accumulation of fact. It means, in short, equipping individuals with the capacity to learn. Education must be premised on the whole man and woman, engaging not only his cognitive activities, but all others—political, affective, playful, and so on. It must include the development of a multitalented human being, optimally equipped for a range of activities—planning, creating, forecasting, decision making, communicating, and many others. The process of education should develop free inquiry and the concomitant self-confidence in imagination and institution that this implies. It should foster plurality and divergence, a capacity for play with ideas, and hospitality toward those of others. It should encourage both associative and analogical thinking. The enormous potentials for inventiveness within a limited framework, as well as the refashioning of ever broader frameworks, must be conveyed.

Particular emphasis must be directed to the teacher, a term that, in its full contemporary connotation, may be a misnomer. Teaching entails reciprocal learning, not one-way communication. In our world of high-velocity change, the teacher who has ceased to learn has ceased to teach. In particular, the problem of the generation gap should be perceived as a reciprocal educational problem. Generations separated by a tremendous experience gap and peoples of extraordinary diversity of previous and future cultural experience will be in continuous contact. This can be either mutually destructive or mutually reinforcing. Teaching and learning can no longer be a linear process from those who know to those who do not know yet, as it has been in the past, but a process by which those who have one form of experience or one current form of knowledge or expertise, wisdom, or inspiration are in mutual communication with each other. In such a system, the teacher of a three-year-old will have to learn first about the experience of that three-year-old, new and unknown to that teacher, because no one will ever have been a three-year-old in just that cultural and situational state. Similarly, the experience of the old, now carelessly disregarded, will be cherished and placed in context as equally precious because, just as the experience of the young today has never existed before, so the experience of each older group will never exist in that form again. The extent to which the experience of each age group differs in a rapidly changing society will have to be recognized in a new design for educational institutions to which individuals will periodically return, or in which they will periodically participate as both learners and teachers. The knowledge and habits of living needed by men and women and children in our changing society will need continuous responsible reviewing, renewing, and innovation.

New environmental needs will require a new technology of learning a new conception of the congruence of learning and living and a full appreciation of the regional and local variations of knowledge diffusion in a multilingual world. We commend to scholars the urgent consideration of these challenges.

4. The media represent a major opportunity for the dissemination of knowledge about decisions affecting man and his environment to the widest audience. We view them as a resource and urge that they be viewed functionally as a public utility of the world community aimed at realizing the most inclusive

interests. We urge that the media be exploited to the fullest as a modality for enhancing the conditions of learning for all mankind. Where the media permit a vaulting over intermediate stages, this should be done. There is no inexorable progression in the adaptation of knowledge for an improvement of life.

Accordingly, we recommend to the plenary conference:

1. Within the framework of WAAS the establishment of an effective worldwide institution to monitor the flow of knowledge relevant to man and his environment in transition. This institution would provisionally be called the Global Knowledge Center. Its membership policy would seek to transcend the transient political divisions of political leaders. It would perform among others the following functions:

—establish goals and priorities for the processing of knowledge relevant to global decisions.

—establish contact with existing international regional and national centers of knowledge, creating thereby a functional grid for the collection, processing, and dissemination of knowledge. To this end, GKC might seek consultative status with appropriate international intergovernmental agencies,

—encourage the establishment of new knowledge centers that it deems necessary,

—recommend the allocation of funds for the establishment and maintenance of such centers as are deemed appropriate to the most inclusive community goals,

—publish timely reports of the world knowledge system as well as substantive levels of knowledge in all areas relevant to decisions of global impact,

—foster an ongoing study of the world knowledge process within the framework of the World Academy and regional and national universities and, in particular, encourage innovations in gathering, processing, and disseminating knowledge.

2. That an interim committee be formed and be charged with the drafting of plans and a statute for GKC and that the Committee return its report and recommendations at the next conference.

3. That a number of crucial problems, some of immediate and some of prospective importance, be initially studied by WAAS or under its auspices and that these same studies, relevant to the global knowledge process, be thereafter transferred to GKC or one of its associated institutions. Among these problems and projects are:

—the monitoring and correlation of different developments in a number of countries in order to speed up the entire process of the advancement of knowledge and to avoid the waste of valuable funds for work that has already been done,

—studies of cognitive processes that might facilitate learning.

—the optimal integration of hardware and software,

—the complex of matters relating to population control; the development of contextual and conceptual demographic models,

—new techniques for the storage and delivery of different types of data and knowledge.

4. That governments and international organizations increase significantly the funds for research about the global knowledge process and about specific areas of substantive knowledge. In particular, we recommend that the distinction between basic and applied research be suspended, because the two interlock and are equally important. The content of the terms is often parochial: what is basic and what applied shifts with perspective, with time, and with the problems encountered

REPORT OF THE WORKING GROUP
ON DECISION PROCESSES
CONFLICT RESOLUTION AND THE CONTROL OF WAR
THE ORGANIZED PLANET
HUMAN RIGHTS AND INDIVIDUAL PARTICIPATION

Rapporteurs: Burns H. Weston, Francis Wolf

The very profound papers and comments that have been presented to this Conference have done much to delimit and clarify the problems that confront mankind in securing and protecting the earth's ecosystem. Many of the specific injuries to our natural environment, their causes, and their potential technical solutions have been outlined with precision and eloquence. Some of the more important general policies for the world community have been urged and explained with conviction and intensity.

It was the understanding of this Working Group that its principal responsibility was to recommend ways in which the world community might better organize itself for clarifying and implementing the common interests of all mankind in relation to the environmental problems that have concerned this Conference. The Working Group deemed it necessary, first, to account for some of the complex considerations that underlay the assignment and, second, to suggest some of the modalities by which mankind might be able to edge closer toward improved decision processes. A recommendation to the World Academy of Art and Science for a continuing commission of inquiry concluded its deliberations.

I

A. The Working Group notes at the outset that there is increasing awareness of the nature and urgency of ecological problems throughout much of the world (as in this Conference) and important public and private initiatives in many countries and at the international level (as in the scheduled 1972 Stockholm Conference on the Human Environment). But there has been little effort toward and less achievement of a comprehensive and homogeneous formulation of all the related problems concerning our natural environment that would readily facilitate the detailed clarification of relevant community policies, the description of past experience in the management of the problems involved, the study of the factors that have affected prior successes and failures, the projection of future conditions, and the invention and evaluation of appropriate solutions. The Working Group therefore recommends that all enlightened individuals and groups, such as have been represented at this Conference, continue to share their human and material resources at all public and private levels (from local to global) for the purpose of achieving a more precise and workable formulation of the cluster of interrelated problems that brought this Conference together.

B. The Working Group notes also that the basic community policies that relate to environmental and associated problems embrace both minimum order (in the sense of the minimization of violence and coercion) and optimum order (in the sense of the greatest production and widest distribution of all values). The emerging aspiration of most of mankind is not so much for the mere conservation of some resources as for an effectively concerted and constructive employment of all resources in the greatest protection and widest distribution of all values. Peace is at stake, both in the major sense that people who are deprived of the basic values of life are inexorably driven to the use of force and in the minor sense that even weather resources, for example, may be employed for attack. Beyond the minimum goal of peace, the whole well-being, even survival, of peoples is fundamentally at stake. In addition to the better identification of specific environmental problems, therefore, there is urgent need for comprehensive and precise clarification, from a perspective that identifies with the whole of mankind, of the basic policies that can guide us to the solution of such problems.

C. The Working Group would note, however, that merely to detail the many problems of our environment, and even to clarify the general policies relevant to their solution, is only a beginning, albeit indispensable, toward the ecological planning, development, and control that are so desperately needed. The critical next steps include the identification and improvement of those processes of authoritative decision that can be made to contribute to the effective implementation of the common interests of the peoples of the world in securing and protecting the constructive enjoyment of their shared environmental base. In the opinion of the Working Group, the processes of decision that require modernization and adaptation in this vital connection are the basic constitutive processes of all of man's different interpenetrating communities, from local to global. Of course, some ecological problems may be economically managed at local or national levels. But when proper account is taken of the maze of interdependencies that prevail among both the organic and inorganic features of the natural environment and between such features and the institutions and practices by which man seeks to satisfy his many social, psychological, and bodily demands, the aggregate of problems is global in impact and thus requires global as well as local remedial measures. To be effective, improvement in decision process must therefore extend to the constitutive processes of all the component communities of the globe, as well as to the global order itself, with all sectors of participants objectively informed about and appropriately coöpted into the solution of the problems in issue at the local, national, and international levels. This effort requires the full support and assistance not only of governments but of university research centers, scientists and the representatives of professional societies, industrial organizations and other such nongovernmental groups.

D. The Working Group further notes that a remedial approach to man's natural environment that is global in scale, involving inquiry into all of man's interpenetrating processes of community decision, necessarily implies a conception of law—in particular, international law—that is more functional than structural in nature. Despite surviving popular myths that international law is not true law or there is very little international law, the Working Group finds that the world arena today exhibits a comprehensive constitutive process that, in its basic features, is comparable both in form and effect to the constitutive processes of our more mature national societies. This process—reflecting a fusion of authority with control (the appropriate reference of "law") that is to be found in the secular empirical expectations of community members about who is to make what decisions, in respect of whom, in accordance with what criteria, and by what procedures—identifies appropriate decision makers, clarifies basic community policies, establishes necessary structures of authority, allocates bases of power for sanctioning purposes, specifies procedure for the making of various kinds of decisions, and secures performance of all the different decision functions necessary to the creation and maintenance of law (intelligence, promotion, prescription, invocation, application, termination, and appraisal). This is not to suggest that the existing global process of constitutive decision, or the governmental structures through which it operates, is today adequate to the task of securing man's basic goals in relation to his environment. The Working Group would insist, however, that we do have today considerable knowledge about, and skills in, the management of decision processes generally, and about our improving global constitutive process specifically, to be genuinely encouraged that more rational and efficient processes of authoritative decision can be constituted and managed for the better securing of basic environmental goals.

II

Recommendations for securing and maintaining our natural environment could be made with respect to every phase of global decision process: Principles and procedures could be designed for making participation in that process both more representative (ensuring that all who are affected by environmental decisions have a voice in making such decisions) and responsible (ensuring that all who affect community environmental policy are made subject to such policy). Clarifications could be sought to eliminate the simple-minded dichotomy between "national" and "international" interests, to demonstrate that the most enduring interests of any nation are the inclusive interests it shares with other communities. New structures of authority could be suggested for many different community levels, with appropriate openness in access and compulsory attendance for participants charged with transgression of community policy. For economy and specificity in illustration, however, the Working Group limits its suggestions for possible improvements to each of the seven decision functions mentioned above, primarily at the international level but with occasional supplementation at the national and local levels, and in relation to nongovernmental participation as well.

The Intelligence Function

The intelligence function comprises mainly the gathering, processing, and dissemination of information relevant to decision making. With the postwar proliferation of international governmental and nongovernmental organizations and the emergence of new techniques and technologies of observation and communication, this function holds a potential for truly massive inquiry into the causes and cures of environmental problems.

Recognizing that few of the many environmental perils that now face mankind can be adequately handled without facilitation of the performance of the intelligence function at all levels of social interaction, the Working Group ecommends, as a matter of overriding general concern, the deliberate and systematic undertaking of all measures that may be seen to enhance the efficiency, dependability, comprehensiveness, and availability of information and information processes pertinent to the clarification and understanding of ecological problems. In this general connection, the Working Group recommends that special attention be given to the fullest possible utilization and coordination of existing intelligence-serving techniques and technologies (e.g., polls, data banks, satellite observation), and, where appropriate, the creation and development of new intelligence-serving techniques and technologies.

Recognizing that existing services do not adequately serve the urgent need for information relevant to the establishment of a physically and psychologically habitable environment, the Working Group recommends, in particular:

—*at the international governmental level,* (a) the creation of new global and regional machinery with facilities, funding, and skills adequate for the comprehensive and dependable gathering, processing, and dissemination of ecological information, to be made freely and widely available to people everywhere, and (b) the strengthening of the intelligence functions of existing global and regional agencies that are charged with responsibility for given environmental problems through additional facilities, funding, and skills (e.g., the United Nations Secretariat and General Assembly organs, ILO, FAO, UNESCO, WHO, WMO, IAEA, ICAO, etc); and

—*at the national and local governmental levels,* (a) the improvement of existing environmental-information-gathering, processing, and dissemination functions of professional associations and research institutes specialized to the technical understanding of various aspects of the natural environment and, where appropriate, the creation of new such services, and (b) the widespread sharing of private ecological expertise with all other individuals and groups, public and private.

The Promotion Function

The promotion function comprises both simple recommendation and active advocacy and, as such, is designed to transform individual and group demands into authoritative community prescriptions. With the increasing availability and openness of specialized education, the burgeoning of contemporary communication media, and the rising democratization of participation in political processes, it is of course an operation of great significance to those who would protect the world's environment.

Recognizing the ease and skill with which environmental demands can be formulated and propagated and, therefore, the potential for harm, as well as good that inheres in the promotion process, the Working Group recommends, as a matter of overriding general concern, the promotion of common rather than special interests, the use of persuasive rather than coercive promotion strategies, and healthy skepticism rather than dogmatic conviction about the rationality of one's environmental proposals in the face of future unknown conditions.

Recognizing that mankind has but recently begun to awaken to the need for comprehensive and concerted ecological control and that, in any event, enlightened prescription (i.e., lawmaking) is largely dependent on enlightened promotion, the Working Group recommends, in particular:

—at the international, national and local governmental levels, (a) the improvement of existing agencies and, where appropriate, the creation of new machinery for the purpose of continuously re-evaluating and reporting on ecological information received, (b) the increased use of the mass media and other channels of communication for the purpose of mobilizing élite groups and broader audiences to a "world view" appreciation of the fundamental interdependencies of the earth's ecological system and of threats thereto, (c) the subsidization, coöption, and other encouragement of educational and research programs for the purpose of enlightening and mobilizing public opinion about the establishment and maintenance of a rational global environment, and (d) the wholehearted support by all interested parties at all levels of the 1972 Stockholm Conference on the Human Environment; and

—at the nongovernmental level, (a) the support of qualified specialists, research centers, and institutions of learning to devise new curricula, texts, and other educational tools that will foster greater citizen enlightenment about beneficial and injurious environmental use and practices, (b) opening, by the mass media and other channels of communication, of free time and space for the purpose of allowing concerned specialists and other qualified persons to influence élite and wider audiences regarding ecological problems, and (c) the self-mobilization of scientific and other professional associations, pressure groups, and political parties in attacks on environmental deterioration and in constructive recommendation of ameliorative practices.

The Prescription Function

Broadly put, the prescription function refers to lawmaking, to the designation of policies and the creation of expectations of authority and control that will sustain the policies designated. As such, it cuts across all institutional lines (i.e., legislatures, courts, administrative agencies, etc.) and includes even the outcomes of unorganized decision (i.e., "customary law").

Recognizing that individual and group willingness to comply with designated environmental policies is in large measure conditioned by the manner in which such policies are prescribed, it is of course essential that lawmaking about the environment itself conform to certain basic public-order goals. Accordingly, as a matter of overriding general concern, the Working Group recommends that all environmental policy formulations be comprehensive and contextual in their general or specific orientation, consistent with the basic goals of the plurilateral or bilateral community to which they are addressed, and effective in their communication of expectations of authority and control to target audiences.

Although there already exists substantial national legislation for dealing with environmental problems (e.g. conservation of natural resources, urbanization, etc.) and important international standards in this area (e.g. the conventions about the oceans and outer space and the ILO conventions on such matters as radiation protection, industrial safety and hygiene, underground work, etc.), clearly, many other prescriptions are needed to control overpopulation, the plundering of the earth's resources, and the precipitate mechanization of many of life's processes. Accordingly, the Working Group recommends, in *particular:*

—at the international governmental level, (a) the formulation and adoption of a set of environmental standards that might take the form of a World

Environment Code (perhaps including much of the above-mentioned and extant convention standards), setting forth policies carefully articulated to promote the world's physical and psychological environment, (b) the continued and accelerated use of the lawmaking machinery of the United Nations, its specialized agencies, and environmental conferences, including the 1972 Stockholm Conference on the Human Environment, for the purpose of achieving multilateral agreement and standardization on a variety of ecological problems, and (c) the adoption by all international organizations and conferences concerned with environmental protection of measures designed to stimulate responsive national legislation consistent with world ecological priorities;

—*at the national and local governmental levels,* (a) the formulation and adoption of comprehensive environmental legislation after due consultation with technical experts and qualified representatives of the national and international communities, and (b) the public encouragement of private rule making that will conform to basic community policies concerning the environment; and

—*at the nongovernmental level,* the formulation and adoption by private individuals and groups, of self-governing measures that will conform to basic community policies concerning the environment, thus creating expectations that will facilitate compliance with publicly prescribed environmental norms.

The Invocation Function

The invocation function concerns setting into motion authoritative processes of decision for the purpose of securing the application of community norms. Although some arenas remain closed to some claimants (as in the International Court of Justice to private persons, for example), most persons do have opportunity, either in personal or surrogate capacity, to complain about the violation of environmental protection prescriptions.

It is recognized on the one hand, that environmental assaults can affect most grievously the life of every human being and, on the other, that the efficient administration of justice may sometimes require limitations on jurisdictional access, and it is apparent that an appropriate balance must be struck among the policies that regulate the invocation function. Accordingly, as a matter of general overriding concern, the Working Group recommends that the extent to which claimants may be authorized access should be determined by reference to context, that there should be a presumption of free access whenever and wherever possible, and that, in any event, where access is authorized, prompt sanctions should be applied to avoid irremedial deprivations.

Recognizing, however, that there remains an overwhelming need to hold environmental violators publicly accountable, the Working Group recommends, in particular:

—*at the national and local governmental levels,* (a) the fostering of free and direct access, to the widest extent possible, for all environmentally damaged individuals and groups, to impartial judicial and quasi-judicial arenas of decision under procedures comparable to existing international and regional human rights mechanisms, and (b) the establishment of an environmental ombudsman under the auspices of the United Nations to whom access may be had irrespective of United Nations membership:

—*at the national and local governmental levels,* (a) the fostering of free and direct access to the widest extent possible for all environmentally damaged individuals and groups to all national and subnational processes of decision, (b) the establishment of environmental ombudsmen (or commissions of inquiry) at all levels of national life, and (e) the encouragement and safeguarding of all noncoercive forms of expression for protest against violations of national and local environmental protection policies; and

—*at the nongovernmental level,* (a) the coordination of efforts by individuals, private associations, and pressure groups in their invocation of national and local environmental protection policies and (b) the establishment of procedures requiring that the management of private associations (particularly those engaged in commerce and industry) be held accountable to their memberships for the violation of community environmental policies.

The Application Function

Broadly put, the application function refers to law enforcement; that is, to the transformation of authoritatively prescribed policies into controlling events. As such, like the prescription function, which it closely parallels, it cuts across all institutional lines (legislatures, courts, administrative agencies, etc.) and similarly includes even the outcomes of unorganized decision ("customary law").

Recognizing that the full—even partial—realization of environmental policies, not to mention public confidence in the legal process itself, is fundamentally dependent on the successful enforcement of prescribed community norms, again we must be concerned for the promotion of basic constitutive policies. Accordingly, as a matter of overriding general concern, the Working Group recommends that the application process be prompt and efficient in initiation and resolution, comprehensive and realistic in its specific orientation, and constructive and consistent in its final conclusions.

Recognizing that international enforcement of environmental policies is probably more effective today than is commonly realized and that, until recently, national and local enforcement of environmental policies probably has been less effective than is popularly believed, the Working Group feels that it is nonetheless self-evident that there is need for much improvement in the processes of norm application at all levels of social interaction. Accordingly, the Working Group recommends, in particular:

—*at the international governmental level* (organized arenas), (a) the consideration of specialized adjudicative machinery regionally and functionally environmental in focus and, in any event, technically qualified, interdisciplinary, cross-culturally representative, and open to direct access by private individuals and groups (paralleling, for example, the World Bank's new Centre for the Settlement of International Investment Disputes), (b) the repeal of those restrictions, such as the Connolly Amendment, which make the International Court of Justice and other adjudicative bodies virtually powerless to resolve environmental controversies, and (c) the adoption, wherever feasible, of compulsory arbitration procedures in ways that are compatible with the "counciliation" arrangements set forth in the recent Draft Convention on the Law of Treaties;

—*at the international governmental level* (unorganized arenas), greater democratization of foreign-office-to-foreign-office procedures aimed at resolving environmental conflicts and, in any event, an increase in technically qualified and interdisciplinary foreign office staffing;

—*at the national and local governmental levels,* (a) wider use of expert witnesses, court-appointed masters, and technically equipped bodies specialized in the resolution of environmental conflicts and (b) the highest possible co-ordination of applicative procedures and programs between and among national and subnational governments and agencies; and

—*at the nongovernmental level,* the encouragement of individuals and private associations, particularly those engaged in commerce and industry, to make explicit how and to what degree they will comply with, and assure adherence to, community environmental policies.

The Termination Function

The termination function puts an end to prescriptions and to arrangements under prescriptions when they have ceased to serve their purpose or have become inequitable. In general, the termination function is performed in the same structures and by the same procedures as the prescription function. Occasionally however, there is resort to unilateral decision, sometimes even by naked power, with special deprivation of some parties. Hence, ameliorating the attendant losses is a unique problem of termination for which provision must be made.

The importance of the termination function at the international level is especially evident because of the relative absence of prescriptive organs. During the past 25 years it has been performed mainly by the General Assembly of the United Nations, as is demonstrated by the great number of territorial communities that, with the help of the United Nations, have been released from prior arrangements. The Working Group would recommend continued resort to the United Nations and, in general, recourse to the same structures of authority and to the same procedures that were recommended for the prescription function whenever environmental crises may so require.

At the national level, the Working Group recommends, again, recourse to the same structures of authority and to the same procedures that were recommended for the prescription function and, by these devices, to put an end to arrangements that may be seen as detrimental to environmental protection.

The Appraisal Function

The appraisal function monitors and evaluates past decision processes in terms of the degree to which in fact they secure the community goals for which they have been established. It is thus a part, but a very specialized part, of the intelligence function first mentioned, which permits an audit of the success and failures of past practices in realizing perceived goals.

Recognizing that international and national environmental policies are certain not to succeed without the undertaking of the appraisal function, the Working Group recommends: (a) the establishment and maintenance of self-monitoring and self-appraisal systems on a continuous basis by all international and, national organizations and bodies that deal with environmental problems and (b) the establishment of outside monitoring and appraisal systems comprised of qualified experts charged with the responsibility of periodically reviewing the progress made by all organizations and bodies in quest of ecological protection.

Continuing Commission for Inquiry

The recommendations hereinabove made are obviously only illustrative and tentative. An immense amount of work by specialists from many disciplines would be required to make them otherwise.

Among the commissions for inquiry that the World Academy of Art and Science might establish in relation to environmental and associated problems, one should be specialized to decision processes. Such a commission should be interdisciplinarily holistic in character and establish necessary liaison with all the more important governmental and nongovernmental bodies concerned with the general problems under investigation.

1. We are unanimous in our conviction that humanity is at the danger point of its evolution. In a world supposedly rendered impervious to war by atomic weapons, we are nevertheless engaged in wars and live amid events that narrow the spectrum of our options. A number of continuous problem areas have emerged whose symptoms are increasingly obvious to all men. Examples of these are the progressive degradation and poisoning of the environment and the major disparities between the rich and the poor nations, which mean that a large proportion of the world's population has little possibility of decent conditions of life. Furthermore, explosive population increases threaten famine in many parts of the world. Other signs are the sterility and frustration of life, especially urban life; the alienation of an increasing number of individuals who realize they have no say in decisions that greatly affect their lives; increase in crime and delinquency, and a constant devaluation of human personality. These problems seem to accompany affluence and technological development in the advanced countries and are accompanied by a loss of confidence in the effectiveness of our institutions and political parties, and with the failure of religion and its institutions to provide a coherent point of view. They are evident in other forms in the developing countries.

These critical problem areas are global and seem to arise in all industrialized and nonindustrialized societies, whatever the social or economic system. They are, by nature, exceedingly complex and are interactive to the extent that it is increasingly impossible to think in terms of clearly defined problems and clearly defined solutions.

We are convinced that this highly intermeshed cluster of problems cannot be tackled by our present motives, attitudes, approaches, political and organizational processes, and institutions; that mere modification of these will bring only marginal improvements of little or no relevance and that unless a completely new approach is taken quickly, the result can only be disastrous to the human species.

2. We feel that a solution must be sought through a global, systemic, and normative approach, based on new values that take account of the potentialities and rights of the individual and of the species itself. This will require a substantial discontinuity in the current momentum of society and a new approach to education. Change limited to economic or even socioeconomic policy seems to us insufficient for the creation of a new reality.

3. We feel that it is necessary to start, in terms of a perspective of about 20 years, to explore the main desirable characteristics of society and plan now toward the achievement of such a society.

4. The goals of tomorrow cannot be meaningfully visualized in terms of today. By as wide participation as possible, the basis of a new ethic must be established whose values will determine the individual goals and detailed nature of the society to be thus constructed. Such an ethic is essentially the definition of "the good" in light of the new situation and not of present conceptions of their linear extension. Many fundamental questions will have to be asked early; for example, should the world of the future be based on economic growth through industrialization, as now? A necessary prelude to consensus on goals would be the definition and institutionalization of the means

and methods through which a process of continuous convergence attuned to changing situations can be achieved. Thus, in the setting of policies and strategies, maximum flexibility must be ensured to allow change and dynamic adaptation of the policy-setting process itself. Participation of those cultures, nations, and social and professional groups that today have little or no say in the shaping of the future must be made possible. Only by such means can the wealth of creative, original, and innovative capacities that are suppressed or neglected today contribute to widening the range of possible alternatives.

5. An essential step will be to define a new bill of rights and obligations of man in 1990, of which the first article will be the right of each individual in the world to a guaranteed decent share of the world's goods and services, irrespective of his occupation or contribution to society. This must ensure his food and shelter and opportunities for health, education, and cultural development.

6. Achievement of this will be possible only through the formulation of global population policies that take into account the future of the species as a whole, and if the enormous energies and resources now wasted on armaments are liberated and put back into the productive cycle.

7. Achievement of such goals can be only through a radical change in political structures and institutions. A world organization will be essential, necessitating abolition or substantial modification of national sovereignty, required in any case to prevent war and permit disarmament, as well as to facilitate a rational distribution of goods and to generate a world revenue necessary for the creation and maintenance of many services that will be required on a global scale.

8. Such an organization would lead to a rational world pattern of agricultural and industrial production and distribution.

9. The above changes will necessitate new concepts of regionalism. The global approach means much centralized consent, planning, and research; nevertheless, decisions will have to be made at regional and local levels as well, depending on the issues, to encourage the maximum of participation. Although there will be much uniformity of structure and method within a world system, diversity will be a quality requirement, because different cultures represent different methods of coping with the environment, both natural and man-made.

10. The political and institutional changes required to bring about the type of world desired for 1990 can be achieved only by public appreciation of the nature of the present predicament of mankind and of the possible plans to meet it. People of knowledge and vision have a particular responsibility to inform and influence both governments and the public.

11. A number of immediate steps are recommended:

(a) Research should be started immediately to define more clearly and in detail the critical problems facing man, with emphasis on their interactions as well as the interdisciplinary and transnational nature of the problems.

(b) Groups should be set up, at first composed of scientists and other independent individuals and later by governments and transitional entities, to develop policies towards the achievement of 1990 in which (i) new world, national, and regional goals will be formulated and (ii) interacting and integrating strategies of economic, technological, educational, and social policy will be developed.

(c) Government centers should be set up immediately for the assessment of social and cultural as well as economic consequences of alternative technological developments. There should also be a world center for the development of methodologies for such assessments as well as to ensure dissemination of the national results and to act as a world early warning point on environmental and social threats.

(d) We think that further economic growth in the advanced countries is necessary but should be undertaken not as an end in itself, but as a mechanism to provide the resources for worldwide social development. A new basis of growth measurement is required that incluudes social and quality elements in addition to economic factors.

12. The foregoing is predicated on an understanding of the fact that today we have at best very primitive methods for infusing our goals with operative meaning and relevance. Hence, we recognize that one of the most urgent tasks that confronts us is the development of methodologies for the setting of goals that are valid and legitimate within the context of a dynamic evolutionary reality. This task is the central responsibility of the emerging social science.

DINNER

Boris Pregel, *presiding*

PREGEL: I would like to welcome you to this evening's banquet. We do not have a very long program, but before we introduce the three major speakers, I would like to make a few remarks.

Until the last 30 years, man's whole history has been lived in the shadow of scarcity, of insufficient food, of inadequate shelter. This scarcity has been a measure of the availability of energy. Even the present relative abundance in the United States has not meant sufficient food or adequate shelter for those of our people at the bottom of the socioeconomic ladder. We retain in our vocabulary the phrase "poverty-stricken," and we all understand its connotations.

How many of us have used the phrase "leisure-stricken," or given thought to its implications? Leisure may soon be one of the major problems with which individuals, families, and governments will be faced. The average industrial work week will probably soon be no more than 20 hours. In the not-distant future, it may very well be less. What kind of problems will this create? We can see at once that the problems of leisure will be different for different types of persons in different environments.

In the cities, which are already becoming atavistic anachronisms, our chief concern should perhaps be with those classes of workers who are to be displaced by automation—the production workers with specialized skills and the clerical workers whose jobs will be taken over by machines. In the city of New York alone there are several hundred thousand potential displaced persons. With their jobs gone, deprived of status, their earning power depreciated or lost entirely, these people will be downgraded to the status of unskilled labor, the most dangerous and explosive segment of society. Unless we wish to see violence, crime, and vice increase, we must find ways to forestall these undesirable results.

Retraining these displaced specialists will be imperative. However, the acquisition of new skills and new jobs will not solve all the problems of the leisure-stricken, who will still be faced with the threat of boredom. Our people, particularly the present generation, have not been prepared through education or culture to spend leisure in a socially desirable or dignified manner. We can predict with relative accuracy what use will be made of the goods an economy of abundance will provide for us. We cannot so readily predict the uses to which the new leisure will be put. This will be a problem for all segments of society, individuals, families, and governments. Culture cannot bear the initial burden of preparing us for leisure. Our culture has valued money more than music, industry more than art, food more than leisure, labor more than leisure. This, incidentally, is a good example of the way in which technical developments outrun social progress. It will become a responsibility of government to educate our people to use their free time constructively. The choice is clear, but we must act immediately. It will not be easy to educate large masses of people in one lifetime or one generation, and I fear this is all the time we have. This is a problem that should be faced now, with courage and initiative, lest the tremendous new energies we are about to unleash be turned ignorantly against us. An unpleasant fact remains: that although famine is no threat to the United

States, it is a continuing and increasing danger in some of the less fortunate nations. The energy supplies available in the world today are sufficient to provide for every inhabitant of the globe the minimum necessities of food, clothing, and shelter. That poverty and famine any longer exist anywhere is the result of defective and obsolete systems of distribution. The turmoil of the emergent nations and the spiritual distress of the West, the anxiety we exhibit, the gloom we feel, can be attributed, I suggest, to our failure to develop in the social field techniques as advanced as those adopted in scientific areas.

Distribution of wealth is essentially a social mechanism. In the last few years we have lived in a world where the distribution of food, clothing, and shelter was governed by a railroad timetable. Today we live in a world of giant supersonic jet airliners. We are unhappy with our performance because we lack the courage to act as if we lived in a jet age. We cling, instead, to the railroad tracks and the slow schedule of the past. We must abandon the past and embrace the future. Technically, we have moved into the 21st century, and we can no longer afford to cling to 19th-century thinking—that of the era of the railroad. This is, of course, a problem for space men, economists, sociologists.

As an engineer, I feel impelled to point out that distribution of essentials is a worldwide problem and requires a worldwide solution. Worldwide availability of energy will make necessary not only worldwide distribution, but a world plan for production. The new world of abundance will be, inescapably, one world.

The scientists who have made this one world possible, who have made it inevitable, have perceived this truth. They are men educated chiefly in schools and universities in which the old concepts of an economy of scarcity and a world divided were accepted as unchanging truths, hallowed by ages. The fact that the scientists could change not only their minds but their hearts leads me to expect and hope that similar changes can be brought about in the thinking of statesmen. I believe mankind will survive, that there will be no nuclear war. The masses of the world fear nuclear destruction. The beneficial aspects of the new discoveries are less apparent to them, especially because many of the applications are so technically complex, and their implications so revolutionary, that it is difficult for the common man to grasp them. Therefore, it becomes the duty of the scientific community, the press, the other communications media, to bring to the masses a fuller understanding of the good life that lies ahead.

This also must be done on a global scale. Along with an understanding of the promise the future holds, the world's peoples must be warned of the trials that are certain to accompany our transition from an economy of scarcity to one of abundance. Some of these troubles will seem, at times, all but unbearable. The day of abundance already has dawned on us, but the situation has not changed in the expected way because a lot of important events took place that the foretellers of the future did not take into consideration. The explosion of the population, the explosion of automation, the explosion of mass production, the explosion of consumption, produced an accelerated deterioration of our ecology, a deterioration that may become irreversible. Cars, trucks, and planes, which appeared on the roads and in the sky in fantastic quantities, polluted the air of our cities. The empty bottles, beer cans, and plastic containers cluttered our landscape, our garbage cans, and our garbage-processing plants. Our chemical factories poisoned and polluted our lakes and rivers. And what is still more important, the system of distribution of goods did not really ameliorate, and the poor countries became poorer and the rich countries became richer.

I have to repeat that, because of our scientific knowledge and our technical ability, we are already able to produce whatever would be necessary to keep the whole human race in satisfactory material condition, but we have not moved at all to solve the problems of distribution nor made a real effort to coordinate world production in a meaningful way. It seems that people and governments react only after the catastrophe takes place: nobody is prepared to sacrifice his narrow and parochial interest. It seems that everyone expects a miracle, but there are no miracles in our times, and unless some conditions unexpectedly produce pressure, an unbearable pressure, on the governing bodies of many countries simultaneously, and an authoritative scientific body convinces people of the real dangers they will encounter in a short time, I see a great danger in the future.

This means that efforts have to be continued. Those who are responsible have to continue to do whatever is possible in order to establish a general plan and elaborate all the necessary recommendations and suggestions to prevent the extinction of our species. We hope that this effort will be successful and that our conference will be one of the important steps in the right direction.

THE POSITIVE POWER OF SCIENCE

Glenn Seaborg

To speak to such a distinguished gathering, and one that over the past few days has immersed itself in matters of such importance, is indeed a challenge. This has been a conference with a difficult and most urgent task—that of considering "the environmental and social consequences of science and technology" and, in the light of these considerations, attempting to "formulate policy goals and strategies for the needs of man." Of course, the kind of thinking that has been responsible for initiating this conference and organizing its direction and content has been long overdue. Most of us, I am sure, realize that, whatever has been accomplished during these deliberations and discussions, we cannot be too self-congratulatory. There is a great segment of the public that would tell us we are "too late with too little." And a significant part of that group will go even further and say that we of the scientific community and our work are at the root of much of man's ills today.

Nevertheless, I am not here tonight to add to the despair and guilt of our times. I am not one who happens to believe that we can degrade the environment less by degrading man more. We are not going to save the earth merely by despising ourselves as its inhabitants any more than we are going to build a better society by belittling ourselves as individuals.

I think it is unfortunate—almost calamitous—that in the often sincere effort to establish a better perspective on our past errors and a sense of proportion about our present powers there are many who feel they have to go to the extremes of demeaning man, of denying reason and downgrading science. Tonight I plan to defend all three. I particularly want to speak up in behalf of science and urge that it rise to the crisis that faces it and man. In a sense, it will be the success or failure of science—science epitomizing man's extended perception and power—that may well determine his survival as a species. But it is not science as we have used it in the past that will do the job. It must be science as part of a new human philosophy; as part of a new age of enlightenment that we are entering. And it is both the past shortcomings and successes of science that are moving us into the age. Let me explain.

What we are seeing today in all our social upheavals, in all our alarm and anguish over an environmental feedback and, in general, the apparent piling of crisis upon crisis to an almost intolerable degree, is not a forecast of doom. It is the birthpangs of a new world. It is the period of struggle in which we are making the physical transition from man to mankind—a mankind that will be an organic as well as a spiritual whole on this earth. I see this transition as a natural evolutionary process, a continuation of the growth and growing complexity of life on this planet. I do not believe that this growth is malignant in nature. It will not destroy itself by devouring its host or poisoning itself in its own waste. Neither will it self-destruct after delivering its message. Rather, it will self-adjust through listening to and responding to that message—one that for all the static surrounding it is coming through quite clear.

I would like to touch on several ideas related to my admittedly broad and optimistic outlook for the future. In each case I hope to bring in the relationship of science and technology to the more general idea. First, let me dwell on the subject of growth. This is something that until recently was a major value but of which we now seem to be developing a deadly fear. Both that growth and today's reactions to it, I believe, are natural and necessary. There seem to be many people today who, with no mean amount of eloquent hindsight, deplore the fact we did not, long before now, predict our population growth and our growing productivity, with its accompanying waste, and somehow forecast our current environmental dilemma. Unfortunately, most great minds of the past foresaw only small segments of the evolving problem. And the values of the past were centered about unlimited growth because this seemed to be to the advantage of the individual and his society. In a world of seemingly endless physical frontiers, where the exploration and exploitation of these frontiers by new human creativity generated and fulfilled new human needs and values, few if any could think in terms of limits, balance and stability.

Many of us who today talk so glibly about our use of the atmosphere and oceans as "sinks" and who have embraced ecology almost as fervently as we have become embarrassed with economics, tend to overlook or forget the facts and feelings of the most recent past. It does little good to make scapegoats of our ancestors or of one or another segment of our society for the crises we face today. Our environmental crisis in particular could not have been theorized or accepted in the abstract before. It was an experiment that had to be lived in conjunction with the other problems of human growth that have evolved and which we must now move on to resolve. I think it was inevitable that man had to grow to this point. He is now entering what must be a period of tremendous maturity. And just as there is written within the genetic code some incipient biological mechanism that stops physical growth at a certain point in life, so I believe there is within our evolving mankind a well-coded message that is now being released. Before I go on to discuss some of what that message means, let me backtrack to discuss the role of science and technology in the growth to which I have referred.

It is obvious that for some people science and technology are among the best scapegoats of the time. They are said to be the cause of most of our ills today. By conquering disease and extending life they have been responsible for an explosion of population. By increasing productivity and raising living standards they have been responsible for depleting resources and polluting nature. By expanding knowledge and emphasizing efficiency, they have been responsible for deflating myths and diminishing man. And by placing enormous power in the hands of man they have brought him to the brink of his own destruction. The list of accusations is endless, and it is not fashionable today to attempt to answer the charges or put the matter in perspective. It is more fashionable to dwell on man's relationship to the natural world, to lament that he is not more like the animal life from which he descended, and to wish for his return to a simpler, perhaps more primitive, existence.

This approach may be more fashionable, it may even contain a certain amount of wisdom we should heed, but it is not the whole message we should be hearing. The major part of that message, I believe, should tell us not to deny our dependence on nature or, on the other hand, to deny our differences from what we left behind in our evolution. It is this recognition of what we are—with all its potential as well as its shortcomings—and the emphasis on what we must and can become that are important. The fact is that in this transitional period from tribal man to a truly organic mankind, and to a world in which we can live in harmony with each other and in balance with our global environment, we need a new level of excellence in our science and technology and a new degree of integration between them.

There is, no doubt, a great deal of pain and shock involved in this transitional period, for we are breaking one set of long-established natural bonds and forming new ones. The whole process of change produces shock, reaction, and readjustment. There is always a tendency to revert, to flee from the new and challenging, before new understanding and confidence allow us to move ahead. We are living through such times. We are experiencing what Alvin Toffler refers to as "future shock," and it is often difficult to sort out our movement and its direction. For example, take note of the action and reaction of our youth as it resists some change with an antirational thrust—and often a flight from reality—while at the same time demanding realization of a new level of idealism that can only be achieved through change, change employing the highest form of rationality.

A similar dichotomy and flux exist in our confrontation with environmental problems. Many feel the need for simplicity, limits, and balance. Yet we know that what we must do to accomplish the goals we express in these terms entails to some degree the mastery of a greater complexity, new growth, and a dynamic rather than static type of balance. We really do not want to freeze the world as it is now or go back to "the good old days." What we want, if I read correctly the ideals that so many of you—and so many deeply thinking and concerned people around the world—feel today, is a world far different than we have ever known it. Today—at this conference and at similar meetings that are being held and organized—we are struggling, fiercely but fruitfully, I think, to clarify our conception of that world and work toward its realization. We must not weaken or lose heart in this struggle.

What will be the outcome when we begin to succeed? What will the evolution of this new mankind mean? And what will be some of the manifestations that it is taking place and succeeding? Perhaps most important, we will see the elimination of war as an attempt to resolve human differences. It will not be only that war becomes untenable as a form of such resolution. Neither will it be only that through the wisest and fullest application of science and technology we will eliminate most of the physical insecurity and want at the root of war. What may be most significant as both a cause and effect in establishing world peace will be a sublimation of man's territorial instincts, and the aggressiveness that is tied to them, to a new feeling, one of the communality of man in the possession of the entire earth. Men are already in some measure sharing the earth through international travel, communication, and exchange of resources. As this sharing is enhanced by a parallel releasing of the age-old bond of fear of scarcity and adjustments in the economic system we have built to institutionalize that bond, we will begin to see the true meaning of the brotherhood of man materialize. And as this happens, the tribal loyalty that Arthur Koestler has seen as the root of much of man's conflict will be broken and shift to a new global loyalty—a loyalty of man to all his fellowmen.

Concurrently with this establishment of world peace— and, again, as both a cause and effect of it—will be the closing of the chasms between the peoples of the world. Aurelio Peccei, Barbara Ward, and many others have warned us that we cannot live in a world growing apart in the rate of development of its peoples. What I think we must, and will, see is a new concerted effort to raise the standard of living and productivity of the underdeveloped areas of the world while readjusting the growth of what many feel are becoming "over-developed" areas, harming themselves and others by some unwise management of their power and affluence. Anyone fully attuned to the problems of the world today can feel the need for this effort and readjustment. I think it is vitally important that we in scientific and technological fields do everything we can to encourage and work with those social and political forces that recognize this need and are trying to fulfill it. I will have more to say about such cooperative efforts in a moment.

Another manifestation of our evolving mankind will be the reduction, and eventual elimination, of environmental pollution. The organic mankind of which I have spoken could exist but momentarily on this earth if it were to act as a parasite or cancer. It must learn to exist as an integral and con-tributing part of the earth that up to now has supported it unquestioningly. This can be achieved only by the formulation and application of a whole new scientific outlook and new ecological-technological relationship. This relationship must be based on a nonexploitive, closed-cycle way of life that is difficult to conceive of in terms of the way we live today. We will have to achieve what René Dubos has referred to as a "steady-state world." We will have to think and operate in terms of tremendous efficiencies. We will have to work with natural resources, energy, and the dynamics of the biosphere as a single system, nurturing and replenishing nature as she supports and sustains us. Such a system can be operated at various levels. A steady-state world does not have to be one in which mankind merely subsists and waits for natural evolution to take place. In fact, a steady-state world would be a challenge and stimulus to man's creative evolution—which I believe we should not deny is a natural process and which may be the highest form of natural evolution. Perhaps the organic global mankind that I have portrayed will be the acme of physical evolution on this planet. And I will not try to speculate beyond this point.

I have no doubt that many people envision the concept of such a complex, efficient, and organic mankind as a nightmare—an anthill civilization in which individuals are mere automatons or mindless cells in a emotionless body. I do not agree with such thinking. What I see evolving is quite the opposite of this. It is a world in which the sphere of freedom of action and choice, individual creativity and sensitivity, is enlarged by the growth and application of knowl-edge and by greater efficiency and organization. These elements buy time and provide freedom. It is ignorance, confusion, and waste that enslave and eventually destroy.

Of course, new values and greater education must accompany the transition to this type of civilization. That is why I believe that the age of enlightenment we must enter must be one that combines scientific understanding with a new humanistic philosophy. We need both now to survive and grow. What will be the role and direction of science in achieving this new age? I believe this conference offers us a good indication of both. But let me briefly sketch what I feel has been the movement of science and where it is going today.

Science, we might say, has become a victim of its own success. Or to put it more precisely, it has become a victim of its own single-minded success. I think this has happened in several ways. First, in going from the broad and general philosophy from which it originated into a growing number of more precise disciplines—each becoming more productive the narrower its focus became—science traded off wisdom for knowledge and, to some extent, knowledge for information. In recent years this process has been reversed, and we are now seeing the growth of interdisciplinary sciences and a striving for a more all-encompassing grasp of the physical world and even broader relationships, such as you have been exploring in this conference. This type of growth is essential if science is to be the guiding force, as it must be, behind our evolving mankind. Science must grow stronger by continuing to nourish and improve its individual disciplines. We need the specific knowledge they offer. At the same time, it must grow wiser through its correlation of knowledge. And it must be able to transmit its wisdom in the most effective way to society.

We have a tremendous task before us in humanizing the focus and feeling of science while at the same time organizing and rationalizing the forces of humanity. In recent years we have not been too successful in either of these directions. That is the reason why we have been faced, and are faced today, with a decline in the prestige of science, an antirational reaction on the part of many of our disillusioned youth, talk of the "eroding integrity of science," and even a feeling of guilt and despair in much of the scientific community. We must move away from all this. We must work toward a unification of the scientific spirit and a restoration of our self-confidence as well as a new degree of respect for science on the part of those who have lost faith and hope in it.

Let me offer some specific proposals as to ways we might accomplish this. I think we should establish more international interdisciplinary conference—such as the outstanding one we have been attending here—and more organizations that integrate our various disciplines, within and outside of the sciences. I believe these conferences and organizations should bring together—for positive, constructive exploration, discussion, and action—participants of varied interests, opinions, and talents: visionaries and realists, environmentalists and technologists, ecologists and economists, theorists and activists. But the purpose of these meetings should not be that of many of the "confrontations" we are witnessing today. We should seek not the degraded power of polarization but the more beneficial strength of unity—that achieved through recognizing and working toward common goals. In this regard I would like to see those scientists who in recent years have done a great service to man by calling attention to his environmental problems now contribute an even greater service by joining their colleagues in a concerted effort to solve those problems.

I think that out of such conferences as we have held here should come concrete programs or ideas that can be acted on, as well as the broader policy-setting type of recommendations. We must give our activists something constructive to act on and encourage the idea that many small positive measures can add up to a significant force. They can develop an important momentum and a spirit that in time can become overwhelming in its total effect.

I think that the information and programs generated at these conferences must be brought to the attention of the public and the world's political leaders more successfully. We have been very unsuccessful in communicating with the public, in bridging the gap between the Two Cultures. Now we must not only communicate, we must involve. We should particularly encourage the participation of youth in scientific and technical activities. Merely to decry their alienation, to speak of their immaturity or their unrealistic, "nonnegotiable" approach to achieving their ideals, is pointless; more than that, it is disastrous. We must at all levels engage them in the realities of life, not in order to blunt their ideals or enthusiasm, but for the purposes of capturing what is good and constructive in them, of harnessing their energy and creativity, of growing with them.

If some sparks must fly between the gap of our generations, let us not use them to ignite conflagrations but rather to fire an engine of human progress. We in the scientific community in particular need our young people working with us, and it is one of the tragedies of our time that so many of them have become cynical about the accomplishments and prospects of science. I believe we can win many of them back, especially by showing them how effective we can be in working toward the solution of our environmental ills. We must prove to them that science and technology are among man's most creative and constructive forces when they are used by creative and constructive men.

Finally, I think that in bringing together the many forces I have referred to tonight, and in emphasizing the importance of their working together, we must establish the leadership and goals to direct and sustain their efforts. Never before has the world had such a desperate need for greatness, for inspiration, for a vision. The cynics today will tell us that any vision we might have now would be a delusion. But I cannot agree. I feel as it says in Proverbs: "Where there is no dream, the people perish." Let us create that dream then and work to achieve it—not only that man shall live, but that mankind shall be born.

IN CONCLUSION: IMPLEMENTING CONFERENCE OBJECTIVES

Boris Pregel, Harold Lasswell, John McHale

A recurring theme of the Working Groups and the plenary discussions of the Conference was the importance for science and society of continuing activities that link scientific knowledge with public policy. The contributions of the scientific community must be a regular feature of life if the social consequences and policy implications of knowledge are to be kept visible, and if the involvement of scientists and of science with society is to be widely understood.

The theme of continuity implies much more than that public policy should take scientific knowledge into account or that qualified persons and professional associations should give occasional attention to public issues. Analysts of the decision process of both governmental and private organizations are aware of the fact that, sooner or later, the makers and executors of policy are unescapably affected by some of the findings of scientific investigation.

The critical words in the previous sentence are "sooner or later" and "some of the findings of scientific investigation." In our interdependent world we can no longer afford the time gaps that have customarily separated the development of new knowledge in many fields from the effective awareness of social consequences and policy implications. Whatever aspect of society and environment was under consideration at the Conference, it became painfully evident that no adequate connection had as yet been made between many scientific findings and public action.

If gaps are to be closed, our established ways of doing things must be drastically renovated and improved. It is necessary to move from episodic and piecemeal methods to continuing and inclusive connections between knowledge and policy. The distinctive institutions of the scientific community need to be better adapted to the task of motivating scientists to participate in decision processes of the larger community. The specialized governmental and political institutions of society need to be more cognizant of the knowledge pertinent to the realization of human values. The institutions of education and public information require drastic revision if civic and public order are to become more adequate to the urgencies of the time.

Members of the Conference who examined the decision process at the national, transnational, and subnational levels called attention to two decision functions where scientists and scientific knowledge can have optimum impact. These functions are referred to as the "intelligence" or "planning" and the "appraisal" or "reviewing" phases of the total process of policy formation and execution. By the intelligence function is meant the gathering, processing, and dissemination of the information required for enlightened action by participating individuals and agencies in the policy process. It is unnecessary to labor the point that physical, biological, and cultural scientists have indispensable contributions to make to the mobilization of knowledge pertinent to the clarification of the goals and the options open to the body politic. The function of appraisal is closely connected with intelligence or planning. It is concerned, however, with past policy, seeking to provide an answer to the question whether, and to what degree, public policies have succeeded or failed. It is inconceivable that a realistic judgment of such matters can be made without mobilizing scientific knowledge of the options that were available to decision makers.

Throughout the discussion of "society and environment in transition" members of the Conference continually stressed the critical importance of time. An ever-present dimension of the ecological system of which man is part is the temporal interdependency of its component elements. No one doubted that failures of mobilization *in time* were at the root of the degree to which the safety, health, and positive well-being of man are imperiled by polluted air, toxic food, radioactive waste, and demoralizing noise. Commentators agreed that the Conference would be worthwhile in the degree that it was an act in a series of acts designed to increase present awareness of the achievements and limitations of past policy, and the options available for future action.

In view of the opinions formulated at the Conference, the American Division of the World Academy of Art and Science and the American Geographical Society have decided to give high priority to the task of instituting a *continuing review of public policy toward environment*. Such a Review is intended to provide a means by which the scientific resources of society are mobilized for the benefit of official and unofficial participants in the policy process at every level of jurisdiction. It will continually remind the scientific community of their obligations as citizens of the larger community with whom their future is intermingled. The Review will demonstrate to the general public that scientists are actively concerned with the survival of man and with the quality of life. By strengthening the effectiveness of the *civic* order the Review will contribute to the effectiveness of the *public* order as an instrument of goals at once rational and realistic.

The Review of public policy toward environment may serve as a prototype of the way in which our institutions can be overhauled or redesigned to meet the needs of man. It is to be hoped that the reviewing procedure will be systematically extended to every sector of public policy at the national, transnational, and subnational level. Besides the appraisal function, the intelligence or planning function is particularly adapted for the distinctive skill and outlook of scientists and scientific societies.

We assume that individual scientists will find themselves motivated to go beyond occasional participation in appraisal and intelligence functions and devote themselves in varying degree to *promotional* activities intended to mobilize support for action through party and pressure group channels. No doubt some professional men will engage in legislative and other *prescribing* activities. Presumably many members of the scientific community will continue to serve as officials charged with full responsibility for administrative operations (the provisional *invocation* of prescriptions in concrete situations; the final *application* of prescriptions; the *termination* of obsolescent prescriptions).

The scope of the appraising function can be made explicit by considering the contemplated Review of public policy toward environment. The Review is a product of committees that focus on the various policy-environmental sectors. Each assessment will necessarily include (1) *a statement of assumptions about the goals of public policy* toward environment. The reviewers take responsibility for the formulation of goals that it utilizes. No doubt each committee will clarify and lend weight to its statements by relating them to pertinent authoritative pronouncements (such as the language of treaties,

Looking back at the New York Conference, one cannot fail to be impressed once more by the willingness of eminent men and women from many nations and disciplinary backgrounds to take time out from their personal commitments to engage in a common quest for deeper understanding. If we succeed in navigating the "transitions" of today and tomorrow, it will be in no small measure the result of the vision and energy of those who are committed to the permeation of public policy with the relevant knowledge. The goals of man, though remarkably different in detail, are remarkably similar in commitment to the search for a relationship of man to man and to nature that goes beyond simple survival to the release of creativity that continually redefines and improves the quality of life.